IMMEDIATE EARLY GENES IN SENSORY PROCESSING, COGNITIVE PERFORMANCE AND NEUROLOGICAL DISORDERS

This book is dedicated to Daniel
(R.P. and L.A.T.)

IMMEDIATE EARLY GENES IN SENSORY PROCESSING, COGNITIVE PERFORMANCE AND NEUROLOGICAL DISORDERS

Edited by

RAPHAEL PINAUD, Ph.D.
*Department of Neurobiology, Duke University Medical Center,
Durham, NC, USA*

and

LIISA A. TREMERE, Ph.D.
*Department of Neurobiology, Duke University Medical Center,
Durham, NC, USA*

🐴 Springer

Library of Congress Control Number: 2006923379

ISBN-10: 0-387-33603-6   e-ISBN-10: 0-387-33604-4
ISBN-13: 978-0-387-33603-9

Printed on acid-free paper.

Printed in the United States of America.

9 8 7 6 5 4 3 2 1

springer.com

Contents

Preface

A thorough comprehension of normal and aberrant activities of the central nervous system (CNS) relies directly our ability to understand the anatomical and functional properties of neurons and the brain circuits in which they participate. Studies on the roles and response profiles of immediate early genes (IEGs) have shed significant light into a number of properties of CNS neurons and networks. IEGs represent a set of genes that are expressed without new protein synthesis, in a rapidly and transiently fashion following synaptic input and, therefore, constitute the first genomic response to sensory input. The most widely studied IEGs in neurobiology encode transcription factors, as is the case of *c-fos* and NGFI-A (a.k.a *zif268*, *egr-1*, *krox-24* and *zenk*), however, efforts from several research groups in the field have identified more recently members of this class of gene that also encode proteins with varied physiological roles such as growth factors, cytoskeletal proteins and signaling molecules, to name a few. Thus, either by regulating the expression of target genes, as in the case of transcription factors, or by direct actions of "effector" protein products, IEGs are well positioned to reactively and prospectively shape CNS function.

The dependence of IEG expression on synaptic activity has been, for almost two decades now, a valuable tool for researchers within virtually all fields of neuroscience, given that the detection of the products that result from their expression allows for a detailed mapping of neuronal activity patterns, with extremely high spatial resolution. The use of IEG expression as a mapping tool for neuronal activity has significantly improved our understanding of a large variety of critical processes during development and adulthood, in normal and abnormal activities of the CNS. More recently the development of modern technologies that allow for direct and selective interference with gene expression has yielded significant information on the specific roles of IEG products in neuronal physiology, brain network functioning and behavior.

The central goal of this book was to compile the most updated information on how the expression of IEGs is impacted by external (environmental) and internal (physiological) information and how IEG expression has contributed to our understanding of the anatomical and functional organization, as well as the aberrant function of brain systems. In addition, a main goal of this book was to review the current knowledge on the specific roles of IEG products to neuronal physiology and the regulation of behavior. In order to generate as complete and engaging a work as possible, we invited several leading authors in the field to join us in this project and lend us their expertise. Our goal was to place readers who were new to this subject at the crossroads of several avenues of work. For investigators who are active in IEG-related research, we wanted to assemble an easy reference for the breadth of discovery that is advancing in several directions in this field.

We would like to take this opportunity to thank our many contributors and collaborators on this book. Assembling a book requires enormous generosity of spirit on the part of all participants. We consider ourselves extremely lucky to have had the opportunity to work along side scientific leaders in the various fields reviewed in this volume. We greatly appreciated your incredible efforts and cooperation both in terms of content and timelines, as well as for your endless patience and enthusiasm throughout the editorial process. It is our hope that you will be as pleased as we are with the final product.

In addition to thanking those who contributed intellectually to this work, we must also extend our sincerest thanks to the amazing team that works behind the scenes at Springer in order to materialize the final product and bring it to the marketplace. We are greatly indebted to Joe Burns and Marcia Kidston, who provided advice and administrative help throughout this project. We extend our sincerest thanks to our partner Sheri Campbell whose expertise, timely help, endless energy and commitment to quality was essential to the generation of this book. We also thank Arūnas Ūsaitis for a superb work in the generation of book proofs and Marc Palmer for his outstanding work on the development of the cover art. Finally we thank all members of our families whose lay interest, enthusiasm and support gave us continued energy to assemble this work.

Raphael Pinaud, Durham, NC, USA
Liisa A. Tremere, Durham, NC, USA

May, 2006

Contributors

ANTONIO ARMARIO
Institut de Neurociències and Unitat
de Fisiologia Animal (Facultat
de Ciències)
Departament de Biologia Cellular,
de Fisiologia i d'Inmunologia
Universitat Autònoma de Barcelona
Barcelona, Spain

RACHEL CAMERON
Signal Transduction Laboratory
Department of Pharmacology and
Clinical Pharmacology
The University of Auckland
Auckland, New Zealand

R. WILLIAM CURRIE
Department of Anatomy and
Neurobiology
Dalhousie University
Halifax, NS, Canada

SABRINA DAVIS
Laboratoire de Neurobiologie de
l'Apprentissage, de la Mémoire
et de la Communication
CNRS UMR 8620
Université Paris-Sud
Orsay, France

SAMUEL DEURVEILHER
Department of Anatomy and
Neurobiology
Dalhousie University
Halifax, NS, Canada

MIKE DRAGUNOW
Signal Transduction Laboratory
Department of Pharmacology and
Clinical Pharmacology
National Research Centre for Growth
and Development
The University of Auckland
Auckland, New Zealand

ISIDRO FERRER
Institut de Neuropatologia, Servei
Anatomia Patològica
Hospital Universitari de Bellvitge
Universitat de Barcelona
Barcelona, Spain

ROBERT K. FILIPKOWSKI
Nencki Institute
Warsaw, Poland

ANTONIO F. FORTES
Department of Neuroscience
University of Minnesota
Minneapolis, MN, USA

JOHN F. GUZOWSKI
Department of Neurobiology &
Behavior and Center for the
Neurobiology of Learning and
Memory
University of California,
Irvine, CA, USA

LESZEK KACZMAREK
Nencki Institute
Warsaw, Poland

EWELINA KNAPSKA
Nencki Institute
Warsaw, Poland

SHANELLE W. KO
Department of Physiology
University of Toronto
Toronto, ON, Canada

SERGE LAROCHE
Laboratoire de Neurobiologie de
    l'Apprentissage, de la Mémoire
    et de la Communication
CNRS UMR 8620
Université Paris-Sud
Orsay, France

MEGAN LIBBEY
Laboratoire de Neurobiologie de
    l'Apprentissage, de la Mémoire
    et de la Communication
CNRS UMR 8620
Université Paris-Sud
Orsay, France

CLAUDIO V. MELLO
Neurological Sciences Institute
Oregon Health & Science University
Beaverton, OR, USA

MONIQUE MONTAG-SALLAZ
Neurogenetics Research Group
Leibniz Institute for Neurobiology
Magdeburg, Germany

DIRK MONTAG
Neurogenetics Research Group
Leibniz Institute for Neurobiology
Magdeburg, Germany

RAPHAEL PINAUD
Department of Neurobiology
Duke University Medical Center
Durham, NC, USA

BERTA PUIG
Institut de Neuropatologia, Servei
    Anatomia Patològica
Hospital Universitari de Bellvitge
Universitat de Barcelona
Barcelona, Spain

GABRIEL SANTPERE
Institut de Neuropatologia, Servei
    Anatomia Patològica
Hospital Universitari de Bellvitge
Universitat de Barcelona
Barcelona, Spain

KAZUE SEMBA
Department of Anatomy and
    Neurobiology
Dalhousie University
Halifax, NS, Canada

PER SVENNINGSSON
Section for Molecular
    Neuropharmacology
Department of Physiology and
    Pharmacology
Karolinska Institute
Stockholm, Sweden

THOMAS A. TERLEPH
Department of Biology
Sacred Heart University
Fairfield, CT, USA

LIISA A. TREMERE
Department of Neurobiology
Duke University Medical Center
Durham, NC, USA

XIAOQUN ZHANG
Section for Molecular
    Neuropharmacology
Department of Physiology and
    Pharmacology
Karolinska Institute
Stockholm, Sweden

1

The Use of Immediate Early Genes as Mapping Tools for Neuronal Activation: Concepts and Methods

THOMAS A. TERLEPH[a] and LIISA A. TREMERE[b]

[a]Department of Biology, Sacred Heart University, Fairfield, CT, USA
[b]Department of Neurobiology, Duke University Medical Center, Durham, NC, USA

1. Introduction

This chapter provides an overview of the immediate early genes (IEGs) most commonly used as tools for neuronal activity mapping in the vertebrate brain: *c-fos*, *zif268* (also known as *krox-24*, NGFI-A, *zenk*, and *egr-1*), and more recently *arc*. In addition, the most commonly employed techniques for the visualization of products that result from IEG expression (mRNA and protein) will be discussed, as well as advantages and potential pitfalls of these methods. We conclude the chapter by guiding the reader through the general topics in each chapter of this volume. One such topic is the many important roles that the protein products encoded by IEGs play in central nervous system physiology. The diversity of approaches and experimental designs described in this chapter highlight how powerful a tool IEGs have proven to be.

1.1. Immediate Early Genes: A Focus on zif268, c-fos and arc

Immediate early genes (IEGs) represent a class of genes that respond rapidly and transiently to a variety of cellular stimuli, ranging from stimulus-induced neuronal depolarization to chemical treatment (reviewed in Hughes and Dragunow, 1995; Kaczmarek and Chaudhuri, 1997; Herdegen and Leah, 1998; Pinaud, 2005). The expression of these genes occurs in the presence of protein synthesis inhibitors, indicating that they do not require new protein synthesis or the activation of any other inducible gene prior to their activation (Hughes and Dragunow, 1995; Chaudhuri, 1997; Clayton, 2000). In the absence of sensory stimulation, as in the case of sensory deprivation, the brain generally expresses low to moderate

R. Pinaud, L.A. Tremere (Eds.), Immediate Early Genes in Sensory Processing, Cognitive Performance and Neurological Disorders, 1–10, ©2006 Springer Science + Business Media, LLC

basal levels of most IEGs, although some IEGs, such as *zif268*, are expressed at high basal levels (Wallace et al., 1995; Kaczmarek and Chaudhuri, 1997; Herdegen and Leah, 1998; Pinaud et al., 2002). Different types of stimulus paradigms may also influence which IEGs are predominantly expressed. The IEGs most commonly used to map neuronal activity, *c-fos* and *zif268*, have had their expression time-course systematically studied. These IEGs exhibit a similar time-course: peak-mRNA levels are detectable around 30 minutes after stimulation onset. Highest protein levels occur between 1.5 and 2 hours post-stimulus, but may be detectable as soon as 5–10 minutes after stimulation (Mello and Ribeiro, 1998; Bisler et al., 2002; Chaudhuri and Zangenehpour, 2002).

Although IEGs encode proteins that have various functions, including growth factors and signal transduction molecules, the most commonly studied IEGs, including the *c-fos* and *zif268*, encode transcription factors that influence neuronal physiology by regulating the expression of downstream target genes, typically referred to as late-response genes, or late genes (LGs) (Herdegen and Leah, 1998; O'Donovan et al., 1999; Tischmeyer and Grimm, 1999; Pinaud, 2005). Other IEGs, however, encode proteins that directly influence cellular function, as in the case of *homer1a* and *arc* (reviewed in Lanahan and Worley, 1998); these genes are often referred to as 'effector' IEGs, and are more extensively discussed in later chapters of this volume (e.g. Chapters 2, 4, 9, 11).

The total number of IEGs constituting a single neuron's response to stimulation lies in the range of tens to hundreds (Sheng and Greenberg, 1990; Nedivi et al., 1993; Lanahan and Worley, 1998). Lanahan and Worley (1998) reported that from approximately 30–40 genes making up an IEG neuronal response, 10–15 encoded regulatory transcription factors, while the remainder encoded effector proteins. The events that lead to IEG induction are largely coordinated by intracellular calcium (Ca^{2+}) influx either as a result of the activation of NMDA-type glutamate receptors, which exhibit a Ca^{2+} conductance, or the activation of voltage sensitive Ca^{2+} channels (reviewed in Finkbeiner and Greenberg, 1998; Pinaud, 2005). These increases in intracellular Ca^{2+} levels trigger the series of biochemical cascades that culminate in the induction of IEGs (Pinaud, 2005). As briefly described above, IEG expression leads to the rapid accumulation of gene-specific transcripts that, in the case of *c-fos* and *zif268*, are subsequently translated into transcriptional regulators. These transcription factors act to influence the expression of target LGs in a rapid and transient manner and may also influence more stable, long-term changes in cellular phenotype (Curran and Morgan, 1987; Curran and Franza, 1988; Pinaud, 2004, 2005).

Of the large numbers of IEGs discovered, only a subset has been used extensively in functional mapping studies. The first, and subsequently most commonly used IEG for neuronal activation mapping is *c-fos* (Morgan et al., 1987; Sagar et al., 1988; Herdegen and Leah, 1998). c-Fos belongs to a large family of leucine-zipper transcriptional regulators that also includes FosB, Fra-1 and Fra-2. This protein has been shown to dimerize with other members of the Fos family or members of the Jun family of leucine-zipper transcription factors, that include c-Jun, JunB and JunD (Morgan and Curran, 1991; Herdegen and Leah, 1998). These homo- or hetero-dimers participate in the assembly of the activator protein-1 (AP-1) transcription factor which, in turn, directly regulates the expression of LGs via a specific DNA consensus present in their promoters (Angel and Karin, 1991; Kobierski et al., 1991; Hughes and Dragunow, 1995; Kaminska et al., 2000).

After *c-fos* was initially cloned, the identification of other IEGs quickly followed. A number of IEGs are often co-induced with *c-fos*, albeit with differences in their expression patterns and kinetics. An IEG that is employed extensively in both mammalian and avian models is *zif268* (Milbrandt, 1987; Sukhatme et al., 1988; Kaczmarek and Chaudhuri, 1997; Herdegen and Leah, 1998; Mello et al., 2004; Pinaud, 2004, 2005). Like *c-fos*, the expression of this gene is highly sensitive to membrane depolarization. *zif268* encodes a zinc-finger transcription factor that has high affinity for a specific DNA sequence that is present in hundreds of promoters expressed throughout the central nervous system. This protein functions in regulating key aspects of normal cellular physiology: it regulates the expression of LGs that are involved in a wide range of physiological processes such as neuronal excitability, neurotransmitter release, and metabolic processes (Pospelov et al., 1994; Thiel et al., 1994; Petersohn et al., 1995; Petersohn and Thiel, 1996; Carrasco-Serrano et al., 2000; Wong et al., 2002).

More recently, the activity-regulated, cytoskeleton-associated (*arc*) gene has been used for functional and neuronal mapping studies. Unlike the regulatory IEGs described above, *arc* is an effector IEG. The mRNA encoded by this IEG is rapidly transported to neuronal dendrites in an activity-dependent manner. It is here that local translation takes place, in polyribosomal complexes generally found at the base of dendritic spines (Lyford et al., 1995; Steward et al., 1998; Steward and Worley, 2001a, 2001b). Arc encodes a growth factor and has been shown to associate with F-actin (Lyford et al., 1995). It is these features that have led to the proposal that Arc may be involved in experience-dependent dendritic reconfiguration (Pinaud et al., 2001; Steward and Worley, 2001b; Pinaud, 2004).

A rapid rise in IEG products as a result of stimulation and associated neuronal activation means that IEG responses can often serve as regional activity markers, mapping the responses to specific stimulus inputs and associated behavioral states. Rapid induction means that only a brief stimulation period is needed, permitting the use of IEGs for simple and easily implemented experimental protocols. *c-fos* expression resulting from sensory stimulation is easy to visualize, in part because baseline levels of the expression of this IEG are quite low. As mentioned above, the time course of *c-fos* expression is relatively rapid, with mRNA levels increasing within minutes of stimulation, and protein expression peaking within a few hours. The ability to easily visualize a signal is therefore highly dependent upon the stimulus or behavioral paradigm used, as well as the brain structure being mapped. *zif268* has often been used in conjunction with *c-fos*, and generally shows higher baseline levels (Keefe and Gerfen, 1996; Kaczmarek and Chaudhuri, 1997; Herdegen and Leah, 1998; Pinaud et al., 2002; Velho et al., 2005). High baseline levels allow for bidirectional changes to be assayed. It is important to note, however, that the responses to various pharmacological interventions often differ for these two IEGs (Herdegen and Leah, 1998). In recent years *Arc* has also been used in mapping studies, and often in conjunction with both *c-fos* and *zif268* (Guzowski et al., 2001b; Velho et al., 2005). These IEGs appear to be co-regulated by sensory experience in the same neurons, although there are some differences in their expression profiles in response to the same stimulus (Guzowski et al., 2001a, 2001b; Velho et al., 2005). The expression of these genes constitute a part of the orchestrated wave of gene expression that ensues after sensory stimulation.

1.2. Visualization Techniques

The expression of IEGs is most commonly visualized in brain mapping studies by means of either immunocytochemistry (ICC) or *in-situ* hybridization histochemistry (ISHH). ICC allows for the detection of the proteins encoded by IEGs through the use of specific, often commercially available, antibodies. The detection of transcripts resulting from IEG expression is achieved by the use of ISHH. This technique employs deoxyribo- or riboprobes that are complementary to the original mRNA sequence of interest and generally tagged with either radioactivity or fluorescence (Ziolkowska and Przewlocki, 2002). The probes most commonly used for *in-situ* hybridization experiments are single stranded oligonucleotides (approximately 20–60 nucleotides in length), complementary RNAs (several hundred nucleotides in length), or double-stranded complementary DNAs (Ziolkowska and Przewlocki, 2002).

For ISHH, the mRNA signal is generally visualized via film autoradiography, which does not provide spatial resolution sufficient for analysis at the cellular level (Fig. 1.1). To overcome this limitation, sections that have been hybridized with radioactive probes can be immersed in an emulsion solution, allowing for silver-grain precipitation over cells that express the mRNA of interest. In addition, emulsion-dipped sections are regularly counter-stained with Cresyl-violet for single-cell identification (Fig. 1.2). More recently, fluorescence *in-situ* hybridization has been used, a technique that allows for reliable detection of mRNAs along with preservation of single-cell resolution (Guzowski et al., 2001a; Pinaud et al., 2004) (Fig. 1.3).

The histological labeling of proteins encoded by IEGs by means of ICC involves secondary detection systems and allows for both single-cell identification and subcellular resolution. As c-Fos and Zif268 proteins are transcription factors, ICC methods reveal immunopositive signal from each cell's nucleus (Fig. 1.4). Secondary detection systems often involve the use of fluorescent antibodies, or fluorescence detection systems (e.g., tyramide signal amplification—TSA), as well as permanent chromogen labeling, such as diaminobenzidine (Mello and Ribeiro, 1998; Pinaud et al., 2000, 2002, 2006). As ICC procedures are relatively simple and allow for cellular-level resolution, this methodology has become the most widely used in mapping studies. In addition, double- and even triple-labeling experiments that combine IEG expression with neurochemical markers (e.g., GAD65 expression for the identification of GABAergic neurons) by use of fluo-rescence ISHH and ICC methods, contribute to a more complete understanding of

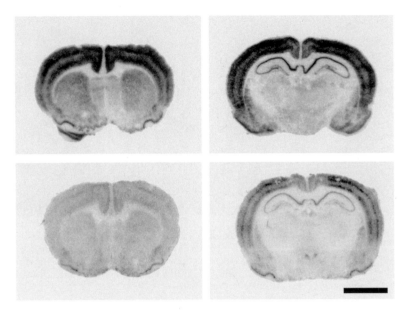

Figure 1.1. Autoradiogram depicting NGFI-A mRNA hybridization in the brains of an animal exposed to an enriched environment (left) and an undisturbed control (right). Sections are presented from rostral to caudal, left to right. Note that NGFI-A is regulated by sensory experience over sensory cortices, striatum and hippocampus. Scale bar = 500 μm. Images courtesy of Dr. Raphael Pinaud.

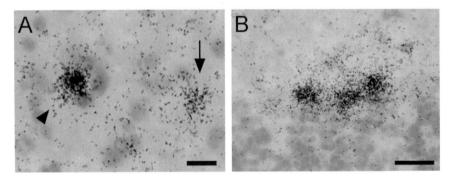

Figure 1.2. Representative examples of emulsion-dipped sections showing accumulation of silver grains over cells expressing mRNA of interest (in this case GABAergic cells). (A) Detailed view of emulsion-dipped section showing differential accumulation of silver grains over GABAergic cells in the songbird NCM, a region of the songbird brain that is analogous to the primary auditory cortex in mammals. Arrowhead depicts a cell with high GAD65 expression while the arrow indicates a cell with low GAD65 mRNA levels. (B) View of emulsion-dipped section showing GAD65 expression in cerebellar Purkinje cells. Scale bars: (A) 10 μm; (B) 25 μm. Images courtesy of Dr. Raphael Pinaud.

the specific population of neurons that participates in a given response (Staiger et al., 2002; Pinaud et al., 2004). Identification of antigens with different subcellular localization, such as Zif268 which exhibits nuclear localization and GABA

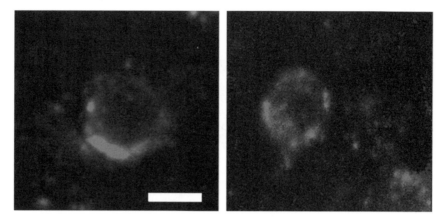

Figure 1.3. Fluorescence in-situ hybridization for NGFI-A/zenk. Representative examples of neurons that express NGFI-A/zenk mRNA in NCM, a region of the songbird brain that is analogous to the primary auditory cortex in mammals, following song stimulation. These images were obtained with confocal microscopy. Note that fluorescence in-situ hybridization allows for single cell resolution. Scale bar = 5 μm. Images courtesy of Dr. Raphael Pinaud.

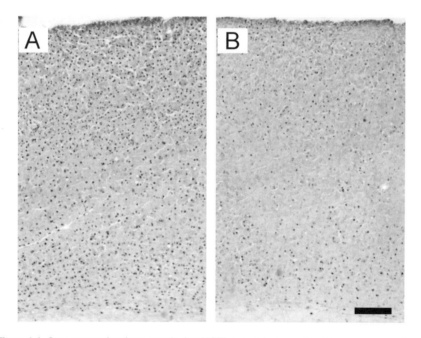

Figure 1.4. Immunocytochemistry reveals that NGFI-A protein is regulated by sensory experience across all layers of the rat primary visual cortex (V1). (A) V1 of an animal exposed to an enriched environment exhibits an enhanced number of NGFI-A positive neurons compared to an animal maintained in an impoverished visual environment—a standard lab home cage (B). NGFI-A protein has a nuclear localization given that it is a transcription factor. In addition, note that immunocytochemistry allows for a single-cell level of resolution. Scale bar = 200 μm. Images courtesy of Dr. Raphael Pinaud.

which is primarily expressed in the cytoplasm, also allows for double-labeling approaches (Pinaud et al., 2004). Such approaches further our understanding of the neurochemical identity of IEG-expressing neurons. Finally, relatively new techniques such as catFISH, which exploit the temporal dynamics of intracellular mRNA trafficking, allow for the visualization of neuronal ensemble responses to two experimental conditions or stimuli (see Chapter 9) (Guzowski et al., 1999; Vazdarjanova and Guzowski, 2004; Burke et al., 2005; Velho et al., 2005).

Specific transcripts and proteins encoded by IEGs have also been detected by using northern- and western-blot approaches, respectively (Ennulat et al., 1994; Mataga et al., 2001; Mokin and Keifer, 2005). Although these techniques offer the same specificity for recognition of either the mRNA or proteins encoded by IEGs, they do not permit the spatial resolution mapping of ISHH and ICC methods.

A limitation that is commonly present in mapping studies that employ immunocytochemical methods is cross-reactivity of the primary antibody of interest with proteins of close homology. In the Fos family, for example, earlier work that employed c-Fos antibodies often detected other members of the Fos family (e.g., Fra-1, Fra-2) (Caston-Balderrama et al., 1998; Van Der Gucht et al., 2000). The recent use of more specific antibodies has helped to reduce this problem (Van Der Gucht et al., 2000). An additional problem in ICC experiments is that sometimes the specificity of an antibody directed against an IEG protein is species-specific and thus not suitable to detect the homologous protein of a different species. For example, initially the identification of c-Fos in avian species was limited by a lack of suitable antibodies that were able to specifically recognize this protein in birds, given that most available antibodies were directed at the rodent protein. This problem was recently addressed by D'Hondt and colleagues who prepared an antibody directed against the chicken c-Fos (D'Hondt et al., 1999). This antibody has been shown to specifically and reliably identify c-Fos in a number of avian species. Similar issues have been addressed in mammalian preparations, such as cats and primates.

1.3. Immediate Early Genes in Sensory Processing, Cognitive Performance and Neurological Disorders

As discussed above, IEG expression has been extensively used to map neuronal activity in a number of different experimental paradigms, and in several animal models. These include sensory stimulation and behavioral manipulations, such as learning experiments. For example, in Part I of this book, Chapters 2, 3 and 4 describe how visual, auditory, gustatory and olfactory experience regulate IEG expression in the vertebrate brain. In addition, Chapters 5 and 6 detail how our understanding of the anatomical and functional organization of the somatosensory system has been furthered by studies that employ IEG expression as an activity marker of touch and pain processing, respectively.

Part II reviews work that has used IEG expression to explore the organization and function of complex systems of higher-order cognitive function. Chapter 7, for example, describes how c-Fos expression has been successfully used to map brain regions involved in the regulation of the sleep-wake cycle. Both regulatory and effector IEGs have also been intensely scrutinized over the past few years, not only to gain insight into the organization of networks that underlie complex brain functions, but also to pinpoint the specific contributions of each of these types of IEGs in such processes. These issues, particularly as they relate to learning and memory formation, are discussed in Chapters 8–10.

Finally, Part III of this volume describes the use of IEGs in a variety of preparations associated with neurological function and dysfunction. Chapter 11 reviews how IEGs have been used to map brain areas involved in behavioral responses to stressors, and Chapter 12 discusses how the activity of transcription factors regulates nerve cell death, survival and the repair of axons. Finally, Chapters 13 and 14 discuss the uses and roles of IEGs in the study of the neurological disorders Alzheimer's and Parkinson's diseases, respectively.

References

Angel P, Karin M (1991). The role of Jun, Fos and the AP-1 complex in cell-proliferation and transformation. Biochim Biophys Acta 1072:129-157.

Bisler S, Schleicher A, Gass P, Stehle JH, Zilles K, Staiger JF (2002). Expression of c-Fos, ICER, Krox-24 and JunB in the whisker-to-barrel pathway of rats: time course of induction upon whisker stimulation by tactile exploration of an enriched environment. J Chem Neuroanat 23:187-198.

Burke SN, Chawla MK, Penner MR, Crowell BE, Worley PF, Barnes CA, McNaughton BL (2005). Differential encoding of behavior and spatial context in deep and superficial layers of the neocortex. Neuron 45:667-674.

Carrasco-Serrano C, Viniegra S, Ballesta JJ, Criado M (2000). Phorbol ester activation of the neuronal nicotinic acetylcholine receptor alpha7 subunit gene: involvement of transcription factor Egr-1. J Neurochem 74:932-939.

Caston-Balderrama AL, Cameron JL, Hoffman GE (1998). Immunocytochemical localization of Fos in perfused nonhuman primate brain tissue: fixation and antisera selection. J Histochem Cytochem 46:547-556.

Chaudhuri A (1997). Neural activity mapping with inducible transcription factors. Neuroreport 8:v-ix.

Chaudhuri A, Zangenehpour S (2002). Molecular activity maps of sensory function. In: Immediate Early Genes and Inducible Transcription Factors in Mapping of the Central Nervous System Function and Dysfunction (Kaczmarek L, Robertson HA, eds), pp. 103-145. Amsterdam: Elsevier Science B.V.

Clayton DF (2000). The genomic action potential. Neurobiol Learn Mem 74:185-216.

Curran T, Morgan JI (1987). Memories of fos. Bioessays 7:255-258.

Curran T, Franza BR, Jr. (1988). Fos and Jun: the AP-1 connection. Cell 55:395-397.

D'Hondt E, Vermeiren J, Peeters K, Balthazart J, Tlemcani O, Ball GF, Duffy DL, Vandesande F, Berghman LR (1999). Validation of a new antiserum directed towards the synthetic c-terminus of the FOS protein in avian species: immunological, physiological and behavioral evidence. J Neurosci Methods 91:31-45.

Ennulat DJ, Babb S, Cohen BM (1994). Persistent reduction of immediate early gene mRNA in rat forebrain following single or multiple doses of cocaine. Brain Res Mol Brain Res 26:106-112.

Finkbeiner S, Greenberg ME (1998). Ca^{2+} channel-regulated neuronal gene expression. J Neurobiol 37:171-189.

Guzowski JF, McNaughton BL, Barnes CA, Worley PF (1999). Environment-specific expression of the immediate-early gene Arc in hippocampal neuronal ensembles. Nat Neurosci 2:1120-1124.

Guzowski JF, McNaughton BL, Barnes CA, Worley PF (2001a). Imaging neural activity with temporal and cellular resolution using FISH. Curr Opin Neurobiol 11:579-584.

Guzowski JF, Setlow B, Wagner EK, McGaugh JL (2001b) Experience-dependent gene expression in the rat hippocampus after spatial learning: a comparison of the immediate-early genes Arc, c-fos, and zif268. J Neurosci 21:5089-5098.

Herdegen T, Leah JD (1998). Inducible and constitutive transcription factors in the mammalian nervous system: control of gene expression by Jun, Fos and Krox, and CREB/ATF proteins. Brain Res Brain Res Rev 28:370-490.

Hughes P, Dragunow M (1995). Induction of immediate-early genes and the control of neurotransmitter-regulated gene expression within the nervous system. Pharmacol Rev 47:133-178.

Kaczmarek L, Chaudhuri A (1997). Sensory regulation of immediate-early gene expression in mammalian visual cortex: implications for functional mapping and neural plasticity. Brain Res Brain Res Rev 23:237-256.

Kaminska B, Pyrzynska B, Ciechomska I, Wisniewska M (2000). Modulation of the composition of AP-1 complex and its impact on transcriptional activity. Acta Neurobiol Exp (Wars) 60:395-402.

Keefe KA, Gerfen CR (1996). D1 dopamine receptor-mediated induction of zif268 and c-fos in the dopamine-depleted striatum: differential regulation and independence from NMDA receptors. J Comp Neurol 367:165-176.

Kobierski LA, Chu HM, Tan Y, Comb MJ (1991). cAMP-dependent regulation of proenkephalin by JunD and JunB: positive and negative effects of AP-1 proteins. Proc Natl Acad Sci USA 88:10222-10226.

Lanahan A, Worley P (1998). Immediate-early genes and synaptic function. Neurobiol Learn Mem 70:37-43.

Lyford GL, Yamagata K, Kaufmann WE, Barnes CA, Sanders LK, Copeland NG, Gilbert DJ, Jenkins NA, Lanahan AA, Worley PF (1995). Arc, a growth factor and activity-regulated gene, encodes a novel cytoskeleton-associated protein that is enriched in neuronal dendrites. Neuron 14:433-445.

Mataga N, Fujishima S, Condie BG, Hensch TK (2001). Experience-dependent plasticity of mouse visual cortex in the absence of the neuronal activity-dependent marker egr1/zif268. J Neurosci 21:9724-9732.

Mello CV, Ribeiro S (1998). ZENK protein regulation by song in the brain of songbirds. J Comp Neurol 393:426-438.

Mello CV, Velho TA, Pinaud R (2004) Song-induced gene expression: a window on song auditory processing and perception. Ann N Y Acad Sci 1016:263-281.

Milbrandt J (1987). A nerve growth factor-induced gene encodes a possible transcriptional regulatory factor. Science 238:797-799.

Mokin M, Keifer J (2005). Expression of the immediate-early gene-encoded protein Egr-1 (zif268) during in vitro classical conditioning. Learn Mem 12:144-149.

Morgan JI, Curran T (1989). Stimulus-transcription coupling in neurons: role of cellular immediate-early genes. Trends Neurosci 12:459-462.

Morgan JI, Curran T (1991). Stimulus-transcription coupling in the nervous system: involvement of the inducible proto-oncogenes fos and jun. Annu Rev Neurosci 14:421-451.

Morgan JI, Cohen DR, Hempstead JL, Curran T (1987). Mapping patterns of c-fos expression in the central nervous system after seizure. Science 237:192-197.

Nedivi E, Hevroni D, Naot D, Israeli D, Citri Y (1993). Numerous candidate plasticity-related genes revealed by differential cDNA cloning. Nature 363:718-722.

O'Donovan KJ, Tourtellotte WG, Millbrandt J, Baraban JM (1999). The EGR family of transcription-regulatory factors: progress at the interface of molecular and systems neuroscience. Trends Neurosci 22:167-173.

Petersohn D, Thiel G (1996). Role of zinc-finger proteins Sp1 and zif268/egr-1 in transcriptional regulation of the human synaptobrevin II gene. Eur J Biochem 239:827-834.

Petersohn D, Schoch S, Brinkmann DR, Thiel G (1995). The human synapsin II gene promoter. Possible role for the transcription factor zif268/egr-1, polyoma enhancer activator 3, and AP2. J Biol Chem 270:24361-24369.

Pinaud R (2004). Experience-dependent immediate early gene expression in the adult central nervous system: evidence from enriched-environment studies. Int J Neurosci 114:321-333.

Pinaud R (2005). Critical calcium-regulated biochemical and gene expression programs involved in experience-dependent plasticity. In: Plasticity in the Visual System: From Genes to Circuits (Pinaud R, Tremere LA, De Weerd P, eds), pp. 153-180. New York: Springer-Verlag.

Pinaud R, Tremere LA, Penner MR (2000). Light-induced zif268 expression is dependent on noradrenergic input in rat visual cortex. Brain Res 882:251-255.

Pinaud R, Penner MR, Robertson HA, Currie RW (2001). Upregulation of the immediate early gene arc in the brains of rats exposed to environmental enrichment: implications for molecular plasticity. Brain Res Mol Brain Res 91:50-56.

Pinaud R, Fortes AF, Lovell P, Mello CV (2006). Calbindin-positive neurons reveal a sexual dimorphism within the songbird analogue of the mammalian auditory cortex. J Neurobiol 66:182-195.

Pinaud R, Tremere LA, Penner MR, Hess FF, Robertson HA, Currie RW (2002). Complexity of sensory environment drives the expression of candidate-plasticity gene, nerve growth factor induced-A. Neuroscience 112:573-582.

Pinaud R, Velho TA, Jeong JK, Tremere LA, Leao RM, von Gersdorff H, Mello CV (2004). GABAergic neurons participate in the brain's response to birdsong auditory stimulation. Eur J Neurosci 20:1318-1330.

Pospelov VA, Pospelova TV, Julien JP (1994). AP-1 and Krox-24 transcription factors activate the neurofilament light gene promoter in P19 embryonal carcinoma cells. Cell Growth Differ 5:187-196.

Sagar SM, Sharp FR, Curran T (1988). Expression of c-fos protein in brain: metabolic mapping at the cellular level. Science 240:1328-1331.

Sheng M, Greenberg ME (1990). The regulation and function of c-fos and other immediate early genes in the nervous system. Neuron 4:477-485.

Staiger JF, Masanneck C, Bisler S, Schleicher A, Zuschratter W, Zilles K (2002). Excitatory and inhibitory neurons express c-Fos in barrel-related columns after exploration of a novel environment. Neuroscience 109:687-699.

Steward O, Worley PF (2001a). A cellular mechanism for targeting newly synthesized mRNAs to synaptic sites on dendrites. Proc Natl Acad Sci U S A 98:7062-7068.

Steward O, Worley PF (2001b). Selective targeting of newly synthesized Arc mRNA to active synapses requires NMDA receptor activation. Neuron 30:227-240.

Steward O, Wallace CS, Lyford GL, Worley PF (1998). Synaptic activation causes the mRNA for the IEG Arc to localize selectively near activated postsynaptic sites on dendrites. Neuron 21:741-751.

Sukhatme VP, Cao XM, Chang LC, Tsai-Morris CH, Stamenkovich D, Ferreira PC, Cohen DR, Edwards SA, Shows TB, Curran T, et al. (1988). A zinc finger-encoding gene coregulated with c-fos during growth and differentiation, and after cellular depolarization. Cell 53:37-43.

Thiel G, Schoch S, Petersohn D (1994). Regulation of synapsin I gene expression by the zinc finger transcription factor zif268/egr-1. J Biol Chem 269:15294-15301.

Tischmeyer W, Grimm R (1999). Activation of immediate early genes and memory formation. Cell Mol Life Sci 55:564-574.

Van Der Gucht E, Vandenbussche E, Orban GA, Vandesande F, Arckens L (2000). A new cat Fos antibody to localize the immediate early gene c-fos in mammalian visual cortex after sensory stimulation. J Histochem Cytochem 48:671-684.

Vazdarjanova A, Guzowski JF (2004). Differences in hippocampal neuronal population responses to modifications of an environmental context: evidence for distinct, yet complementary, functions of CA3 and CA1 ensembles. J Neurosci 24:6489-6496.

Velho TA, Pinaud R, Rodriguez PV, Mello CV (2005). Co-induction of activity-dependent genes in songbirds. Eur J Neurosci (in press).

Wallace CS, Withers GS, Weiler IJ, George JM, Clayton DF, Greenough WT (1995). Correspondence between sites of NGFI-A induction and sites of morphological plasticity following exposure to environmental complexity. Brain Res Mol Brain Res 32:211-220.

Wong WK, Ou XM, Chen K, Shih JC (2002). Activation of human monoamine oxidase B gene expression by a protein kinase C MAPK signal transduction pathway involves c-Jun and Egr-1. J Biol Chem 277:22222-22230.

Ziolkowska B, Przewlocki R (2002). Methods used in inducible transcription factor studies: focus on mRNA. In: Immediate Early Genes and Inducible Transcription Factors in Mapping of the Central Nervous System Function and Dysfunction (Kaczmarek L, Robertson HA, eds), pp. 1-38. Amsterdam: Elsevier B.V.

PART I

Immediate Early Gene Expression as Part of Sensory Processing

2

Regulation of Immediate Early Genes in the Visual Cortex

RAPHAEL PINAUD[a], THOMAS A. TERLEPH[b], R. WILLIAM CURRIE[c]
and LIISA A. TREMERE[a]

[a]*Department of Neurobiology, Duke University Medical Center, Durham, NC, USA*
[b]*Department of Biology, Sacred Heart University, Fairfield, CT, USA*
[c]*Department of Anatomy and Neurobiology, Dalhousie University, Halifax, NS, Canada*

1. Introduction

Light is the fastest and likely the most complex source of physical energy processed by the mammalian central nervous system (CNS). Throughout evolution, most mammals that rely heavily on vision for their normal behaviors have developed an exquisitely elaborate pattern of connectivity and functionality for harnessing, processing and integrating visual information. This process allowed for remarkable environmental adaptation and ecological success for a number of species. The complexity and behavioral relevance of the visual system, and the relative experimental ease associated with research using this modality in laboratory animal models, has arguably placed research in this sensory system in the forefront of contemporary sensory neuroscience research.

Tremendous gains in our understanding of the anatomical and functional organization of the visual system were obtained with the pioneering experiments of David Hubel and Torsten Wiesel, who employed traditional single-unit electrophysiological recordings to describe orientation and directional selectivity, in addition to several other response properties of neurons in the primary visual cortex (V1) of the anesthetized mammal (Hubel and Wiesel, 1962, 1968, 1970, 1972; Wiesel and Hubel, 1963; Hubel et al., 1977). These experiments paved the way for subsequent research using awake animals (e.g., Evarts, 1968; Wurtz, 1969) and eventually for ensemble recordings in awake, behaving mammals (e.g., Nicolelis et al., 1993, 2003; Fanselow et al., 2001; Gail et al., 2003, 2004).

Despite the large strides obtained through the use of multi-site, multi-electrode recordings in awake animals, one of the main limitations of this methodology is the low spatial resolution and the invasiveness associated with recording both chronically and acutely. Histological methodologies have supplemented data obtained with electrophysiological recordings by allowing for evaluation of the global brain activity patterns that result from experimental manipulations

R. Pinaud, L.A. Tremere (Eds.), Immediate Early Genes in Sensory Processing, Cognitive Performance and Neurological Disorders, 13–33, ©2006 Springer Science + Business Media, LLC

(e.g., dark-rearing or light stimulation). With such manipulations, high spatial resolution can be obtained, whereby virtually the entire brain can be mapped for activity, at the cost of low temporal resolution. At the forefront of visual system research, the activity of the mitochondrial enzyme cytochrome oxidase, and of 2-deoxyglucose, have consistently been the most employed histological activity markers, allowing for generation of global response profiles of the visual cortex following sensory stimuli. These methodologies have permitted the successful anatomical delineation of ocular dominance columns in V1, as well as the eye-specific interdigitated organization of retinal projections in the lateral geniculate nucleus (LGN), in primates and cats (reviewed in Sokoloff, 1981; Horton, 1984; Horton and Hedley-Whyte, 1984; Tieman, 1985; Wong-Riley, 1989; Krubitzer and Kaas, 1990; Gattass et al., 2005).

In the late 1980's, activity-dependent immediate early genes (IEGs) began to be employed as a new generation of markers for neuronal activity. The association between neuronal activation and IEG expression has allowed researchers to map the activation of brain networks through the specific probing of the products (both mRNA and protein) that result from gene induction following a number of stimuli, including sensory input (discussed in most chapters in this volume and in Kaczmarek and Chaudhuri, 1997; Herdegen and Leah, 1998; Kaczmarek and Robertson, 2002). Importantly, this methodology allows for the investigation of large-scale activity patterns, with a single-cell level of resolution, and the use of awake animals that have experienced minimal interference with their "natural" behaviors. In addition, high spatial resolution can be achieved with this approach, given that multiple brain areas within the same animal can be studied. One major disadvantage of this methodology is the low temporal resolution, since IEG expression reflects a cumulative effect of activity that occurs in a time-scale ranging from minutes to tens of minutes, depending on the detection methods used (Chapter 1 and Herdegen and Leah, 1998; Kaczmarek and Robertson, 2002; but see Chapter 9 in this volume). In mammals, IEG expression has been successfully used to map neuronal activation within virtually all stations of the ascending visual system, including the retina, the lateral geniculate nucleus (LGN) and lower- and higher-order visual cortical areas, including the V1 and the infero-temporal cortex (IT), respectively, to name a few examples (Chaudhuri and Cynader, 1993; Chaudhuri et al., 1995; Okuno and Miyashita, 1996; Kaczmarek and Chaudhuri, 1997; Okuno et al., 1997; Miyashita et al., 1998; Kaczmarek et al., 1999; Arckens et al., 2000; Pinaud et al., 2002b, 2003a, 2003b; Pinaud, 2004, 2005; Arckens, 2005; Montero, 2005; Pinaud and Tremere, 2005; Soares et al., 2005). This methodology has provided insights into the anatomical and functional properties of visual stations and has proven to be superior to the standard histological markers of activity, as discussed above, given that it offers better spatial and temporal resolution.

The goal of this chapter is to discuss how our understanding of the functional organization of visual cortical circuits was advanced by the use of IEGs as markers for neuronal activation. We will describe and discuss the expression profiles of the most commonly used IEGs in studies of the visual cortex, including the transcription factors encoded by NGFI-A (also known as *egr-1*, *krox-24*, *zenk* and *zif-268*), *c-fos*, *c-jun*, *junB* and *junD*, as related to the AP-1 transcription factor, and the effector IEG *arc*. These expression profiles will be described from the results of a variety of deprivation and stimulation paradigms. Space constraints do not permit an extensive review of this topic; however, several reviews have recently been published and provide additional resources for the reader (Kaczmarek and Chaudhuri, 1997; Herdegen and Leah, 1998; Montero, 2005; Pinaud, 2005; Pinaud and Tremere, 2005). These reviews also discuss findings on IEG expression obtained from other stations of the ascending visual pathway, including the retina and LGN. Here, however, key scientific works describing the regulation of IEG responses to visual experience in V1 are presented and discussed, as well as the emerging findings of the roles that IEGs play in the normal physiology of visual cortical neurons. Finally, we will compare the similarities and differences in the expression profiles of these genes and will discuss how their protein products may orchestrate experience-dependent changes in the visual cortical circuitry, based on some of our own findings.

2. Immediate Early Gene Expression

The expression of IEGs has been extensively used as a mapping tool for neuronal activation, given that their fast and transient expression is associated with the neuronal depolarization that follows sensory input, without the requirement of *de-novo* protein synthesis. For these reasons, it has been argued that the expression of IEGs provide the first genomic response to cell stimulation (Worley et al., 1991; Kaczmarek and Robertson, 2002; Pinaud, 2004). Research on the biochemical cascades that lead to IEG induction has revealed that, not surprisingly, different genes are activated through mostly different, and often times multiple, intracellular cascades, although in some cases there is a significant degree of overlap. A common requirement for the expression of the most widely used IEGs is the influx of calcium (reviewed in Ginty et al., 1992; Ghosh et al., 1994; Finkbeiner and Greenberg, 1998; Pinaud, 2005). This influx is often associated with the activation of the NMDA-type of glutamatergic receptors, which have been repeatedly implicated in the experience-dependent enhancement of synaptic efficacy (Collingridge and Lester, 1989; Hollmann and Heinemann, 1994; Mayer and Armstrong, 2004), as well as the activation of voltage-sensitive calcium channels (reviewed in Pinaud, 2005). The requirement of calcium for IEG induction potentially places the protein products encoded by this class of genes in the early stages of genomic responses associated with plastic changes in the CNS circuitry, including that of the visual system.

It is important to mention, however, that under certain conditions and in certain CNS regions, IEG expression appears to be dissociated from neuronal excitability. For example, light-driven activity in the LGN is not paralleled by an upregulation of the IEG NGFI-A in a number of mammalian species, including cats, monkeys and opossums (Arckens et al., 2000; Pinaud et al., 2003b; Soares et al., 2005). The specific reasons for such an uncoupling between electrophysiological activity and IEG expression remain largely unknown but might involve the modulation of critical components of the intracellular cascade that leads to IEG induction. This issue has been discussed in greater detail in other reviews and will not be further extended here (Sharp et al., 1993; Pinaud, 2005). As discussed below, however, the large majority of IEGs reviewed in this chapter are regulated by visual experience in the V1; therefore, they provide a direct read-out of activity patterns that arise following visual input.

In the following paragraphs, we will briefly detail the characteristics, and some of the functional properties of the IEGs discussed in this chapter, followed by a review of the expression profiles of these IEGs in the V1.

2.1. The Fos and Jun Families: AP-1

Members of the Fos family (C-Fos, FosB, Fra-1 and Fra-2) and of the Jun family (c-Jun, JunB and JunD) form homo- or hetero-dimers through "leucine zipper" interactions positioned in the protein's structure; these dimers constitute the AP-1 (activator protein-1) transcriptional regulator (Morgan and Curran, 1991; Herdegen and Leah, 1998). AP-1 exhibits specificity and binds to a well-characterized DNA motif located in the promoter of a number of CNS genes, thereby regulating their expression. The AP-1 transcription factor exerts bidirectional regulatory effects (either up- or down-regulation) of target genes (Angel and Karin, 1991; Kobierski et al., 1991) and its specific effect on target gene regulation has been shown to be influenced by the identity of the members of the Fos and Jun proteins that compose the dimer. For instance, FosB/JunB heterodimers tend to suppress transcriptional activity of target genes given that this dimer binds to its specific cis-element but is inactive. Coversely, Fos/JunD heterodimers display high positive regulatory capabilities upon binding to its regulatory DNA motif (Hughes and Dragunow, 1995; Kaminska et al., 2000; Hess et al., 2004). Subsequent post-translational modifications, such as phosphorylation, are able to promote additional modifications in the transcriptional properties of the AP-1 (Herdegen and Leah, 1998).

2.2. NGFI-A

The protein encoded by the IEG NGFI-A is a transcription factor of the zinc-finger class (Milbrandt, 1987). Finger-like bulges in the structure of the NGFI-A protein are stabilized by zinc and exhibit high affinity for a specific DNA consensus that has been detected in the promoter region of numerous genes expressed in the mammalian nervous system, thereby potentially exerting regulatory effects on their expression. The identities of late-response genes (LGs) regulated by NGFI-A have been under extensive scrutiny and some targets have emerged from studies primarily conducted in in-vitro preparations. These assays have demonstrated, for example, that NGFI-A regulates the expression of the synapsin I, synapsin II and synaptobrevin genes that encode pre-synaptic proteins involved in the regulation of the size of the readily releasable pool of neurotransmitters and in neurotransmitter release (Thiel et al., 1994; Petersohn et al., 1995;

Petersohn and Thiel, 1996). NGFI-A also appears to regulate the expression of ligand-gated ion channels, such as the nicotinic acetylcholine receptor (Carrasco-Serrano et al., 2000), metabolic enzymes, such as monoamine oxidase B (Wong et al., 2002), as well as genes involved in neuronal structural stability, such as neurofilament (Pospelov et al., 1994). It is therefore clear that the transcription factor NGFI-A is well positioned to integrate activity-driven changes on the cell surface, with the genomic machinery that mediates a large variety of cellular processes that range from neuronal excitability and neurotransmitter release to long-term metabolic changes and neurite remodeling, as part of experience-dependent network rewiring (Pinaud, 2004, 2005; Pinaud and Tremere, 2005).

2.3. Arc

Unlike NGFI-A, Fos and Jun members, the protein product encoded by the IEG *arc* does not act as a transcription factor, but rather is a growth factor (Lyford et al., 1995). One of the most interesting features of *arc* is that the mRNA that results from its expression is rapidly translocated to dendrites and translated locally, in polyribosomal complexes often located at the base of dendritic spines, in an activity-dependent manner (Lyford et al., 1995; Steward et al., 1998; Steward and Worley, 2001a, b). These properties have spurred significant interest in this IEG, as it is well positioned to mediate fast, activity-driven modifications in synaptic configuration, especially as it relates to dendritic architecture (e.g., retraction and elongation). These types of modifications have been shown to occur in association with alterations in the patterns of sensory input, as in the case of lesions or an animal's exposure to enhanced sensory drive (Grutzendler et al., 2002; Nimchinsky et al., 2002; Trachtenberg et al., 2002; Pinaud, 2004; Guzowski et al., 2005).

3. IEG Regulation by Visual Deprivation and Stimulation

Below we will review the main findings of the effects of light deprivation and stimulation on IEG expression. Most of the studies on visual deprivation have focused on paradigms of either reversible or permanent light deprivation, including dark isolation, monocular and binocular enucleations, as well as retinal lesions. Subsequently, the impact of sensory stimulation on IEG expression in the visual cortex will be discussed.

Unlike *c-fos* and *c-jun*, that are expressed at relatively low basal levels in the rodent, cat and monkey V1, and *arc*, that is expressed at moderate baseline levels, the basal activity of the IEG NGFI-A has been repeatedly reported to be high in these experimental models as a result of ongoing neuronal activity (Worley et al., 1991; Beaver et al., 1993; Zhang et al., 1995; Kaplan et al., 1996; Kaczmarek and Chaudhuri, 1997; Pinaud et al., 2001). This expression property has provided the grounds for extensive use of NGFI-A in investigations of the effects of light deprivation throughout the mammalian visual system, while the low basal expression of *c-fos*, *c-jun*, *junB* and *arc* has largely prevented the use of these IEGs for this purpose.

3.1. Rodent

Pioneering experiments by Paul Worley and colleagues have demonstrated that NGFI-A mRNA and protein levels are dramatically decreased following intraocular infusion of tetrodotoxin (TTX) and dark adaptation in the rodent visual cortex, suggesting that the basal expression of this IEG is regulated by ongoing sensory input (Worley et al., 1991). These results were corroborated and expanded upon by subsequent work in the rat V1 using monocular eyelid suture, as well as intra-vitreal TTX infusion (Caleo et al., 1999). In these experiments, a marked suppression of NGFI-A basal levels was detected in both monocular and binocular V1, contralateral to the deprived eye. In addition, a minor decrease in NGFI-A expression in the ipsilateral supragranular layers of the binocular region of V1 was reported (Caleo et al., 1999).

While visual deprivation decreased basal NGFI-A expression levels, subsequent exposure of animals to light has been repeatedly shown to rapidly recover the normal, baseline levels of this IEG in the rodent visual cortex (Worley et al., 1991; Kaczmarek and Chaudhuri, 1997; Pinaud et al., 2000). Robust light-induced NGFI-A induction has been detected throughout all layers of the rodent V1 both in immature (during the critical period) as well as in adult animals (Worley et al., 1991; Nedivi et al., 1996; Kaczmarek and Chaudhuri, 1997; Pinaud et al., 2000) (Figs 2.1 and 2.2).

While visual deprivation slightly decreased *c-fos* expression in the rodent visual cortex (Yamada et al., 1999), visual input has been shown to significantly drive the expression of this IEG. For example, Correa-Lacarcel and colleagues (2000) have shown that white light stimulation, as well as various intensities of laser flickering, induces c-Fos in the rat V1. Interestingly, the distribution of c-Fos immunoreactive neurons was changed as a function of the stimulation paradigm: while continuous or high frequency flickering stimuli significantly increased the number of c-Fos immunolabeled neurons, stimulation of animals with middle and low frequency flickering laser light did not significantly alter c-Fos expression in V1 neurons of the rat. In all experimental groups, c-Fos immunopositive neurons were detected in cortical layers II/III, IV and VI, and to a lesser extent in layer V (Correa-Lacarcel et al., 2000). Montero (1995) has shown that exposure of animals to patterned visual stimuli (gratings and dots) triggered a significant increase in the number of c-Fos immunolabeled neurons, primarily in layers IV and VI, and to a lesser extent in the supragranular layers (Montero and Jian, 1995).

To the best of our knowledge, no systematic anatomical characterization of the distribution of *c-jun-*, *junB-* and *junD-* and *arc-*expressing cells, in dark-reared and light stimulated rodents has been conducted in V1. However, Kaminska and colleagues (1996) showed that dark-adaptation followed by light stimulation markedly increased the DNA-binding activity of both AP-1 and NGFI-A. It has also been reported that although FosB and JunD are the two primary components of the AP-1 in dark-reared rats, light stimulation for 2 hours recruits JunB, c-Jun and c-Fos to the formation of this transcription factor (Kaminska et al., 1996). Interestingly, longer stimulation periods (6-24 hours) led to a marked decrease in

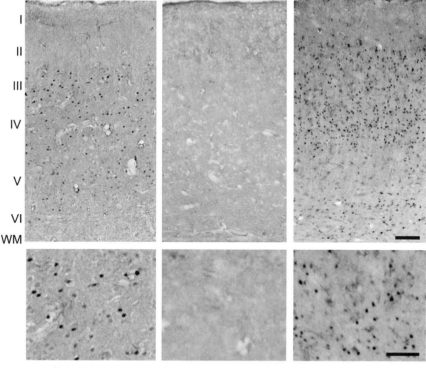

Basal Levels   Light-Deprived   Light-Stimulated

Figure 2.1. Immediate early gene regulation by light experience in the mammalian primary visual cortex (V1). The opossum (*Didelphis aurita*) embodies an ancestral pattern of organization of the central nervous system and, therefore, has been extensively used as a model to study the general principles of the visual system anatomical and functional organization. In the opossum, NGFI-A/Egr-1 is expressed at high basal levels (left column). Light-deprivation (24 hours; middle column) markedly decreases the number of NGFI-A/Egr-1 positive cells, while a 2 hour exposure of ambient light following light-deprivation (right column) elicits a marked upregulation of the protein encoded by this IEG in V1. Very similar results have been obtained with these protocols of light deprivation and stimulation in a number of other mammalian species, including rodents, cats and monkeys, indicating that the regulation of this IEG by light is evolutionarily conserved in the visual cortex of the mammalian lineage. I–VI—cortical layers 1 through 6; WM—white matter. Scale bars (in μm) = 100 (top) and 50 (bottom). Reprinted from Brain Res. Bull., 61, Pinaud et al., Light-induced Egr-1 expression in the striate cortex of the opossum, pp. 139-146, Copyright (2003), with permission from Elsevier.

c-Fos and c-Jun levels, while a sustained expression of JunD, JunB and FosB was detected, suggesting that AP-1 composition, and likely its activity and the identity of its target genes, is regulated by visual activity (Kaminska et al., 1996).

3.2. Cat

As observed in rodents, NGFI-A is expressed at high basal levels in the cat visual cortex, with highest expression detected in cortical layers II/III and VI

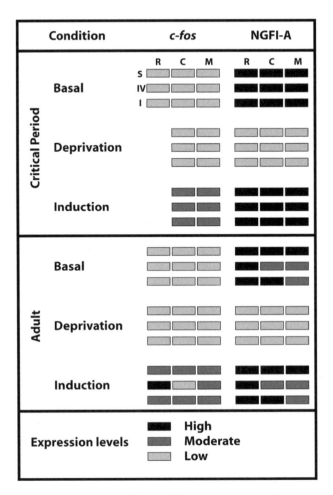

Figure 2.2. Table comparing c-fos and NGFI-A/zif268 expression under different sensory conditions in the rat, cat and monkey visual cortex. The information presented here was compiled from laminar expression profiles of the mRNA and protein products of both genes. In-situ hybridization or immunohistochemical staining levels were assigned one of three levels (low, moderate or high) based on the original histological data and descriptions, as cited in the text and detailed in Kaczmarek and Chaudhuri (1997). These three qualitative levels of staining are represented accordingly by the darkness of the pattern. Each cell in this figure depicts the staining intensity separately for the thalamo-recipient (IV), supragranular (S), and infragranular (I) layers of the rat (R), cat (C), and vervet monkey (M). In most instances of the original table (Kaczmarek and Chaudhuri, 1997), authors took layer VI to represent the infragranular layer because of the consistently poor staining that has been observed in layer V. The laminar expression profiles for all three species are shown for three sensory conditions—basal, visual deprivation and visual stimulation—in adult animals and during the critical period of development. Reproduced with the kind permission of Prof. L. Kaczmarek and reprinted from Brain Res. Rev., 23, L. Kaczmarek and A. Chaudhuri, Sensory regulation of immediate-early gene expression in mammalian visual cortex: implications for functional mapping and neural plasticity, pp. 237-256, Copyright (1997), with permission from Elsevier.

(Zhang et al., 1994; Kaplan et al., 1996). Likewise, protocols of visual deprivation also affected the expression of this IEG in V1 of the cat (Rosen et al., 1992; Zhang et al., 1995) (Fig. 2.2). For example, a marked downregulation of NGFI-A (zif268) mRNA levels in both supragranular and infragranular layers of V1 was described after unilateral sectioning of the optic tract (Zhang et al., 1995). Subsequent work showed that 1-week of dark-rearing significantly downregulated NGFI-A protein levels throughout all cortical layers of the cat V1 (Kaplan et al., 1996). These results differ from those obtained in the rodent V1, where deprivation of visual input has a greater impact on NGFI-A levels in cortical layer IV (Worley et al., 1991), likely reflecting the different cortical architecture of V1 between these two species, as well as dissimilar processing/computational strategies achieved by these two networks.

Kaplan and colleagues (1996) demonstrated that a brief exposure to light after dark-rearing triggers a robust induction of NGFI-A in an age-dependent, cortical layer-specific fashion. Adult cats underwent a substantial increase in NGFI-A protein levels in both supragranular and infragranular layers, with a small number of immunopositive neurons detected in the thalamorecipient layer IV. Conversely, young animals (5-weeks) underwent a marked induction of NGFI-A in all cortical layers (Kaplan et al., 1996) (Fig. 2.2).

c-fos mRNA basal levels were shown to be low and predominantly expressed in the cortical layer VI of the cat striate cortex (Zhang et al., 1994). It has also been reported that c-Fos expression is affected by light experience. Even though basal levels of this IEG are markedly low, dark-rearing has been shown to slightly reduce protein levels in both young and adult animals (Rosen et al., 1992; Kaplan et al., 1996). Similarly, unilateral optic tract sectioning decreased c-fos mRNA levels in layers II/III and V/VI in the cat V1 (Zhang et al., 1995) (Fig. 2.2). c-jun and junD basal expression levels have been shown to be primarily distributed through the supragranular layers and cortical layer VI in area 17 (Zhang et al., 1994). Interestingly, unlike what has been observed for c-fos and NGFI-A, optic tract lesioning was shown to increase c-jun levels, as detected by *in-situ* hybridization, in the infragranular layers of the striate cortex (Zhang et al., 1995). In addition, dark-rearing for 20-weeks, but not 5-weeks, affected the expression levels of *junB*, but not *c-jun*, in the cat V1 (Rosen et al., 1992). Dark-adaptation followed by light stimulation (1 hour) triggered a robust increase in both c-fos and junB (components of the AP-1) mRNA levels in the feline visual cortex (Rosen et al., 1992). This increase in c-fos and junB mRNA levels following light stimulation was shown to be transient, given that exposure of animals to light for two successive days led to a significant decrease in mRNA levels of both genes (but not *c-jun*) (Rosen et al., 1992).

Subsequent detailed anatomical characterization of the distribution of light-induced c-Fos expression revealed that two hours after stimulus onset, the highest density of c-Fos positive neurons was detected in the supragranular layers (II/III) as well as in cortical layer VI (Fig. 2.2). In addition, persistent visual stimulation (6 hours) revealed that c-Fos immunoreactivity levels returned to baseline (control) levels (Beaver et al., 1993). The transient nature of *c-fos* expression following sustained sensory drive has also been described for the IEG *junB* (Rosen et al., 1992); recall that the protein products of both IEGs are major components of the AP-1 transcription factor.

Finally, a systematic comparison between the levels of *c-fos* and NGFI-A during adulthood versus the critical period revealed that light-induced expression of both IEGs occurred at all cortical layers of young animals (5-weeks old), while the highest expression levels of both genes in adult animals were detected in supragranular and infragranular layers, with the lowest expression detected for cortical layer IV (Kaplan et al., 1996) (Fig. 2.2). These data suggest that the discrepant expression profiles of these IEGs during development, as compared to adulthood, may reveal differential activation patterns in V1, perhaps related to various plasticity states, across separate cortical layers (discussed below).

3.3. Monkey

High basal expression of the IEG NGFI-A has also been described for the V1 of both New- and Old-World primates (Chaudhuri and Cynader, 1993; Silveira et al., 1996; Okuno et al., 1997; Soares et al., 2005). Similar to data obtained from cats, the greatest density of NGFI-A positive neurons was detected in cortical layers II/III and VI, with a moderate number of immunolabeled cells in the thalamo-recipient layer IV (IVCb) and cortical layer V (Chaudhuri et al., 1995; Okuno et al., 1997) (Fig. 2.2).

Monocular deprivation, either by means of intraocular TTX infusion, enucleation or eyelid suture, was shown to decrease significantly NGFI-A basal levels in the ocular dominance columns (ODCs) associated with the deprived eye (Chaudhuri and Cynader, 1993; Silveira et al., 1996). Interestingly, the IEG downregulation that occurred as a function of visual deprivation was detected in a very small time window, as early as 1-5 hours after the interference with sensory input (Chaudhuri and Cynader, 1993; Chaudhuri et al., 1995; Silveira et al., 1996).

As discussed above, monocular deprivation in the adult markedly downregulated NGFI-A expression in the ODCs associated with the deprived eye. Interestingly, Silveira and colleagues (1996) have shown that this form of sensory deprivation did not reveal ODCs in 3-month old Cebus monkeys, while 6-month old animals exhibited scantily developed ODCs (Silveira et al., 1996). Conversely, Kaczmarek and colleagues (1999) were able to detect ODCs in the Vervet monkey as early as post-natal day (PD) 6, and robustly at PD 40 and 90 (Kaczmarek et al., 1999) (Fig. 2.2). Although a clear explanation for these seemingly discrepant findings has yet to be systematically revealed—some possibilities are discussed by Kaczmarek and colleagues (1999)—these findings suggest that ODCs, as revealed by NGFI-A expression, are functionally shaped late in the process of development of the visual cortex.

In the vervet monkey, light stimulation followed by dark-rearing triggered a significant increase in NGFI-A expression levels in ODCs associated with the stimulated eye, while minimal levels have been described for the ODCs associated with the deprived eye (Chaudhuri et al., 1995). In this work, NGFI-A immunolabeling appeared to be detected evenly across cortical layers (Chaudhuri et al., 1995) (Fig. 2.2).

The expression of both c-fos mRNA and protein levels has also been investigated in the Vervet monkey V1. It was found that the laminar profile of c-fos mRNA expression undergoes virtually no change with age: animals as young as 6-days of age exhibited high expression in supragranular layers (II/III), layer IVC and VI; this profile was also observed for adult monkeys (Kaczmarek et al., 1999). Immunocytochemistry directed at c-Fos protein yielded a remarkably similar laminar distribution, as compared to *in-situ* hybridization approaches directed at c-fos mRNA (Kaczmarek et al., 1999). Finally, similar findings were recently obtained in the New World monkey *Cebus apella* (Soares et al., 2005).

JunD expression has been assessed in the macaque monkey. Highest expression was detected in the superficial part of layer II and in layer VI, while moderate immunostaining was detected in the remainder of layer II and cortical layers III and IVA. Low JunD expression was detected in layers IVB, IVC and V (Okuno et al., 1997). To the best of our knowledge, *c-jun*, *junB* and *arc* expression have not been assessed in the primate V1.

4. Plasticity in the Visual Cortex and IEG Expression

Little doubt remains that IEG expression depends on sensory input in the mammalian visual cortex. It is also clear that neuronal activity is necessary, but not sufficient, to drive IEG expression in the CNS (Sharp et al., 1993; Arckens et al., 2000; Pinaud, 2004). For example, visual drive reliably triggers activity of LGN neurons, yet, in a number of experimental models, no NGFI-A expression has been detected after visual stimulation in these thalamic neurons in a number of mammalian species (Arckens et al., 2000; Pinaud et al., 2003b; Soares et al., 2005). Given that IEG expression likely relies on specific activity patterns and qualities, substantial interest in the field has been shifting from their use as activity-markers towards studies of their specific roles in cellular physiology. As discussed in a number of chapters in this volume, the proteins encoded by IEGs play a wide variety of roles in the physiology of neurons and neural networks, including memory formation, cell death and trophic regulation, as a few examples. In addition, a group of IEGs, including some discussed in this chapter, appear to play key roles in the induction and/or maintenance of plasticity in neurons. Given that the expression of these IEGs is strongly correlated with neuronal activity patterns that have been associated with plasticity in the CNS, recent studies have focused on these "candidate-plasticity genes" as tools to map circuits undergoing experience-dependent reorganization, as well as the specific roles of these proteins in these processes.

One of the most commonly employed experimental paradigms that reliably triggers plasticity-associated modifications in cortical circuitry is the exposure of animals to an enriched environment (EE). This experimental protocol fundamentally involves exposing freely-behaving animals to a complex sensory setting (related to visual processing: higher contrast, stimuli frequency, depth, colors) and comparing gene expression patterns obtained from these animals with those obtained from control groups, which are often composed of freely-ranging animals that have been exposed to impoverished visual environments (Rosenzweig et al., 1972; van Praag et al., 2000; Pinaud, 2004; Pinaud and Tremere, 2005).

Exposure of animals to an EE triggers dramatic changes in CNS architecture and physiology (reviewed in Rosenzweig et al., 1972; van Praag et al., 2000; Pinaud, 2004; Pinaud and Tremere, 2005). For example, in the visual cortex exposure of animals to an EE significantly increases the dendritic arborization of primary visual cortical neurons and overall neuronal density, as well as mean synaptic disc diameter and synapse-to-neuron ratio, both of which are direct correlates of enhanced synaptic transmission (Volkmar and Greenough, 1972; Bhide and Bedi, 1984, 1985; Beaulieu and Colonnier, 1987, 1988; van Praag et al., 2000). These morphological correlates of visual experience have been shown to translate into enhanced physiological responses in V1 neurons. For example, EE animals exhibit a higher percentage of orientation-selective neurons when compared to animals raised in impoverished visual conditions. Moreover, sharper orientation tuning and increased responsivity to light were reported for EE animals (Beaulieu and Cynader, 1990a, b). Finally, contrast sensitivity and acuity were reported to be enhanced in enriched animals (Beaulieu and Cynader, 1990a, b; Prusky et al., 2000). These findings clearly indicate that exposure of animals to an EE triggers significant anatomical and functional plastic changes in visual cortical circuitry.

In an effort to separate the effects of activity-driven gene expression from those genetic mechanisms associated with the induction of plasticity in the visual cortex, we and others have investigated candidate-plasticity gene expression patterns that result from ambient light-stimulation paradigms, and compared them with those obtained by exposing animals to complex visual environments (Pinaud et al., 2001; Pinaud et al., 2002a). Our working hypothesis was that if IEG expression is associated with the induction and/or regulation of plastic changes in V1, then EE animals will exhibit a different gene expression profile as compared with those animals that experienced simple patterns of visual stimulation (and presumably did not undergo experience-dependent changes in cortical circuitry).

Our questions regarding the relationship of IEG expression to a putative plasticity-inducing set of stimuli were addressed in the following experiment. A population of young adult rats was divided into three experimental groups that were housed in typical home cages of two to three animals per cage. The first experimental group was named the enriched environment group (EE) because each day, at the same hour, these animals were placed into the enriched environment complex. Following 1 hr in the EE complex, animals were returned to their standard housing. The full duration of the experiment was 21 days, so each animal had 21 exposures to the EE. As one control group, we maintained a group of animals in housing conditions that were identical to the home cages used for the EE animals. These animals were simply maintained in parallel with the experimental group with no interference from an investigator; this group was named the undisturbed (UD) control group. We anticipated that the daily handling of EE animals to and from the EE could influence gene expression and cause an overestimation of the EE effect on IEG expression. To capture what proportion of IEG expression could be attributed to this movement of animals and any stress associated with going into and out from the experimental setting we created a second control group called the handling only (HO) group. These subjects were housed in conditions that were identical to both the EE and UD groups. The HO animals were handled at the same time that the EE group underwent relocation to the enriched setting. Unlike EE animals, HO animals were immediately placed back into standard home cages (Pinaud et al., 2001, 2002a; Pinaud, 2004).

In these experiments, we focused our analyses on the expression of the IEGs NGFI-A and *arc*. As briefly described above, both of these IEGs are well positioned to mediate activity-dependent plastic changes in neurons, including those that result from EE exposure. For example, target genes that are regulated by the transcription factor NGFI-A include part of the neurotransmitter release machinery, such as synapsin I, synapsin II and synaptobrevin, as well as structural genes such as neurofilament (Pospelov et al., 1994; Thiel et al., 1994; Petersohn et al., 1995; Petersohn and Thiel, 1996). The protein encoded by the IEG *arc*, on the other hand, is translated in an activity-dependent manner in polyribossomal complexes located in the post-synaptic membrane, often at the base of dendritic spines (Lyford et al., 1995; Steward et al., 1998; Steward and Worley, 2001b). This attribute of *arc* expression potentially places this protein in a good position to orchestrate dendritic reconfiguration as a function of activity (Pinaud et al., 2001; Pinaud, 2004). Finally, the expression of both genes depend on calcium influx associated with NMDA receptor activation, a subtype of glutamatergic receptors that has been repeatedly implicated in paradigms of plasticity and enhanced synaptic efficacy (reviewed in Collingridge and Lester, 1989; Hollmann and Heinemann, 1994; Mayer and Armstrong, 2004).

Both NGFI-A mRNA and protein levels were assessed in V1 of EE, HO and UD animals. As reported previously by other groups, NGFI-A mRNA was detected at high basal levels in UD animals (Wallace et al., 1995; Kaczmarek and Chaudhuri, 1997; Pinaud et al., 2002a). Even though high mRNA levels were detected in all cortical layers except for layer I, the highest signal was observed in cortical layers III and V (discussed below) (Pinaud et al., 2002a). Optical density analysis revealed that the mRNA levels detected in HO controls were not significantly different than those levels observed in UD controls. Conversely, EE animals underwent a significant upregulation of NGFI-A mRNA levels in all cortical layers of V1, when compared to HO and UD controls (Fig. 2.3) (Pinaud et al., 2002a).

Immunocytochemistry directed at NGFI-A protein also showed that exposure to a complex visual environment triggered a marked upregulation of this IEG in all cortical layers of V1, with the exception of layer I, as compared to the protein levels observed in HO and UD animals (Pinaud et al., 2002a) (Fig. 2.4). Similar to our findings obtained with NGFI-A mRNA distribution, the highest increase in

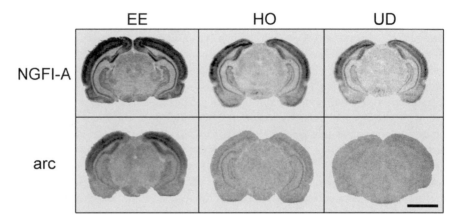

Figure 2.3. Experience-dependent expression of immediate early genes NGFI-A and *arc* in the rodent visual cortex. In-situ hybridization autoradiograms depicting NGFI-A (top row) and arc (bottom row) expression in animals that were exposed to a complex visual environment (EE) for 1 hour/day for a total of 21 days, as compared to the expression levels of animals that were manipulated (HO) or left undisturbed in their home cages (UD). Both NGFI-A and arc are markedly upregulated by exposure to the EE condition, as compared to both control groups. Expression levels for both IEGs are not different across both control groups. Highest induction in response to the EE for both IEGs was detected in cortical layers III and V. Scale bar = 500 μm. NGFI-A, nerve growth factor-induced gene A; arc, activity-regulated cytoskeletal gene; EE, enriched environment group; HO, handled-only group; UD, undisturbed group.

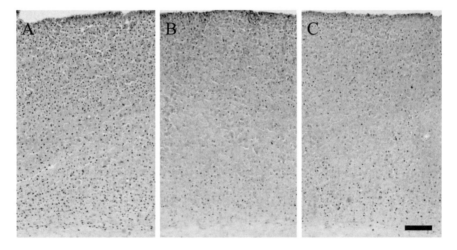

Figure 2.4. NGFI-A protein levels are increased in all cortical layers of the primary visual cortex in rats exposed to an enriched visual environment. Compared to undisturbed (UD) and handled-only (HO) animals, animals exposed to an enriched environment (EE) exhibited an increased number of immunopositive nuclei in cortical layers II/III and VI, while a more modest increase was found in layer IV (A). Both HO (B) and UD (C) controls displayed basal levels of NGFI-A immunoreactivity in all cortical layers, layer IV being the cortical layer with fewest NGFI-A positive nuclei. No NGFI-A positive cells were detected in layer I for all experimental groups. Scale bar = 200 μm.

the number of NGFI-A immunopositive neurons was detected in cortical layers III and V, while the smallest increase in NGFI-A expression was detected in cortical layer IV (discussed below).

Together these findings suggest that it was the complexity of the visual environment—and likely the neuronal activity patterns associated with this experience—rather than simply the light stimulation, that triggered a differential regulation of the IEG NGFI-A in the rodent V1. The underlying notion behind this interpretation considers that both UD and HO groups were fully awake, displayed similar activity levels and were maintained in identical luminance environments, as compared to EE animals and, therefore, were also visually stimulated. Nevertheless, in both HO and UD groups, NGFI-A mRNA and protein levels were substantially different from those obtained in the V1 of EE animals.

In contrast to the findings obtained with NGFI-A, arc mRNA levels were found at moderately low basal levels in the rodent V1 (Pinaud et al., 2001) (Fig. 2.3). Quantitative analysis of arc mRNA levels in HO and UD controls revealed that these groups were not significantly different from each other. Conversely, arc levels in EE animals were markedly increased when compared to both control groups (Pinaud et al., 2001). This upregulation was detected in all cortical layers (Fig. 2.3). Interestingly, and similar to the findings obtained with NGFI-A, the highest arc mRNA levels were detected in cortical layers III and V, with the smallest increases observed in the thalamo-recipient layer IV (discussed below). Together, these findings provided direct evidence for a reliable and significant upregulation of both IEGs following EE exposure. In addition these data suggested that NGFI-A and arc may be part of the machinery involved in the early inducible genomic response associated with experience-dependent plastic changes in the rodent visual cortex, rather than simply echoing neuronal activity.

The significant increase in IEG expression in the V1 of EE animals may therefore be associated with increased levels, or specific patterns, of neuronal activity that result from visual experiences in a complex environment. One interesting possibility for the recruitment of plasticity-associated gene expression programs in V1 is that mechanisms for detecting sub-optimal architecture and processing capabilities are in place, and were activated, in this cortical region following EE exposure. During experience in the EE setting, detection of sensory or information overload may trigger these programs to initiate a potential optimization of cortical architecture and synaptic weights, in order to appropriately process information contained in this new, rich sensory environment. Experimental testing of this hypothesis may shed light on the genetic mechanisms associated with the regulation of plasticity in the CNS.

Alternatively, novelty may play a role in the differential regulation of IEG expression following EE exposure. In fact, other IEGs such as *c-fos* and *c-jun* have been previously demonstrated to be positively regulated by novelty in a number of forebrain areas, including the somatosensory and visual cortices (Papa et al., 1993; Zhu et al., 1995). Interestingly, noradrenergic input, which has been proposed to gate attentional processing as it relates to arousal and behavioral responsivity to novelty, has been shown to be required for the maintenance of basal expression of NGFI-A in the rodent forebrain (Cirelli et al., 1996), as well as for light-induced expression of this IEG in the rat V1 (Pinaud et al., 2000). However, in the EE experiments conducted by our group, novelty is not a probable player in the regulation of IEG expression given that expression levels were assessed in animals that were familiarized with the EE for three weeks (Pinaud et al., 2001, 2002a).

One of the most noticeable effects in our preparations was that the highest expression of both NGFI-A and *arc* was detected in cortical layers III and V, and the smallest increases in gene expression were consistently detected in the thalamo-recipient layer IV of EE animals (Pinaud et al., 2001, 2002a). These findings are intriguing and may be related to electrophysiological results suggesting that cortical layers II/III and V are potentially more plastic, as indicated by the ease of triggering experience-dependent alterations in their functional properties and synaptic strength, as has been observed in long-term potentiation (LTP) and long-term depression (LTD) (Daw et al., 1992; Darian-Smith and Gilbert, 1994; Glazewski and Fox, 1996; Petersen and Sakmann, 2001). On the contrary, these activity-driven changes in synaptic strength were more difficult to trigger in layer IV neurons, suggesting that the thalamo-recipient layer is less plastic than other cortical strata. The potentially low plasticity observed in layer IV may be associated with the stability in cortical map representation, as well as the preservation of fidelity of information transfer at this level (Pinaud, 2004, 2005). Although systematic studies are required to support or refute this hypothesis, should it be supported, the low candidate-plasticity gene expression detected in layer IV following EE exposure may provide a read-out of a decrease in activity for the machinery involved in triggering or regulating experience-dependent network reorganization in V1. Similarly, enhanced expression levels in layers III and V following EE exposure may provide a direct visualization of neurons in V1 that are participating in a neural plastic response triggered by exposure to an enriched visual environment.

References

Angel P, Karin M (1991). The role of Jun, Fos and the AP-1 complex in cell-proliferation and transformation. Biochim Biophys Acta 1072:129-157.

Arckens L (2005). The molecular biology of sensory map plasticity in adult mammals. In: Plasticity in the Visual system: From Genes to Circuits (Pinaud R, Tremere LA, De Weerd P, eds), pp. 181-203. New York: Springer-Verlag.

Arckens L, Van Der Gucht E, Eysel UT, Orban GA, Vandesande F (2000). Investigation of cortical reorganization in area 17 and nine extrastriate visual areas through the detection of changes in immediate early gene expression as induced by retinal lesions. J Comp Neurol 425:531-544.

Beaulieu C, Colonnier M (1987). Effect of the richness of the environment on the cat visual cortex. J Comp Neurol 266:478-494.

Beaulieu C, Colonnier M (1988). Richness of environment affects the number of contacts formed by boutons containing flat vesicles but does not alter the number of these boutons per neuron. J Comp Neurol 274:347-356.

Beaulieu C, Cynader M (1990a). Effect of the richness of the environment on neurons in cat visual cortex. II. Spatial and temporal frequency characteristics. Brain Res Dev Brain Res 53:82-88.

Beaulieu C, Cynader M (1990b). Effect of the richness of the environment on neurons in cat visual cortex. I. Receptive field properties. Brain Res Dev Brain Res 53:71-81.

Beaver CJ, Mitchell DE, Robertson HA (1993). Immunohistochemical study of the pattern of rapid expression of C-Fos protein in the visual cortex of dark-reared kittens following initial exposure to light. J Comp Neurol 333:469-484.

Bhide PG, Bedi KS (1984). The effects of a lengthy period of environmental diversity on well-fed and previously undernourished rats. II. Synapse-to-neuron ratios. J Comp Neurol 227:305-310.

Bhide PG, Bedi KS (1985). The effects of a 30 day period of environmental diversity on well-fed and previously undernourished rats: neuronal and synaptic measures in the visual cortex (area 17). J Comp Neurol 236:121-126.

Caleo M, Lodovichi C, Pizzorusso T, Maffei L (1999). Expression of the transcription factor Zif268 in the visual cortex of monocularly deprived rats: effects of nerve growth factor. Neuroscience 91:1017-1026.

Carrasco-Serrano C, Viniegra S, Ballesta JJ, Criado M (2000). Phorbol ester activation of the neuronal nicotinic acetylcholine receptor alpha7 subunit gene: involvement of transcription factor Egr-1. J Neurochem 74:932-939.

Chaudhuri A, Cynader MS (1993). Activity-dependent expression of the transcription factor Zif268 reveals ocular dominance columns in monkey visual cortex. Brain Res 605:349-353.

Chaudhuri A, Matsubara JA, Cynader MS (1995). Neuronal activity in primate visual cortex assessed by immunostaining for the transcription factor Zif268. Vis Neurosci 12:35-50.

Cirelli C, Pompeiano M, Tononi G (1996). Neuronal gene expression in the waking state: a role for the locus coeruleus. Science 274:1211-1215.

Collingridge GL, Lester RA (1989). Excitatory amino acid receptors in the vertebrate central nervous system. Pharmacol Rev 41:143-210.

Correa-Lacarcel J, Pujante MJ, Terol FF, Almenar-Garcia V, Puchades-Orts A, Ballesta JJ, Lloret J, Robles JA, Sanchez-del-Campo F (2000). Stimulus frequency affects c-fos expression in the rat visual system. J Chem Neuroanat 18:135-146.

Darian-Smith C, Gilbert CD (1994). Axonal sprouting accompanies functional reorganization in adult cat striate cortex. Nature 368:737-740.

Daw NW, Fox K, Sato H, Czepita D (1992). Critical period for monocular deprivation in the cat visual cortex. J Neurophysiol 67:197-202.

Evarts EV (1968). A technique for recording activity of subcortical neurons in moving animals. Electroencephalogr Clin Neurophysiol 24:83-86.

Fanselow EE, Sameshima K, Baccala LA, Nicolelis MA (2001). Thalamic bursting in rats during different awake behavioral states. Proc Natl Acad Sci USA 98:15330-15335.

Finkbeiner S, Greenberg ME (1998). Ca^{2+} channel-regulated neuronal gene expression. J Neurobiol 37:171-189.

Gail A, Brinksmeyer HJ, Eckhorn R (2003). Simultaneous mapping of binocular and monocular receptive fields in awake monkeys for calibrating eye alignment in a dichoptical setup. J Neurosci Methods 126:41-56.

Gail A, Brinksmeyer HJ, Eckhorn R (2004). Perception-related modulations of local field potential power and coherence in primary visual cortex of awake monkey during binocular rivalry. Cereb Cortex 14:300-313.

Gattass R, Nascimento-Silva S, Soares JG, Lima B, Jansen AK, Diogo AC, Farias MF, Botelho MM, Mariani OS, Azzi J, Fiorani M (2005). Cortical visual areas in monkeys: location, topography, connections, columns, plasticity and cortical dynamics. Philos Trans R Soc Lond B Biol Sci 360:709-731.

Ghosh A, Ginty DD, Bading H, Greenberg ME (1994). Calcium regulation of gene expression in neuronal cells. J Neurobiol 25:294-303.

Ginty DD, Bading H, Greenberg ME (1992). Trans-synaptic regulation of gene expression. Curr Opin Neurobiol 2:312-316.

Glazewski S, Fox K (1996). Time course of experience-dependent synaptic potentiation and depression in barrel cortex of adolescent rats. J Neurophysiol 75:1714-1729.

Grutzendler J, Kasthuri N, Gan WB (2002). Long-term dendritic spine stability in the adult cortex. Nature 420:812-816.

Guzowski JF, Timlin JA, Roysam B, McNaughton BL, Worley PF, Barnes CA (2005). Mapping behaviorally relevant neural circuits with immediate-early gene expression. Curr Opin Neurobiol 15:599-606.

Herdegen T, Leah JD (1998). Inducible and constitutive transcription factors in the mammalian nervous system: control of gene expression by Jun, Fos and Krox, and CREB/ATF proteins. Brain Res Brain Res Rev 28:370-490.

Hess J, Angel P, Schorpp-Kistner M (2004). AP-1 subunits: quarrel and harmony among siblings. J Cell Sci 117:5965-5973.

Hollmann M, Heinemann S (1994). Cloned glutamate receptors. Annu Rev Neurosci 17:31-108.

Horton JC (1984). Cytochrome oxidase patches: a new cytoarchitectonic feature of monkey visual cortex. Philos Trans R Soc Lond B Biol Sci 304:199-253.

Horton JC, Hedley-Whyte ET (1984). Mapping of cytochrome oxidase patches and ocular dominance columns in human visual cortex. Philos Trans R Soc Lond B Biol Sci 304:255-272.

Hubel DH, Wiesel TN (1962). Receptive fields, binocular interaction and functional architecture in the cat's visual cortex. J Physiol 160:106-154.

Hubel DH, Wiesel TN (1968). Receptive fields and functional architecture of monkey striate cortex. J Physiol 195:215-243.

Hubel DH, Wiesel TN (1970). The period of susceptibility to the physiological effects of unilateral eye closure in kittens. J Physiol 206:419-436.

Hubel DH, Wiesel TN (1972). Laminar and columnar distribution of geniculo-cortical fibers in the macaque monkey. J Comp Neurol 146:421-450.

Hubel DH, Wiesel TN, LeVay S (1977). Plasticity of ocular dominance columns in monkey striate cortex. Philos Trans R Soc Lond B Biol Sci 278:377-409.

Hughes P, Dragunow M (1995). Induction of immediate-early genes and the control of neurotransmitter-regulated gene expression within the nervous system. Pharmacol Rev 47:133-178.

Kaczmarek L, Chaudhuri A (1997). Sensory regulation of immediate-early gene expression in mammalian visual cortex: implications for functional mapping and neural plasticity. Brain Res Brain Res Rev 23:237-256.

Kaczmarek L, Robertson HA (2002). Immediate Early Genes and Inducible Transcription Factors in Mapping of the Central Nervous System Function and Dysfunction. Amsterdam: Elsevier Science B.V.

Kaczmarek L, Zangenehpour S, Chaudhuri A (1999). Sensory regulation of immediate-early genes c-fos and zif268 in monkey visual cortex at birth and throughout the critical period. Cereb Cortex 9:179-187.

Kaminska B, Kaczmarek L, Chaudhuri A (1996). Visual stimulation regulates the expression of transcription factors and modulates the composition of AP-1 in visual cortex. J Neurosci 16:3968-3978.

Kaminska B, Pyrzynska B, Ciechomska I, Wisniewska M (2000). Modulation of the composition of AP-1 complex and its impact on transcriptional activity. Acta Neurobiol Exp (Wars) 60:395-402.

Kaplan IV, Guo Y, Mower GD (1996). Immediate early gene expression in cat visual cortex during and after the critical period: differences between EGR-1 and Fos proteins. Brain Res Mol Brain Res 36:12-22.

Kobierski LA, Chu HM, Tan Y, Comb MJ (1991). cAMP-dependent regulation of proenkephalin by JunD and JunB: positive and negative effects of AP-1 proteins. Proc Natl Acad Sci USA 88:10222-10226.

Krubitzer LA, Kaas JH (1990). Cortical connections of MT in four species of primates: areal, modular, and retinotopic patterns. Vis Neurosci 5:165-204.

Lyford GL, Yamagata K, Kaufmann WE, Barnes CA, Sanders LK, Copeland NG, Gilbert DJ, Jenkins NA, Lanahan AA, Worley PF (1995). Arc, a growth factor and activity-regulated gene, encodes a novel cytoskeleton-associated protein that is enriched in neuronal dendrites. Neuron 14:433-445.

Mayer ML, Armstrong N (2004). Structure and function of glutamate receptor ion channels. Annu Rev Physiol 66:161-181.

Milbrandt J (1987). A nerve growth factor-induced gene encodes a possible transcriptional regulatory factor. Science 238:797-799.

Miyashita Y, Kameyama M, Hasegawa I, Fukushima T (1998). Consolidation of visual associative long-term memory in the temporal cortex of primates. Neurobiol Learn Mem 70:197-211.

Montero V (2005). Attentional activation of cortico-reticulo-thalamic pathways revealed by Fos imaging. In: Plasticity in the Visual System: From Genes to Circuits (Pinaud R, Tremere LA, De Weerd P, eds), pp. 97-124. New York: Springer-Verlag.

Montero VM, Jian S (1995). Induction of c-fos protein by patterned visual stimulation in central visual pathways of the rat. Brain Res 690:189-199.

Morgan JI, Curran T (1991). Stimulus-transcription coupling in the nervous system: involvement of the inducible proto-oncogenes fos and jun. Annu Rev Neurosci 14:421-451.

Nedivi E, Fieldust S, Theill LE, Hevron D (1996). A set of genes expressed in response to light in the adult cerebral cortex and regulated during development. Proc Natl Acad Sci USA 93:2048-2053.

Nicolelis MA, Lin RC, Woodward DJ, Chapin JK (1993). Dynamic and distributed properties of many-neuron ensembles in the ventral posterior medial thalamus of awake rats. Proc Natl Acad Sci USA 90:2212-2216.

Nicolelis MA, Dimitrov D, Carmena JM, Crist R, Lehew G, Kralik JD, Wise SP (2003). Chronic, multisite, multielectrode recordings in macaque monkeys. Proc Natl Acad Sci USA 100:11041-11046.

Nimchinsky EA, Sabatini BL, Svoboda K (2002). Structure and function of dendritic spines. Annu Rev Physiol 64:313-353.

Okuno H, Miyashita Y (1996) Expression of the transcription factor Zif268 in the temporal cortex of monkeys during visual paired associate learning. Eur J Neurosci 8:2118-2128.

Okuno H, Kanou S, Tokuyama W, Li YX, Miyashita Y (1997). Layer-specific differential regulation of transcription factors Zif268 and Jun-D in visual cortex V1 and V2 of macaque monkeys. Neuroscience 81:653-666.

Papa M, Pellicano MP, Welzl H, Sadile AG (1993). Distributed changes in c-Fos and c-Jun immunore-activity in the rat brain associated with arousal and habituation to novelty. Brain Res Bull 32:509-515.

Petersen CC, Sakmann B (2001). Functionally independent columns of rat somatosensory barrel cortex revealed with voltage-sensitive dye imaging. J Neurosci 21:8435-8446.

Petersohn D, Thiel G (1996). Role of zinc-finger proteins Sp1 and zif268/egr-1 in transcriptional regulation of the human synaptobrevin II gene. Eur J Biochem 239:827-834.

Petersohn D, Schoch S, Brinkmann DR, Thiel G (1995). The human synapsin II gene promoter. Possible role for the transcription factor zif268/egr-1, polyoma enhancer activator 3, and AP2. J Biol Chem 270:24361-24369.

Pinaud R (2004). Experience-dependent immediate early gene expression in the adult central nervous system: evidence from enriched-environment studies. Int J Neurosci 114:321-333.

Pinaud R (2005). Critical calcium-regulated biochemical and gene expression programs involved in experience-dependent plasticity. In: Plasticity in the Visual System: From Genes to Circuits (Pinaud R, Tremere LA, De Weerd P, eds), pp. 153-180. New York: Springer-Verlag.

Pinaud R, Tremere LA (2005). Experience-dependent rewiring of retinal circuitry: involvement of immediate early genes. In: Plasticity in the Visual System: From Genes to Circuits (Pinaud R, Tremere LA, De Weerd P, eds), pp. 79-95. New York: Spinger-Verlag.

Pinaud R, Tremere LA, Penner MR (2000). Light-induced zif268 expression is dependent on noradrenergic input in rat visual cortex. Brain Res 882:251-255.

Pinaud R, Penner MR, Robertson HA, Currie RW (2001). Upregulation of the immediate early gene arc in the brains of rats exposed to environmental enrichment: implications for molecular plasticity. Brain Res Mol Brain Res 91:50-56.

Pinaud R, Tremere LA, Penner MR, Hess FF, Robertson HA, Currie RW (2002a). Complexity of sensory environment drives the expression of candidate-plasticity gene, nerve growth factor induced-A. Neuroscience 112:573-582.

Pinaud R, De Weerd P, Currie RW, Fiorani Jr M, Hess FF, Tremere LA (2003a). Ngfi-a immunoreactiv-ity in the primate retina: implications for genetic regulation of plasticity. Int J Neurosci 113:1275-1285.

Pinaud R, Tremere LA, Penner MR, Hess FF, Barnes S, Robertson HA, Currie RW (2002b). Plasticity-driven gene expression in the rat retina. Brain Res Mol Brain Res 98:93-101.

Pinaud R, Vargas CD, Ribeiro S, Monteiro MV, Tremere LA, Vianney P, Delgado P, Mello CV, Rocha-Miranda CE, Volchan E (2003b). Light-induced Egr-1 expression in the striate cortex of the opossum. Brain Res Bull 61:139-146.

Pospelov VA, Pospelova TV, Julien JP (1994). AP-1 and Krox-24 transcription factors activate the neurofilament light gene promoter in P19 embryonal carcinoma cells. Cell Growth Differ 5:187-196.

Prusky GT, Reidel C, Douglas RM (2000). Environmental enrichment from birth enhances visual acuity but not place learning in mice. Behav Brain Res 114:11-15.

Rosen KM, McCormack MA, Villa-Komaroff L, Mower GD (1992). Brief visual experience induces immediate early gene expression in the cat visual cortex. Proc Natl Acad Sci USA 89:5437-5441.

Rosenzweig MR, Bennett EL, Diamond MC (1972). Brain changes in relation to experience. Sci Am 226:22-29.

Sharp FR, Sagar SM, Swanson RA (1993). Metabolic mapping with cellular resolution: c-fos vs. 2-deoxyglucose. Crit Rev Neurobiol 7:205-228.

Silveira LC, de Matos FM, Pontes-Arruda A, Picanco-Diniz CW, Muniz JA (1996). Late development of Zif268 ocular dominance columns in primary visual cortex of primates. Brain Res 732:237-241.

Soares JG, Pereira AC, Botelho EP, Pereira SS, Fiorani M, Gattass R (2005). Differential expression of Zif268 and c-Fos in the primary visual cortex and lateral geniculate nucleus of normal Cebus monkeys and after monocular lesions. J Comp Neurol 482:166-175.

Sokoloff L (1981). Localization of functional activity in the central nervous system by measurement of glucose utilization with radioactive deoxyglucose. J Cereb Blood Flow Metab 1:7-36.

Steward O, Worley PF (2001a). A cellular mechanism for targeting newly synthesized mRNAs to synaptic sites on dendrites. Proc Natl Acad Sci USA 98:7062-7068.

Steward O, Worley PF (2001b). Selective targeting of newly synthesized Arc mRNA to active synapses requires NMDA receptor activation. Neuron 30:227-240.

Steward O, Wallace CS, Lyford GL, Worley PF (1998). Synaptic activation causes the mRNA for the IEG Arc to localize selectively near activated postsynaptic sites on dendrites. Neuron 21:741-751.

Thiel G, Schoch S, Petersohn D (1994). Regulation of synapsin I gene expression by the zinc finger transcription factor zif268/egr-1. J Biol Chem 269:15294-15301.

Tieman SB (1985). The anatomy of geniculocortical connections in monocularly deprived cats. Cell Mol Neurobiol 5:35-45.

Trachtenberg JT, Chen BE, Knott GW, Feng G, Sanes JR, Welker E, Svoboda K (2002). Long-term in vivo imaging of experience-dependent synaptic plasticity in adult cortex. Nature 420:788-794.

van Praag H, Kempermann G, Gage FH (2000). Neural consequences of environmental enrichment. Nat Rev Neurosci 1:191-198.

Volkmar FR, Greenough WT (1972). Rearing complexity affects branching of dendrites in the visual cortex of the rat. Science 176:1145-1147.

Wallace CS, Withers GS, Weiler IJ, George JM, Clayton DF, Greenough WT (1995). Correspondence between sites of NGFI-A induction and sites of morphological plasticity following exposure to environmental complexity. Brain Res Mol Brain Res 32:211-220.

Wiesel TN, Hubel DH (1963). Single-Cell Responses in Striate Cortex of Kittens Deprived of Vision in One Eye. J Neurophysiol 26:1003-1017.

Wong WK, Ou XM, Chen K, Shih JC (2002). Activation of human monoamine oxidase B gene expression by a protein kinase C MAPK signal transduction pathway involves c-Jun and Egr-1. J Biol Chem 277:22222-22230.

Wong-Riley MT (1989). Cytochrome oxidase: an endogenous metabolic marker for neuronal activity. Trends Neurosci 12:94-101.

Worley PF, Christy BA, Nakabeppu Y, Bhat RV, Cole AJ, Baraban JM (1991). Constitutive expression of zif268 in neocortex is regulated by synaptic activity. Proc Natl Acad Sci USA 88:5106-5110.

Wurtz RH (1969). Visual receptive fields of striate cortex neurons in awake monkeys. J Neurophysiol 32:727-742.

Yamada Y, Hada Y, Imamura K, Mataga N, Watanabe Y, Yamamoto M (1999). Differential expression of immediate-early genes, c-fos and zif268, in the visual cortex of young rats: effects of a noradrenergic neurotoxin on their expression. Neuroscience 92:473-484.

Zhang F, Vanduffel W, Schiffmann SN, Mailleux P, Arckens L, Vandesande F, Orban GA, Vanderhaeghen JJ (1995). Decrease of zif-268 and c-fos and increase of c-jun mRNA in the cat areas 17, 18 and 19 following complete visual deafferentation. Eur J Neurosci 7:1292-1296.

Zhang F, Halleux P, Arckens L, Vanduffel W, Van Bree L, Mailleux P, Vandesande F, Orban GA, Vanderhaeghen JJ (1994). Distribution of immediate early gene zif-268, c-fos, c-jun and jun-D mRNAs in the adult cat with special references to brain region related to vision. Neurosci Lett 176:137-141.

Zhu XO, Brown MW, McCabe BJ, Aggleton JP (1995). Effects of the novelty or familiarity of visual stimuli on the expression of the immediate early gene c-fos in rat brain. Neuroscience 69:821-829.

3

Immediate Early Gene Regulation in the Auditory System

CLAUDIO V. MELLO [a] and RAPHAEL PINAUD [b]

[a]Neurological Sciences Institute, Oregon Health & Science University, Beaverton, OR, USA
[b]Department of Neurobiology, Duke University Medical Center, Durham, NC, USA

1. Introduction

Immediate early genes (IEGs) are genes whose expression in the brain is sensitive to the activation state of neuronal cells (see other chapters in this book and the reviews in Kaczmarek and Robertson, 2002). Although the exact relationship between neuronal activation and IEG expression is not entirely understood, the analysis of induced expression of IEGs has been extremely useful in the identification and study of brain regions activated by specific sensory stimuli or behavioral conditions. This is the case because IEG studies are non-invasive, allowing the mapping of brain activation without interfering with the animal's ability to behave or respond to the stimulus being presented. In addition, although IEG expression studies generally lack temporal resolution (but see the chapter by Guzowski in this volume), this type of analysis allows for the mapping of global patterns of activation with cellular resolution. Furthermore, some IEGs have been linked to neuronal plasticity (Guzowski et al., 1999; Jones et al., 2001) and their activation may indicate sites where experience-dependent changes take place in the brain. The present chapter discusses the use of IEG expression analysis to study brain regions and pathways involved in the processing of auditory stimuli in vertebrates (for previous reviews on related topics, see Chaudhuri, 1997; Clayton, 2000; Chaudhuri and Zangenehpour, 2002; Mello, 2002a, 2004; Mello et al., 2004).

2. IEGs are Induced by Auditory Stimulation

When songbirds hear song, the IEG *zenk*—the songbird homologue of *egr-1* (a.k.a. *zif-268*, *NGFI-A* and *krox-24*)—is induced in discrete areas of the songbird brain, most markedly in the caudomedial telencephalon (Mello et al., 1992;

R. Pinaud, L.A. Tremere (Eds.), Immediate Early Genes in Sensory Processing, Cognitive Performance and Neurological Disorders, 35–56, ©2006 Springer Science + Business Media, LLC

Mello and Clayton, 1994). *zenk* encodes a zinc-finger transcriptional regulator linked to neuronal plasticity (Milbrandt, 1987; Lemaire et al., 1988; Sukhatme et al., 1988; Christy and Nathans, 1989; Jones et al., 2001). The phenomenon of song-induced *zenk* expression represents one of the first and clearest examples of IEG induction by a specific and naturally occurring sensory stimulus of behavioral relevance, rather than by direct electrical stimulation of the brain or by administration of convulsivant agents. Tract-tracing and electrophysiological studies have determined that the caudomedial telencephalic areas of songbirds that show a marked *zenk* induction after song stimulation are high-order regions of the avian central auditory pathways, with important roles in song perceptual processing and potentially in the memorization of song (Mello, 2004; Mello et al., 2004). Other studies have shown that stimulation with various types of auditory stimuli induces *zenk* and other IEGs in parts of the ascending auditory pathway and in auditory telencephalic regions of non-oscine avian species, frogs and mammals (see below). Altogether, such studies have demonstrated the utility of using IEGs for the identification and functional analysis of areas involved in auditory perceptual processing, memorization and discrimination. Below we first briefly review the general organization of auditory pathways in vertebrates. We then discuss how the analysis of IEG induction by auditory stimulation has influenced our understanding of the anatomical and functional organization of the auditory system. Stronger emphasis is given to the paradigms that have been the focal points of study in the field, e.g. analysis of expression of *zenk* in the songbird telencephalon and of *c-fos*, another inducible IEG regulated by auditory input, in the rodent brain. Notice that throughout the chapter we use the recently revised avian brain nomenclature (Reiner et al., 2004).

3. Auditory Pathways in Vertebrates

Several aspects of the overall organization of the ascending auditory pathways are conserved among vertebrates (Butler and Hodos, 1996) and are represented schematically in Fig. 3.1. In general, auditory information gains access to the brain through the auditory branch of the VIII cranial nerve, and ascends through a series of pontine and midbrain nuclei before reaching the auditory thalamus. The nuclei that constitute this ascending pathway include cochlear, olivary, lemniscal and midbrain nuclei. Species-specific specializations along this pathway reflect adaptations present in some specific animal groups, such as nuclei involved in sound localization in the barn owl (Knudsen, 1999, 2002; Knudsen et al., 2000) and in sonar navigation in bats (Harnischfeger et al., 1985; Suga, 1989). In all species investigated, robust projections have been described from the auditory midbrain (the central nucleus of the inferior colliculus in mammals, the nucleus mesencephaliculs lateralis pars dorsalis, or MLd, in birds) to the auditory thalamus (medial geniculate nucleus in mammals and nucleus ovoidalis, or Ov, in birds). Projections originating in the auditory thalamus then reach thalamo-recipient zones in the telencephalon (layer IV, or granular cell layer, in the auditory cortex of mammals, and field L2 in birds). The auditory telen-cephalic (pallial) areas have different organizations in different animal classes,

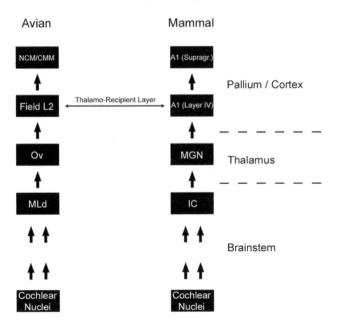

Figure 3.1. Basic organization of the avian and mammalian ascending auditory pathway. The box diagrams illustrate the main stations of the ascending auditory pathway in birds (left column) and mammals (right column). Both diagrams are highly schematic and represent the presence of multiple parallel projections at the level of the brainstem converging onto the midbrain, one main thalamic station, a telencephalic pallial/cortical thalamo-recipient layer, and intra-telencephalic/cortical connections. Abbreviations: MLd, dorsal part of the lateral mesencephalic nucleus; Ov, nucleus Ovoidalis; NCM, caudomedial nidopallium; CMM, caudomedial mesopallium; IC, inferior colliculus; MGN, medial geniculate nucleus; A1, primary auditory cortex.

e.g., cortico-laminar in mammals and nuclear in birds and reptiles (Jarvis et al., 2005). In general, however, such areas are thought to play important roles in the perceptual processing and discrimination of complex auditory stimuli such as vocal communication signals, as well as in associative learning involving auditory stimuli (see below).

4. IEG Induction in Early Stations of the Auditory Pathway

This level of the auditory system has been most studied in the rat. It has been reported that direct electrical stimulation of the cochlea induces the expression of the IEGs *c-fos* and, more recently, *egr-1* (*zenk*), as assessed by immunocytochemistry (ICC), in several nuclei of the ascending auditory pathway, namely the anteroventral and posteroventral cochlear nuclei, the lateral superior olive, the medial dorsal nucleus of the trapezoid body, the dorsal and ventral nuclei of the lateral lemniscus, and the central nucleus of the inferior colliculus (Vischer et al., 1994, 1995; Zhang et al., 1996; Saito et al., 1999, 2000; Illing and Michler, 2001; Illing et al., 2002). IEG expression in these nuclei shows some topography according to the cochlear area stimulated. Such studies indicate that nuclei in the ascending auditory pathway are capable of exhibiting an IEG response upon the activation of this pathway via electrical stimulation of the cochlea. Because the kind and intensity of stimulation used in several of these studies may lead to actual changes in the physiological properties of the auditory pathway, it is possible that the ensuing IEG activation is related to such changes (Illing et al., 2002). Overall, these IEG induction studies have been useful in the context of evaluating the effect of cochlear implant devices.

With regards to IEG induction in nuclei of the auditory pathway following actual auditory stimulation, the results have been more controversial. The early studies that initiated this field of research indicated that auditory stimulation with pure tones induces c-Fos protein expression in a tonotopic fashion in the dorsal cochlear nucleus and in the inferior colliculus of the mouse (Ehret and Fischer, 1991; Brown and Liu, 1995), rat (Friauf, 1992; Rouiller et al., 1992; Friauf, 1995) and cat (Adams, 1995). However, the results of these and subsequent studies using ICC to reveal IEG protein expression have been quite variable in terms of the effective stimuli and stimulation conditions, as well as whether other auditory brainstem nuclei and cell groups exhibit an IEG response upon auditory stimulation. For instance, c-Fos expression in the cochlear nuclei of rats is most pronounced if the auditory stimulus used is novel, and declines in parallel with adaptation to repeated stimulation (Keilmann and Herdegen, 1996; Kandiel et al., 1999). In other cases, the expression of an IEG like c-Fos is highly modulated after manipulations that result in plastic alterations of the auditory circuitry, such as unilateral tympanotomy (Hillman et al., 1997), partial deafening by unilateral cochlear lesions in rats (Riera-Sala et al., 2001) or stimulation with tinnitus-inducing noise in the gerbil (Wallhausser-Franke et al., 2003). Surprisingly, after such manipulations, c-Fos expression can increase in some, and decrease in other, nuclei of the auditory pathway.

To a large extent this issue has been clarified in more systematic studies that indicate that *in-situ* hybridization for mRNA detection is a more reliable method to reveal IEG induction and modulation in auditory brainstem nuclei than ICC (Luo et al., 1999; Saint Marie et al., 1999). These studies have revealed that a *c-fos* mRNA induction response occurs in a tonotopic fashion to the simple presentation of tones in all cochlear nuclei, the medial nucleus of the trapezoid body, the lateral superior olive, the dorsal nucleus of the lateral lemniscus, and the inferior colliculus of the rat, and that this response is also present without an evident tonotopy in other lemniscal nuclei (Luo et al., 1999; Saint Marie et al., 1999). The fact that a c-Fos response is apparently weaker or absent in many of these nuclei, as assessed by ICC, indicates important differences in sensitivity and/or specificity between methods or, alternatively, that *c-fos* mRNA undergoes post-transcription regulation in these nuclei. Importantly, the tonotopic organization revealed by *c-fos* mapping along the auditory pathway, regardless of the detection method, is critically dependent upon stimulus intensity. Both the number and spatial distribution of labeled cells change with increasing intensity, such that at high intensities a clear tonotopic organization of expression of the *c-fos* gene is disrupted (Saint Marie et al., 1999). Another important issue is that relatively high basal levels of c-Fos protein have sometimes been reported in the auditory pathways even in the absence of auditory stimulation (Cody et al., 1996). This expression is likely to relate, at least in some cases, to the slow time course of c-Fos protein induction and inactivation as compared to changes in mRNA levels, potentially masking the response to subsequent stimulation. Because of the fast kinetics of mRNA synthesis and degradation, this problem is much less severe when using *in-situ* hybridization methods (Saint Marie et al., 1999).

In songbirds, preliminary evidence using ICC initially suggested that the expression of ZENK protein might be induced in cochlear nuclei upon presentation of conspecific song stimuli (Mello, C.V., pers. comm.), but the signal was very weak compared to the strong activation observed in other brain areas (see below). Subsequent experiments utilizing *in-situ* hybridization analysis for *zenk* gene expression have not detected significant levels of *zenk* mRNA in cochlear nuclei (Mello, C.V., pers. comm.), suggesting that IEG induction does not occur at this brain level in birds. Alternatively, only stimuli that have particular behavioral relevance are effective, or only IEGs other than *zenk* are inducible at this level of the auditory pathway, but further investigation is needed to settle the issue.

5. IEG Induction in the Auditory Midbrain and Diencephalon

As opposed to the preceding nuclei of the auditory pathway, there is clear evidence that IEGs are induced by auditory stimuli in the midbrain of all species examined to date, including frogs, birds, bats, mice, rats, and cat. Significant induction of the *c-fos* and/or *zenk* genes has been observed for various stimuli, ranging from simple tones to behaviorally relevant stimuli such as species-specific vocalizations. To the extent that IEG induction is indicative of plasticity related events, these observations suggest that the midbrain is a site within the auditory pathway where experience-dependent changes are highly prevalent.

In the tungara frog, *egr-1* (*zenk*) expression in the auditory midbrain has been studied in detail in males, in the context of the response to mate calls (Hoke et al., 2004). This study has revealed different activation patterns in the different subdivisions that constitute the torus semicircularis, the amphibian equivalent of the auditory part of the inferior colliculus. However, the results also indicate that simple acoustic features are apparently not represented in any one of these subdivisions. Moreover, analysis of the expression patterns of all the subdivisions of the torus indicate that, considered as a group, these regions have a higher ability to discriminate among acoustic stimuli than any single subdivision alone. These findings are consistent with an important role of the auditory midbrain in the recognition and discrimination of species–specific vocalizations, in particular those associated with reproductive behavior. Preliminary evidence indicates that *egr-1* is also expressed in the torus semicircularis (as well as in other brain areas) of teleosts (Burmeister and Fernald, 2005), but whether this expression is triggered by auditory stimulation has not yet been determined.

In songbirds (canaries, zebra finches), auditory stimulation results in strong *zenk* induction in the MLd, the avian equivalent of the central nucleus of the inferior colliculus of mammals (Mello and Clayton, 1994; Mello and Ribeiro,

1998). This effect has also been observed with species-specific vocalizations in the other avian vocal learners, i.e. parrots and hummingbirds (Jarvis and Mello, 2000; Jarvis et al., 2000), as well as in doves, a vocal non-learner avian group (Terpstra et al., 2005), but the stimulus specificity of this response has not yet been investigated at this brain level.

The mapping of c-Fos expression in the inferior colliculus (IC) after auditory stimulation has consistently revealed a robust response in different mammalian species, including rat, mouse, bat, and cat (Ehret and Fischer, 1991; Friauf, 1992; Qian and Jen, 1994; Adams, 1995). Studied with tonal stimuli, this response has a clear tonotopic organization, visualized as continuous isofrequency bands of labeled neurons that extend across the central nucleus and parts of the dorsal cortex. Some studies above have also suggested that multiple and separate tonal representations resulting from intracollicular projections originating in the central nucleus of the IC occur in more lateral and medial parts of the IC, but this remains to be confirmed electrophysiologically. In the big brown bat, binaural stimulation leads to significant expression of c-Fos at most if not all levels of the ascending auditory pathway (Qian and Jen, 1994; Qian et al., 1996). Interestingly, monaural stimulation results in an increase in the number of resulting c-Fos labeled cells as compared with binaural stimulation, a result that can be attributed to the removal of an inhibitory influence over neurons that receive bilateral inputs (Qian and Jen, 1994). While this effect occurs in both subcortical and cortical structures, the increase is seen in contralateral nuclei of the ascending pathway up to the IC, and in ipsilateral nuclei from the IC upwards, indicating the existence of a predominantly inhibitory crossed influence and a particular role of the IC in interhemispheric coordination. Another fact worth mentioning with regards to the IC is that a significant modulation of c-Fos expression is seen in this area in mice primed for audiogenic seizures by early exposure to high intensity noise, suggesting an active involvement of the IC in the priming of this seizure modality (Kai and Niki, 2002; for further discussion of this issue, see also Chaudhuri and Zangenehpour, 2002).

In contrast to the midbrain, IEG induction by auditory stimulation is not obvious in the thalamus. In songbirds, *zenk* induction in Ov, the avian homologue of the medial geniculate of mammals, is conspicuously absent upon auditory stimulation with conspecific song (Mello and Clayton, 1994; Mello and Ribeiro, 1998). Given that Ov is an obligatory station in the avian ascending auditory pathway (Karten, 1967, 1968; Brauth et al., 1987; Vates et al., 1996), the lack of *zenk* induction suggests that the expression of IEGs has been uncoupled from neuronal activation in this nucleus. A possible functional significance for this phenomenon is that the thalamus might serve primarily as a relay of auditory information from the midbrain to the telencephalon, rather than being a site of active neuronal plasticity. Whether the lack of an induction response upon activation of the auditory thalamus is a general characteristic of different classes of IEGs or whether it occurs in different species remains to be determined.

In the hypothalamus, significant modulation in *zenk* expression has been observed in specific hypothalamic nuclei (the suprachiasmatic nucleus, the posterior tuberculus and the lateral hypothalamus) of the tungara frog in the context of exposure to behaviorally relevant stimuli, i.e. conspecific vocal communication signals, as compared to non-salient stimuli or expression levels in unstimulated controls (Hoke et al., 2005). Interestingly, *zenk* mRNA levels in these nuclei correlate with the expression levels in distinct auditory midbrain and thalamic nuclei, suggesting possible functional relationships among such nuclei and/or potential pathways through which the ascending auditory system could modulate autonomous and endocrine centers in the hypothalamus. Such organization, if correct, would suggest that the influence of the auditory system on hypothalamic function can bypass processing centers higher up in the auditory pathway. Indeed, anatomical evidence for such projections has been described in some avian species (Durand et al., 1992; Cheng and Zuo, 1994).

6. IEG Induction in the Avian Telencephalon

In birds, the marked induction of *zenk* mRNA (Fig. 3.2A, B) and protein (Fig. 3.2C–F) that occurs upon auditory stimulation with song localizes primarily to discrete pallial areas of the telencephalon, namely the caudomedial nidopallium (NCM) and caudomedial mesopallium (CMM) (Mello and Clayton, 1994). Together with the primary thalamo-recipient zone field L2, these areas comprise a caudomedial auditory lobule thought to be involved in various aspects of auditory processing and discrimination (Mello, 2004). In addition to NCM and CMM, the act of hearing song induces *zenk* in the more laterally located field L subfields L1 and L3 (Mello and Clayton, 1994; Mello and Ribeiro, 1998). *zenk* induction by auditory stimulation in the areas above appears to be a general characteristic of the avian brain, as it has been observed in representative species of all vocal learning avian orders (songbirds such as canaries and finches, budgerigars— a representative parrot species, and hummingbirds), as well as in doves, a vocal non-learner group (Mello et al., 1992; Mello and Clayton, 1994; Jarvis and Mello, 2000; Jarvis et al., 2000; Mello, 2002a; Terpstra et al., 2005). In songbirds, *zenk* induction also occurs in the shelf and cup regions adjacent to nuclei HVC and robustus arcopallialis, or RA (Mello and Clayton, 1994; Mello and Ribeiro, 1998), the latter representing nuclei of the direct motor pathway of the songbird's song control system (Nottebohm et al., 1976, 1982). Areas equivalent to the shelf and cup regions are also present and show *zenk* induction in response to song stimulation in budgerigars and hummingbirds (Jarvis and Mello, 2000; Jarvis et al., 2000). Equivalent areas, respectively the dorsal nidopallium, or Nd, and the ventromedial nucleus of the intermediate arcopallium, or Aivm, have been described in vocal non-learners such as chicken and pigeon, (Wild et al., 1993; Metzger et al., 1998), but IEG induction in these areas in response to auditory stimulation has not been investigated in detail in these birds.

Tract-tracing studies in zebra finches have revealed that the telencephalic areas showing *zenk* induction upon song auditory stimulation altogether comprise a pathway whose connectivity bears considerable resemblance to that of the auditory cortex of mammals, including a thalamo-recipient zone, intratelencephalic

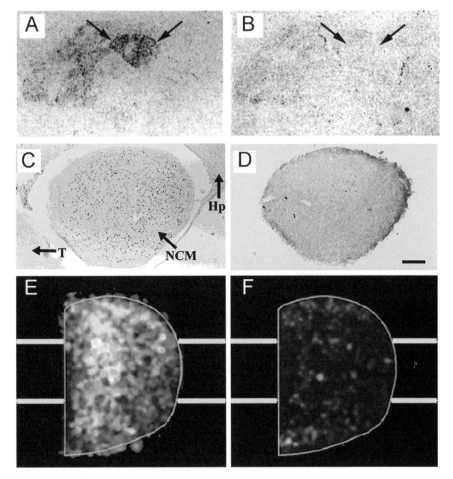

Song-Stimulated   Silence

Figure 3.2. Induction of zenk mRNA and protein levels in the songbird forebrain following exposure to song. (A) *In-situ* hybridization autoradiogram of a section of an adult male zebra finch exposed to 45 minutes to recorded conspecific song. (B) Unstimulated control. These sections correspond to the parasagittal plane 250 μm lateral to the medial surface of the brain. (C) ZENK expression, revealed by immunocytochemistry, in the caudomedial telencephalon. Notice the presence of numerous ZENK-labeled nuclei after song stimulation in the caudomedial nidopallium (NCM) of a zebra finch from the hearing-only group. (D) Unstimulated control. Scale bar = 500 μm. (E) Map of ZENK expression in the NCM of a canary resulting from presentation of a playback of conspecific songs. (F) Unstimulated control. Adapted from Mello et al. (1992), Mello and Ribeiro (1998) and Ribeiro et al. (1998). For color details see Ribeiro et al. (1998).

projections, and areas that originate descending projections upon thalamic and midbrain nuclei of the ascending auditory pathway (Vates et al., 1996; Mello et al., 1998). In addition to *zenk*, induced expression upon auditory stimulation with conspecific song has also been demonstrated in some or all of the above areas for the IEGs *c-fos*, *c-jun* and *arc* (Nastiuk et al., 1994; Bolhuis et al., 2000; Bailey et al., 2002; Velho et al., 2005), the former two encoding transcription factors (Sonnenberg et al., 1989) and the latter encoding an early effector gene linked to synaptic plasticity (Guzowski et al., 2000) and thought to act at the level of recently activated synapses (Lyford et al., 1995; Steward et al., 1998; Steward and Worley, 2002). These different IEGs show a remarkable degree of cellular co-localization following song stimulation, suggesting their participation in a coordinated program of gene expression in song-activated neuronal cells (Velho et al., 2005).

The mapping data obtained with IEG expression in songbirds is consistent with previous electrophysiological studies demonstrating that field L plays an important role in auditory processing in birds. In songbird species such as starlings, selective responses of field L to complex stimuli such as vocalizations have long been known and studied (Muller and Leppelsack, 1985). More recently, interest has been rekindled in the study of the role of field L in the processing of conspecific song stimuli (Gehr et al., 1999, 2000; Sen et al., 2001; Grace et al., 2003; Hsu et al., 2004). The *zenk* induction studies, however, have drawn particular attention to areas that are direct projection targets of field L, namely NCM and both the medial and lateral subdivisions of the CM, respectively CMM and CLM (Vates et al., 1996). From an anatomical standpoint, these areas may be considered analogous to supragranular layers of the auditory cortex. Electrophysiological recordings in various songbird species have confirmed that these areas show responses to song stimulation, but also that their responses differ significantly from those recorded in field L, being typically less phasic and showing higher selectivity for complex stimuli (Chew et al., 1995, 1996; Stripling et al., 1997; Gentner and Margoliash, 2003; Hsu et al., 2004; Terleph et al., 2005). Most interestingly, these caudomedial areas show an experience-dependent, song-specific, decrease (habituation) of their electrophysiological responses (Chew et al., 1995; Stripling et al., 1997). In addition, the long-term maintenance of song-specific habituation has been shown to depend on novel gene expression in song-responding neurons in NCM (Chew et al., 1995). This phenomenon has been suggested as a possible contributing mechanism for the memorization and perceptual discrimination of songs. The other major telencephalic field L target, CM, has also received considerable attention recently. Its lateral portion, or CLM, is a major candidate for providing auditory input to the song control system (Vates et al., 1996). In starlings, behavioral training in an associative paradigm results in increased selectivity for the trained songs in song-responsive units within CM, indicating that experience-dependent plasticity can have a significant impact on the response properties of this area (Gentner and Margoliash, 2003). In sum, the available evidence indicates that the auditory areas revealed by *zenk* expression play an important role in the perceptual processing and potentially in the memorization of auditory stimuli that are of behavioral relevance.

Considerable effort has been dedicated to characterizing the properties of *zenk* induction in NCM. This interest was initially triggered because *zenk* induction in this area was found to be highest for conspecific song and lower for heterospecific or non-song auditory stimuli (Mello et al., 1992). This finding indicated that neuronal circuits and/or units in NCM are tuned to acoustic features present in conspecific songs. In fact, a detailed study in the canary has shown that the *zenk* expression patterns are tuned to acoustic features present in individual component syllables of the song (Ribeiro et al., 1998). That study was particularly successful due to the simple acoustic structure of many of the individual syllables that constitute the canary song. Such syllables have a tonal quality, resembling pure tones of various frequencies. Using a quantitative mapping analysis (Cecchi et al., 1999), rostral NCM was found to be organized in a frequency dependent fashion, with high frequencies mapping ventrally and low frequencies mapping dorsally, in a gradient fashion (Fig. 3.3A). However, this map is not strictly tonotopic, as it is not clearly revealed when using artificial tones that presumably lack some of the subtle but characteristic features of natural syllables such as frequency and amplitude modulations (Fig. 3.3B, C). That study also revealed that complex syllables containing multiple frequencies, produced in succession or simultaneously, result in complex patterns that cannot be explained as the linear summation of the maps resulting from individual frequencies. Overall, the analysis of *zenk* maps in the canary NCM indicate that the activation patterns of this brain area contains sufficient information for the bird to discriminate and classify the various syllables present in the song, suggesting this area as a possible site for a syllabic auditory representation (Ribeiro et al., 1998). These findings illustrate well how the use of IEG expression analysis to generate global maps of brain activation in awake behaving animals can contribute important insights to our understanding of the representation of complex auditory stimuli.

The studies discussed above indicate that NCM is essentially an auditory processing area. Interestingly, however, *zenk* expression analysis has revealed that the activation of neuronal circuits in NCM can be modulated by the context in which the auditory stimulus occurs. For instance, the induced expression of *zenk* in NCM can be modified by coupling the song to another modality in an associative learning paradigm (e.g., a shock avoidance task in canaries), which results in a change of the behavioral relevance of the song (Jarvis et al., 1995). Similarly, modulation of induced ZENK protein expression has been observed in zebra finches when coupling the auditory stimulus (song or calls) with a behaviorally relevant visual stimulus such as seeing another conspecific individual (Avey et al., 2005; Vignal et al., 2005). A comparable modulation can also be seen with manipulations in the sound source and intensity and by coupling the song with light stimuli (Kruse et al., 2004), or when comparing the response to auditory stimulation in unrestrained vs. restrained birds (Park and Clayton, 2002), but not by simultaneous presentation of background noise (Vignal et al., 2004). Another important factor that can exert a modulatory effect on *zenk* expression levels in various species is the quality and behavioral relevance of the song stimulus (Gentner et al., 2001; Eda-Fujiwara et al., 2003; Maney et al., 2003; Phillmore et al., 2003). Overall, such effects are consistent with the possibility that modulatory neurotransmitters known to mediate the effects of alertness and arousal, such as norepinephrine, also exert a modulatory effect on the

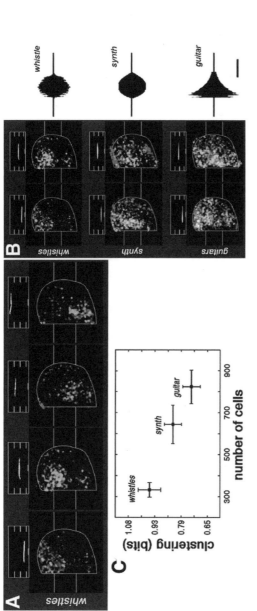

Figure 3.3. Syllabic auditory representation in the canary brain. (A) Panels show maps of ZENK expression in NCM resulting from presentation of whistles of increasing frequencies; the corresponding sonograms (frequency versus time plots) are shown above each map. Scale bars represent 100 msec. (B) Tonotopy in NCM depends on natural features of the stimulus. ZENK expression maps resulting from natural whistles (top), synthetic whistles (middle) and guitar notes (bottom) are compared. As one moves away from natural whistles toward artificial stimuli of corresponding frequencies, ZENK patterns become less clustered both in spatial distribution and range of labeling intensity. Stimulus amplitude envelopes are indicated on the right column. (C) Clustering decreases and number of ZENK-labeled cells per section increases progressively from natural whistles to synthetic whistles and to guitar notes. Plotted are averages of the clustering indices of the whistle maps, synthetic and guitar stimuli presented in (B). Clustering in bits; error bars represent SEM. Adapted from Ribeiro et al. (1998). For color details see Ribeiro et al. (1998).

auditory processing circuits in NCM and on their ensuing *zenk* response. Indeed, preliminary evidence indicates that the *zenk* response in NCM is modulated by noradrenergic transmission (Ribeiro, 1999; Ribeiro and Mello, 2000), as has also been observed for stimulation-driven IEG expression in the mammalian brain (Cirelli et al., 1996; Pinaud et al., 2000). It is also worth mentioning here that GABAergic interneurons have recently been shown to be a major component of song responsive circuits in NCM and CMM (Pinaud et al., 2004), suggesting that GABA neurotransmission plays a major role in song auditory processing and possibly in the ensuing IEG response to song in these areas.

Auditory experience has been shown to deeply affect the *zenk* induction response to song in NCM. In zebra finches, for example, *zenk* expression levels are greatest in response to novel song, decreasing as a song becomes familiar in consequence of repeated presentations (Mello et al., 1995), as also observed for electrophysiological responses in NCM (habituation). Similarly, a marked effect of recent auditory experience has been described in starlings, in this case measured as the number of cells positive for ZENK protein (Sockman et al., 2002, 2005). The effect of auditory experience has also been studied in the context of song learning. For example, ZENK protein expression in the NCM of adult male zebra finches in response to stimulation with tutor song (the song to which the birds were exposed to as juveniles during the song learning period) correlates positively with the extent to which the birds copied the tutor song (Bolhuis et al., 2000, 2001; Terpstra et al., 2004). This observation suggests that NCM may be a storage site for perceptual memories of the songs heard during the song learning period. Interestingly, the *zenk* response to song does not occur at early ages posthatch, but rather develops in concert with the onset of the song learning period (about 4 weeks posthatch), indicating a clear effect of age on the maturation of the *zenk* response (Jin and Clayton, 1997; Stripling et al., 2001). Other developmental studies have revealed potential interactions among factors such as sex, age and auditory experience as early modulators of the ZENK and c-Fos protein responses to song in the NCM and CMM of zebra finches and other avian species (Bailey and Wade, 2003; Phillmore et al., 2003; Hernandez and MacDougall-Shackleton, 2004; Bailey and Wade, 2005).

The perceptual processing of song and the formation of long-lasting song auditory memories are requisites for two important aspects of the life history of songbirds: (1) the acquisition of a song auditory template to be used for vocal learning, and (2) the individual recognition that is used in territorial defense and mate selection (Catchpole and Slater, 1995; Kroodsma and Miller, 1996). The results discussed in the previous paragraphs suggest that the auditory processing and potential memorization of songs that take place in NCM could play an important role both in song perception and in vocal learning. One of the next challenges is to determine whether and how the perceptual processing in NCM influences the physiology of the brain nuclei directly involved in song production and learning. It will also be important to investigate the relationship between auditory processing and non-vocal aspects of the behavioral response to song.

In contrast to the marked expression of *zenk* and other IEGs in the telen-cephalic areas above, a conspicuous characteristic of the IEG response to auditory stimulation in birds is the lack of induction in the primary thalamo-recipient zone, field L2 (Mello and Clayton, 1994; Mello and Ribeiro, 1998; Velho et al., 2005). This area represents the main entry site of auditory information into the telencephalon (Karten, 1968; Vates et al., 1996), and its neuronal cells are therefore necessarily activated during song auditory stimulation, yet no IEGs (including *zenk*, *c-fos*, *c-jun* and *arc*) are induced in this area in songbirds or any other avian group examined to date, including parrots, hummingbirds and doves (Jarvis and Mello, 2000; Jarvis et al., 2000; Terpstra et al., 2005), even after administration of a strong depolarizing agent like the convulsivant metrazole (Mello and Clayton, 1995). Therefore, it seems that IEG induction has been uncoupled from neuronal activation in L2. A possible functional significance for such uncoupling is that L2 might not undergo active experience-dependent plasticity, but rather serve primarily as a relay of auditory inputs to higher order areas in the telencephalon. There are several possible mechanisms through which this uncoupling could be achieved, ranging from downregulation of components in the signaling pathway required for IEG induction (e.g. NMDA receptors, kinase signaling components or constitutive transcription factors such as CREB) to more direct transcriptional regulation of IEG promoters through modulation of the levels of DNA methylation or histone acetylation.

Similarly to what occurs in field L2, song auditory stimulation does not induce the expression of any known song-inducible IEG in nuclei of the brain system that regulates song behavior (Mello and Clayton, 1994; Jarvis and Nottebohm, 1997; Mello and Ribeiro, 1998; Velho et al., 2005). This includes both the nuclei of the direct motor pathway for vocal-motor control and the anterior forebrain pathway involved in song learning. Also similarly to L2, this finding is some-what surprising because evoked electrophysiological responses to song auditory stimulation have been recorded within song control nuclei (see discussions on this topic in (Mello, 2002a, b; Mello et al., 2004)). Interestingly, however, song evoked responses within the song system are highly dependent on the stimulus, showing an overall preference for stimulation with the bird's own song, and on the bird's state, being highly modulated by sleep, arousal and neuromodulatory transmitters (Margoliash, 1986; Doupe and Konishi, 1991; Dave et al., 1998; Schmidt and Konishi, 1998; Nick and Konishi, 2001, 2005a, b; Cardin and Schmidt, 2003, 2004; Shea and Margoliash, 2003). The fact that song-evoked responses tend to be much less pronounced during alertness raises concern as to their relevance to song auditory perception. Although further investigation is needed to clarify this issue, the results obtained with IEG expression mapping are consistent with, at most, a limited role of the song control nuclei in the perceptual processing of birdsong.

7. IEG Induction in the Mammalian Auditory Cortex

In comparison with the brainstem auditory pathways in mammals, or with the auditory telencephalon in birds, there have been fewer studies of IEG induction in auditory cortical structures. Some of the first such studies were performed in gerbils, where c-Fos protein was shown to be induced in a tonotopic fashion in the primary auditory cortex by few (single or triple) presentations of narrow band frequency modulated tone bursts (Scheich and Zuschratter, 1995; Zuschratter et al., 1995). The resulting pattern consisted of a narrow band of c-Fos expressing cells extending through cortical layers II-VI and generally according to known tonotopic maps defined by electrophysiology or with the metabolic marker 2-deoxy-glucose, albeit with a higher (cellular) resolution than the latter. Such a pattern was not observed in the adjacent anterior and caudal auditory fields, where labeling was sparse and not frequency-dependent. Interestingly, with increasing stimulus repetition, the resulting expression pattern of c-Fos spread across cortical auditory fields, its distribution becoming less tonotopic in both the primary auditory cortex and adjacent fields. A similar spreading of c-Fos expression was obtained with increasing amounts of stimulation using conditioning (aversive) tones (Scheich and Zuschratter, 1995; Scheich et al., 1997). These results gave an indication that both spectral features as well as the behavioral relevance of the stimulus, as indicated by the acquisition of an auditory conditioned response, can determine the patterns of mammalian cortical activation revealed by the expression of an IEG.

More recent studies performed in rats have attempted to use gene expression (c-Fos) mapping to tease apart the contributions of different cortical fields to auditory processing *per se* versus the influence of behavioral or contextual value of the auditory stimulus. In one case, analysis was performed in animals trained on an auditory discrimination paradigm requiring a response based on a previous association with a food reward (Carretta et al., 1999). The results revealed higher c-Fos expression in parts of the auditory cortex, particularly the secondary field, of trained animals as compared with animals receiving similar stimulation without training or with unstimulated animals, whereas expression in other auditory brain areas did not differ between trained and stimulated groups. In another study, c-Fos expression was analyzed in various cortical fields in response to different batteries of novel and familiar sound stimuli of various spectral and temporal characteristics (Wan et al., 2001). A small but significant enhancement was observed in the auditory association cortex when comparing novel over familiar stimuli, but not in the inferior colliculus or primary auditory cortex. Sakata and colleagues examined the effects of tonal stimuli with varying behavioral relevance in the context of performance of an auditory discrimination task (using visual discrimination in the same animals as an internal control). The results demonstrate that an enhancement of c-Fos expression occurs in the primary auditory cortex when the auditory stimuli are presented during the discrimination task as compared to a purely perceptual context. Furthermore, this enhancement appears to involve a population of excitatory but not inhibitory neurons, suggesting a possible cellular target for contextual modulation of auditory cortical activation (Sakata et al., 2002).

Another set of studies has made use of different natural vocalizations in mice as stimuli. One study performed with adult females investigated c-Fos expression resulting from presentation of synthetic ultrasonic vocalizations from pups (Fichtel and Ehret, 1999). By comparing two contexts that differ in the behavioral relevance of the calls (using experienced mothers vs. virgins that, respectively, respond or not to such calls) this study revealed a differential activation of both the ultrasonic field and of the secondary auditory cortex (respectively lower and higher in the mothers), but not of the primary or anterior cortical fields. In addition, whereas the patterns elicited in primary auditory cortex in the virgin females—i.e., a context where the stimulus lacked behavioral relevance—were diffuse, without much evidence of a tonotopic distribution, they were well ordered in a frequency-dependent fashion in the mothers. These observations suggest that the activation of cortical auditory fields in a stimulus-specific pattern only occurs in the context of behaviorally relevant stimulation (e.g., stimuli that trigger a significant behavioral response).

A subsequent study analyzed natural wriggling calls of pups, which also trigger maternal behaviors in mice, as compared to similar but modified calls that lack the behavioral relevance of the natural calls (Geissler and Ehret, 2004). Analysis of the resulting c-Fos expression patterns provided evidence for a representation in the primary auditory cortex that reflects the spectral composition of the stimuli in terms of component frequencies. However, evidence for a differential cortical activation that reflects the behavioral relevance of the stimulus (e.g., whether or not a behavioral response is elicited) was observed in the secondary, but not primary, auditory cortex. In addition, this study revealed interesting interhemispheric differences: stronger activation was observed in the left secondary cortex and exclusive activation in the left dorsal field. The latter represents a small region of the dorsal belt of the auditory cortex that contains neurons with high sensitivity to duration and time-dependent spectral integration and that has been previously implicated in the processing of complex sounds such as vocal communication signals.

Expression analysis of c-Fos has also been used as a complement to electrophysiological studies of the auditory cortex of bats (Jen et al., 1997), a group where auditory signals play a very significant role in both echolocation and vocal communication. About 20% of the neurons in the primary auditory cortex of the big brown bat express c-Fos in response to auditory stimulation. Interestingly, a significant increase in c-Fos expression is seen only ipsilaterally upon unilateral ear plugging, indicating that the input to the primary auditory cortex originates mainly in the contralateral ear (Jen et al., 1997).

8. Overview and Future Perspectives

The use of IEG expression analysis has contributed significantly to our understanding of the functional organization of the auditory system. Some general themes have emerged from a comparison among studies in various organisms:

(1) Induced IEG expression analysis provides refined maps of auditory induced activation within the auditory system. Overall, IEG expression maps are in

good accordance with previous studies utilizing electrophysiological recordings or metabolic markers such as 2-deoxy-glucose, although such maps offer more global accounts of brain activation as well as higher spatial resolution in comparison with other methods.

(2) IEG studies greatly facilitate the determination of the neurochemical identity of the neurons involved in auditory processing. Such studies have provided substantial support for a role of GABAergic interneurons in auditory function, as well as a prominent role of modulatory neurotransmitters.

(3) The activation of the various subcortical centers along the ascending auditory pathway seems to primarily correlate with various aspects of the spectral and temporal composition of the stimulus. It is possible that the feature analysis occurring at these levels of the auditory system may be sufficient for the appropriate sensory discrimination across stimuli, regardless of their behavioral or contextual relevance.

(4) The activation patterns of pallial/cortical areas are highly modulated by the behavioral and/or contextual relevance of the stimulus. The processing occurring at this level of the auditory system seems to play critical roles in the analysis of complex stimuli such as those used for vocal communication, as well as in the formation and or storage of associations that confer behavioral or contextual value to auditory stimuli. The evidence in this regard derives from groups as diverse as rodents and birds, indicating that such a role for the higher reaches of the auditory system may be highly conserved among vertebrates. This, in turn, suggests that several aspects of the detailed microcircuitry organization of pallial auditory fields may also be conserved, despite the significant differences observed across animal groups (e.g. nuclear vs. layered layouts).

References

Adams JC (1995). Sound stimulation induces Fos-related antigens in cells with common morphological properties throughout the auditory brainstem. J Comp Neurol 361:645-668.

Avey MT, Phillmore LS, MacDougall-Shackleton SA (2005). Immediate early gene expression following exposure to acoustic and visual components of courtship in zebra finches. Behav Brain Res 165:247-253.

Bailey DJ, Wade J (2003). Differential expression of the immediate early genes FOS and ZENK following auditory stimulation in the juvenile male and female zebra finch. Brain Res Mol Brain Res 116:147-154.

Bailey DJ, Wade J (2005). FOS and ZENK responses in 45-day-old zebra finches vary with auditory stimulus and brain region, but not sex. Behav Brain Res 162:108-115.

Bailey DJ, Rosebush JC, Wade J (2002). The hippocampus and caudomedial neostriatum show selective responsiveness to conspecific song in the female zebra finch. J Neurobiol 52:43-51.

Bolhuis JJ, Zijlstra GG, den Boer-Visser AM, Van Der Zee EA (2000). Localized neuronal activation in the zebra finch brain is related to the strength of song learning. Proceedings of the National Academy of Sciences USA 97:2282-2285.

Bolhuis JJ, Hetebrij E, Den Boer-Visser AM, De Groot JH, Zijlstra GG (2001). Localized immediate early gene expression related to the strength of song learning in socially reared zebra finches. Eur J Neurosci 13:2165-2170.

Brauth SE, McHale CM, Brasher CA, Dooling RJ (1987). Auditory pathways in the budgerigar. I. Thalamo-telencephalic projections. Brain, Behavior and Evolution 30:174-199.

Brown MC, Liu TS (1995). Fos-like immunoreactivity in central auditory neurons of the mouse. J Comp Neurol 357:85-97.

Burmeister SS, Fernald RD (2005). Evolutionary conservation of the egr-1 immediate-early gene response in a teleost. J Comp Neurol 481:220-232.

Butler AB, Hodos W (1996). Comparative Vertebrate Neuroanatomy: Evolution and Adaptation. New York, NY: Wiley-Liss.

Cardin JA, Schmidt MF (2003). Song system auditory responses are stable and highly tuned during sedation, rapidly modulated and unselective during wakefulness, and suppressed by arousal. J Neurophysiol 90:2884-2899.

Cardin JA, Schmidt MF (2004). Noradrenergic inputs mediate state dependence of auditory responses in the avian song system. J Neurosci 24:7745-7753.

Carretta D, Herve-Minvielle A, Bajo VM, Villa AE, Rouiller EM (1999). c-Fos expression in the auditory pathways related to the significance of acoustic signals in rats performing a sensory-motor task. Brain Res 841:170-183.

Catchpole CK, Slater PJB (1995). Bird Song: Biological Themes and Variations. Cambridge, UK.: Cambridge University Press.

Cecchi GA, Ribeiro S, Mello CV, Magnasco MO (1999). An automated system for the mapping and quantitative analysis of immunocytochemistry of an inducible nuclear protein. Journal of Neuroscience Methods 87:147-158.

Chaudhuri A (1997). Neural activity mapping with inducible transcription factors. Neuroreport 8:iii-vii.

Chaudhuri A, Zangenehpour S (2002). Molecular activity maps of sensory function. In: Immediate Early Genes and Inducible Transcription Factors in Mapping of the Central Nervous System Function and Dysfunction (Kaczmarek L, Robertson HA, eds). Amsterdam: Elsevier.

Cheng MF, Zuo M (1994). Proposed pathways for vocal self-stimulation: met-enkephalinergic projections linking the midbrain vocal nucleus, auditory-responsive thalamic regions and neurosecretory hypothalamus. Journal of Neurobiology 25:361-379.

Chew SJ, Vicario DS, Nottebohm F (1996). A large-capacity memory system that recognizes the calls and songs of individual birds. Proceedings of the National Academy of Sciences USA 93:1950-1955.

Chew SJ, Mello C, Nottebohm F, Jarvis E, Vicario DS (1995). Decrements in auditory responses to a repeated conspecific song are long-lasting and require two periods of protein synthesis in the songbird forebrain. Proceedings of the National Academy of Sciences USA 92:3406-3410.

Christy B, Nathans D (1989). DNA binding site of the growth factor-inducible protein Zif268. Proceedings of the National Academy of Sciences USA 86:8737-8741.

Cirelli C, Pompeiano M, Tononi G (1996). Neuronal gene expression in the waking state: a role for the locus coeruleus. Science 274:1211-1215.

Clayton DF (2000). The genomic action potential. Neurobiol Learn Mem 74:185-216.

Cody AR, Wilson W, Leah J (1996). Acoustically activated c-fos expression in auditory nuclei of the anaesthetised guinea pig. Brain Res 728:72-78.

Dave AS, Yu AC, Margoliash D (1998). Behavioral state modulation of auditory activity in a vocal motor system. Science 282:2250-2254.

Doupe AJ, Konishi M (1991). Song-selective auditory circuits in the vocal control system of the zebra finch. Proceedings of the National Academy of Sciences USA 88:11339-11343.

Durand SE, Tepper JM, Cheng MF (1992). The shell region of the nucleus ovoidalis: a subdivision of the avian auditory thalamus. Journal of Comparative Neurology 323:495-518.

Eda-Fujiwara H, Satoh R, Bolhuis JJ, Kimura T (2003). Neuronal activation in female budgerigars is localized and related to male song complexity. Eur J Neurosci 17:149-154.

Ehret G, Fischer R (1991). Neuronal activity and tonotopy in the auditory system visualized by c-fos gene expression. Brain Res 567:350-354.

Fichtel I, Ehret G (1999). Perception and recognition discriminated in the mouse auditory cortex by c-Fos labeling. Neuroreport 10:2341-2345.

Friauf E (1992). Tonotopic Order in the Adult and Developing Auditory System of the Rat as Shown by c-fos Immunocytochemistry. Eur J Neurosci 4:798-812.

Friauf E (1995). C-fos immunocytochemical evidence for acoustic pathway mapping in rats. Behav Brain Res 66:217-224.

Gehr DD, Hofer SB, Marquardt D, Leppelsack H (2000). Functional changes in field L complex during song development of juvenile male zebra finches(1). Developmental Brain Research 125:153-165.

Gehr DD, Capsius B, Grabner P, Gahr M, Leppelsack HJ (1999). Functional organisation of the field-L-complex of adult male zebra finches. Neuroreport 10:375-380.

Geissler DB, Ehret G (2004). Auditory perception vs. recognition: representation of complex communication sounds in the mouse auditory cortical fields. Eur J Neurosci 19:1027-1040.

Gentner TQ, Margoliash D (2003). Neuronal populations and single cells representing learned auditory objects. Nature 424:669-674.

Gentner TQ, Hulse SH, Duffy D, Ball GF (2001). Response biases in auditory forebrain regions of female songbirds following exposure to sexually relevant variation in male song. Journal of Neurobiology 46:48-58.

Grace JA, Amin N, Singh NC, Theunissen FE (2003). Selectivity for conspecific song in the zebra finch auditory forebrain. J Neurophysiol 89:472-487.

Guzowski JF, McNaughton BL, Barnes CA, Worley PF (1999). Environment-specific expression of the immediate-early gene Arc in hippocampal neuronal ensembles. Nature Neuroscience 2:1120-1124.

Guzowski JF, Lyford GL, Stevenson GD, Houston FP, McGaugh JL, Worley PF, Barnes CA (2000). Inhibition of activity-dependent arc protein expression in the rat hippocampus impairs the maintenance of long-term potentiation and the consolidation of long-term memory. Journal of Neuroscience 20:3993-4001.

Harnischfeger G, Neuweiler G, Schlegel P (1985). Interaural time and intensity coding in superior olivary complex and inferior colliculus of the echolocating bat Molossus ater. J Neurophysiol 53:89-109.

Hernandez AM, MacDougall-Shackleton SA (2004). Effects of early song experience on song preferences and song control and auditory brain regions in female house finches (Carpodacus mexicanus). J Neurobiol 59:247-258.

Hillman DE, Gordon CE, Troublefield Y, Stone E, Giacchi RJ, Chen S (1997). Effect of unilateral tympanotomy on auditory induced c-fos expression in cochlear nuclei. Brain Res 748:77-84.

Hoke KL, Ryan MJ, Wilczynski W (2005). Social cues shift functional connectivity in the hypothalamus. Proc Natl Acad Sci USA 102:10712-10717.

Hoke KL, Burmeister SS, Fernald RD, Rand AS, Ryan MJ, Wilczynski W (2004). Functional mapping of the auditory midbrain during mate call reception. J Neurosci 24:11264-11272.

Hsu A, Woolley SM, Fremouw TE, Theunissen FE (2004). Modulation power and phase spectrum of natural sounds enhance neural encoding performed by single auditory neurons. J Neurosci 24:9201-9211.

Illing RB, Michler SA (2001). Modulation of P-CREB and expression of c-fos in cochlear nucleus and superior olive following electrical intracochlear stimulation. Neuroreport 12:875-878.

Illing RB, Michler SA, Kraus KS, Laszig R (2002). Transcription factor modulation and expression in the rat auditory brainstem following electrical intracochlear stimulation. Exp Neurol 175:226-244.

Jarvis ED, Nottebohm F (1997). Motor-driven gene expression. Proc Natl Acad Sci USA 94:4097-4102.

Jarvis ED, Mello CV (2000). Molecular mapping of brain areas involved in parrot vocal communication. Journal of Comparative Neurology 419:1-31.

Jarvis ED, Mello CV, Nottebohm F (1995). Associative learning and stimulus novelty influence the song-induced expression of an immediate early gene in the canary forebrain. Learn Mem 2:62-80.

Jarvis ED, Ribeiro S, da Silva ML, Ventura D, Vielliard J, Mello CV (2000). Behaviourally driven gene expression reveals song nuclei in hummingbird brain. Nature 406:628-632.

Jarvis ED, Gunturkun O, Bruce L, Csillag A, Karten H, Kuenzel W, Medina L, Paxinos G, Perkel DJ, Shimizu T, Striedter G, Wild JM, Ball GF, Dugas-Ford J, Durand SE, Hough GE, Husband S, Kubikova L, Lee DW, Mello CV, Powers A, Siang C, Smulders TV, Wada K, White SA, Yamamoto K, Yu J, Reiner A, Butler AB (2005). Avian brains and a new understanding of vertebrate brain evolution. Nat Rev Neurosci 6:151-159.

Jen PH, Sun X, Shen JX, Chen QC, Qian Y (1997). Cytoarchitecture and sound activated responses in the auditory cortex of the big brown bat, Eptesicus fuscus. Acta Otolaryngol Suppl 532:61-67.

Jin H, Clayton DF (1997). Localized changes in immediate-early gene regulation during sensory and motor learning in zebra finches. Neuron 19:1049-1059.

Jones MW, Errington ML, French PJ, Fine A, Bliss TV, Garel S, Charnay P, Bozon B, Laroche S, Davis S (2001). A requirement for the immediate early gene Zif268 in the expression of late LTP and long-term memories. Nature Neuroscience 4:289-296.

Kaczmarek L, Robertson HA (2002). Immediate Early Genes and Inducible Transcription Factors in Mapping of the Central Nervous System Function and Dysfunction. Amsterdam: Elsevier.

Kai N, Niki H (2002). Altered tone-induced Fos expression in the mouse inferior colliculus after early exposure to intense noise. Neurosci Res 44:305-313.

Kandiel A, Chen S, Hillman DE (1999). c-fos gene expression parallels auditory adaptation in the adult rat. Brain Res 839:292-297.

Karten HJ (1967). The organization of the ascending auditory pathway in the pigeon (*Columba livia*). I. Diencephalic projections of the inferior colliculus (nucleus mesencephali lateralis, pars dorsalis). Brain Research 6:409-427.

Karten HJ (1968). The ascending auditory pathway in the pigeon (*Columba livia*). II. Telencephalic projections of the nucleus ovoidalis thalami. Brain Research 11:134-153.

Keilmann A, Herdegen T (1996). Decreased expression of the c-Fos, but not Jun B, transcription factor in the auditory pathway of the rat after repetitive acoustic stimulation. ORL J Otorhinolaryngol Relat Spec 58:262-265.

Knudsen EI (1999). Mechanisms of experience-dependent plasticity in the auditory localization pathway of the barn owl. J Comp Physiol [A] 185:305-321.

Knudsen EI (2002). Instructed learning in the auditory localization pathway of the barn owl. Nature 417:322-328.

Knudsen EI, Zheng W, DeBello WM (2000). Traces of learning in the auditory localization pathway. Proc Natl Acad Sci USA 97:11815-11820.

Kroodsma DE, Miller EH (1996). Ecology and Evolution of Acoustic Communication in Birds. Ithaca, NY: Cornell University Press.

Kruse AA, Stripling R, Clayton DF (2004). Context-specific habituation of the zenk gene response to song in adult zebra finches. Neurobiol Learn Mem 82:99-108.

Lemaire P, Revelant O, Bravo R, Charnay P (1988). Two mouse genes encoding potential transcription factors with identical DNA-binding domains are activated by growth factors in cultured cells. Proceedings of the National Academy of Sciences USA 85:4691-4695.

Luo L, Ryan AF, Saint Marie RL (1999). Cochlear ablation alters acoustically induced c-fos mRNA expression in the adult rat auditory brainstem. J Comp Neurol 404:271-283.

Lyford GL, Yamagata K, Kaufmann WE, Barnes CA, Sanders LK, Copeland NG, Gilbert DJ, Jenkins NA, Lanahan AA, Worley PF (1995). Arc, a growth factor and activity-regulated gene, encodes a novel cytoskeleton-associated protein that is enriched in neuronal dendrites. Neuron 14:433-445.

Maney DL, MacDougall-Shackleton EA, MacDougall-Shackleton SA, Ball GF, Hahn TP (2003). Immediate early gene response to hearing song correlates with receptive behavior and depends on dialect in a female songbird. J Comp Physiol A Neuroethol Sens Neural Behav Physiol 189:667-674.

Margoliash D (1986). Preference for autogenous song by auditory neurons in a song system nucleus of the white-crowned sparrow. Journal of Neuroscience 6:1643-1661.

Mello C, Nottebohm F, Clayton D (1995). Repeated exposure to one song leads to a rapid and persistent decline in an immediate early gene's response to that song in zebra finch telencephalon. Journal of Neuroscience 15:6919-6925.

Mello CV (2002a). Mapping vocal communication pathways in birds with inducible gene expression. J Comp Physiol A Neuroethol Sens Neural Behav Physiol 188:943-959.

Mello CV (2002b). Immediate-early gene (IEG) expression mapping of vocal communication areas in the avian brain. In: Immediate Early Genes and Inducible Transcription Factors in Mapping of the Central Nervous System Function and Dysfunction (Kaczmarek L, Robertson HA, eds). Amsterdam: Elsevier.

Mello CV (2004). Gene regulation by song in the auditory telencephalon of songbirds. Front Biosci 9:63-73.

Mello CV, Clayton DF (1994). Song-induced ZENK gene expression in auditory pathways of songbird brain and its relation to the song control system. J Neurosci 14:6652-6666.

Mello CV, Clayton DF (1995). Differential induction of the ZENK gene in the avian forebrain and song control circuit after metrazole-induced depolarization. J Neurobiol 26:145-161.

Mello CV, Ribeiro S (1998). ZENK protein regulation by song in the brain of songbirds. Journal of Comparative Neurology 393:426-438.

Mello CV, Vicario DS, Clayton DF (1992). Song presentation induces gene expression in the songbird forebrain. Proceedings of the National Academy of Sciences USA 89:6818-6822.

Mello CV, Velho TA, Pinaud R (2004). Song-induced gene expression: a window on song auditory processing and perception. Ann N Y Acad Sci 1016:263-281.

Mello CV, Vates GE, Okuhata S, Nottebohm F (1998). Descending auditory pathways in the adult male zebra finch (*Taeniopygia guttata*). Journal of Comparative Neurology 395:137-160.

Metzger M, Jiang S, Braun K (1998). Organization of the dorsocaudal neostriatal complex: a retrograde and anterograde tracing study in the domestic chick with special emphasis on pathways relevant to imprinting. Journal of Comparative Neurology 395:380-404.

Milbrandt J (1987). A nerve growth factor-induced gene encodes a possible transcriptional regulatory factor. Science 238:797-799.

Muller CM, Leppelsack HJ (1985). Feature extraction and tonotopic organization in the avian auditory forebrain. Exp Brain Res 59:587-599.

Nastiuk KL, Mello CV, George JM, Clayton DF (1994). Immediate-early gene responses in the avian song control system: cloning and expression analysis of the canary c-jun cDNA. Molecular Brain Research 27:299-309.

Nick TA, Konishi M (2001). Dynamic control of auditory activity during sleep: correlation between song response and EEG. Proc Natl Acad Sci USA 98:14012-14016.

Nick TA, Konishi M (2005a). Neural auditory selectivity develops in parallel with song. J Neurobiol 62:469-481.

Nick TA, Konishi M (2005b). Neural song preference during vocal learning in the zebra finch depends on age and state. J Neurobiol 62:231-242.

Nottebohm F, Stokes TM, Leonard CM (1976). Central control of song in the canary, Serinus canarius. Journal of Comparative Neurology 165:457-486.

Nottebohm F, Kelley DB, Paton JA (1982). Connections of vocal control nuclei in the canary telencephalon. Journal of Comparative Neurology 207:344-357.

Park KH, Clayton DF (2002). Influence of restraint and acute isolation on the selectivity of the adult zebra finch zenk gene response to acoustic stimuli. Behav Brain Res 136:185-191.

Phillmore LS, Bloomfield LL, Weisman RG (2003). Effects of songs and calls on ZENK expression in the auditory telencephalon of field- and isolate-reared black capped chickadees. Behav Brain Res 147:125-134.

Pinaud R, Tremere LA, Penner MR (2000). Light-induced zif268 expression is dependent on noradrenergic input in rat visual cortex. Brain Res 882:251-255.

Pinaud R, Velho TA, Jeong JK, Tremere LA, Leao RM, von Gersdorff H, Mello CV (2004). GABAergic neurons participate in the brain's response to birdsong auditory stimulation. Eur J Neurosci 20:1318-1330.

Qian Y, Jen PH (1994). Fos-like immunoreactivity elicited by sound stimulation in the auditory neurons of the big brown bat Eptesicus fuscus. Brain Res 664:241-246.

Qian Y, Wu M, Jen PH (1996). Tracing the auditory pathways to electrophysiologically characterized neurons with HRP and Fos double-labeling technique. Brain Res 731:241-245.

Reiner A, Perkel DJ, Bruce LL, Butler AB, Csillag A, Kuenzel W, Medina L, Paxinos G, Shimizu T, Striedter G, Wild M, Ball GF, Durand S, Guturkun O, Lee DW, Mello CV, Powers A, White SA, Hough G, Kubikova L, Smulders TV, Wada K, Dugas-Ford J, Husband S, Yamamoto K, Yu J, Siang C, Jarvis ED (2004). Revised nomenclature for avian telencephalon and some related brainstem nuclei. J Comp Neurol 473:377-414.

Ribeiro S, Mello CV (2000). Gene expression and synaptic plasticity in the auditory forebrain of songbirds. Learning and Memory 7:235-243.

Ribeiro S, Cecchi GA, Magnasco MO, Mello CV (1998). Toward a song code: evidence for a syllabic representation in the canary brain. Neuron 21:359-371.

Ribeiro SP, R.; Mello, C. (1999). Noradrenergic modulation of song-induced ZENK expression in the zebra finch brain. In: 29th Annual Meeting of the Society for Neuroscience, p. 348-344.

Riera-Sala C, Molina-Mira A, Marco-Algarra J, Martinez-Soriano F, Olucha FE (2001). Inner ear lesion alters acoustically induced c-Fos expression in the rat auditory rhomboencephalic brainstem. Hear Res 162:53-66.

Rouiller EM, Wan XS, Moret V, Liang F (1992). Mapping of c-fos expression elicited by pure tones stimulation in the auditory pathways of the rat, with emphasis on the cochlear nucleus. Neurosci Lett 144:19-24.

Saint Marie RL, Luo L, Ryan AF (1999). Effects of stimulus frequency and intensity on c-fos mRNA expression in the adult rat auditory brainstem. J Comp Neurol 404:258-270.

Saito H, Miller JM, Altschuler RA (2000). Cochleotopic fos immunoreactivity in cochlea and cochlear nuclei evoked by bipolar cochlear electrical stimulation. Hear Res 145:37-51.

Saito H, Miller JM, Pfingst BE, Altschuler RA (1999). Fos-like immunoreactivity in the auditory brainstem evoked by bipolar intracochlear electrical stimulation: effects of current level and pulse duration. Neuroscience 91:139-161.

Sakata S, Kitsukawa T, Kaneko T, Yamamori T, Sakurai Y (2002). Task-dependent and cell-type-specific Fos enhancement in rat sensory cortices during audio-visual discrimination. Eur J Neurosci 15:735-743.

Scheich H, Zuschratter W (1995). Mapping of stimulus features and meaning in gerbil auditory cortex with 2-deoxyglucose and c-Fos antibodies. Behav Brain Res 66:195-205.

Scheich H, Stark H, Zuschratter W, Ohl FW, Simonis CE (1997). Some functions of primary auditory cortex in learning and memory formation. Adv Neurol 73:179-193.

Schmidt MF, Konishi M (1998). Gating of auditory responses in the vocal control system of awake songbirds. Nature Neuroscience 1:513-518.

Sen K, Theunissen FE, Doupe AJ (2001). Feature analysis of natural sounds in the songbird auditory forebrain. J Neurophysiol 86:1445-1458.

Shea SD, Margoliash D (2003). Basal forebrain cholinergic modulation of auditory activity in the zebra finch song system. Neuron 40:1213-1226.

Sockman KW, Gentner TQ, Ball GF (2002). Recent experience modulates forebrain gene-expression in response to mate-choice cues in European starlings. Proc R Soc Lond B Biol Sci 269:2479-2485.

Sockman KW, Gentner TQ, Ball GF (2005). Complementary neural systems for the experience-dependent integration of mate-choice cues in European starlings. J Neurobiol 62:72-81.

Sonnenberg JL, Rauscher FJ, Morgan JI, Curran T (1989). Regulation of proenkephalin by Fos and Jun. Science 246:1622-1625.

Steward O, Worley P (2002). Local synthesis of proteins at synaptic sites on dendrites: role in synaptic plasticity and memory consolidation? Neurobiol Learn Mem 78:508-527.

Steward O, Wallace CS, Lyford GL, Worley PF (1998). Synaptic activation causes the mRNA for the IEG Arc to localize selectively near activated postsynaptic sites on dendrites. Neuron 21:741-751.

Stripling R, Volman SF, Clayton DF (1997). Response modulation in the zebra finch neostriatum: relationship to nuclear gene regulation. Journal of Neuroscience 17:3883-3893.

Stripling R, Kruse AA, Clayton DF (2001). Development of song responses in the zebra finch caudomedial neostriatum: role of genomic and electrophysiological activities. J Neurobiol 48:163-180.

Suga N (1989). Principles of auditory information-processing derived from neuroethology. J Exp Biol 146:277-286.

Sukhatme VP, Cao XM, Chang LC, Tsai-Morris CH, Stamenkovich D, Ferreira PC, Cohen DR, Edwards SA, Shows TB, Curran T, et al. (1988). A zinc finger-encoding gene coregulated with c-fos during growth and differentiation, and after cellular depolarization. Cell 53:37-43.

Terleph TA, Mello CV, Vicario DS (2006). Auditory topography and temporal response dynamics of canary caudal telencephalon. J Neurobiol. 66:281-292.

Terpstra NJ, Bolhuis JJ, den Boer-Visser AM (2004). An analysis of the neural representation of birdsong memory. J Neurosci 24:4971-4977.

Terpstra NJ, Bolhuis JJ, Den Boer-Visser AM, Ten Cate C (2005). Neuronal activation related to auditory perception in the brain of a non-songbird, the ring dove. J Comp Neurol 488:342-351.

Vates GE, Broome BM, Mello CV, Nottebohm F (1996). Auditory pathways of caudal telencephalon and their relation to the song system of adult male zebra finches. J Comp Neurol 366:613-642.

Velho TA, Pinaud R, Rodrigues PV, Mello CV (2005). Co-induction of activity-dependent genes in songbirds. Eur J Neurosci 22:1667-1678.

Vignal C, Andru J, Mathevon N (2005). Social context modulates behavioural and brain immediate early gene responses to sound in male songbird. Eur J Neurosci 22:949-955.

Vignal C, Attia J, Mathevon N, Beauchaud M (2004). Background noise does not modify song-induced genic activation in the bird brain. Behav Brain Res 153:241-248.

Vischer MW, Hausler R, Rouiller EM (1994). Distribution of Fos-like immunoreactivity in the auditory pathway of the Sprague-Dawley rat elicited by cochlear electrical stimulation. Neurosci Res 19:175-185.

Vischer MW, Bajo-Lorenzana V, Zhang J, Hausler R, Rouiller EM (1995). Activity elicited in the auditory pathway of the rat by electrical stimulation of the cochlea. ORL J Otorhinolaryngol Relat Spec 57:305-309.

Wallhausser-Franke E, Mahlke C, Oliva R, Braun S, Wenz G, Langner G (2003). Expression of c-fos in auditory and non-auditory brain regions of the gerbil after manipulations that induce tinnitus. Exp Brain Res 153:649-654.

Wan H, Warburton EC, Kusmierek P, Aggleton JP, Kowalska DM, Brown MW (2001). Fos imaging reveals differential neuronal activation of areas of rat temporal cortex by novel and familiar sounds. Eur J Neurosci 14:118-124.

Wild JM, Karten HJ, Frost BJ (1993). Connections of the auditory forebrain in the pigeon (Columba livia). J Comp Neurol 337:32-62.

Zhang JS, Haenggeli CA, Tempini A, Vischer MW, Moret V, Rouiller EM (1996). Electrically induced fos-like immunoreactivity in the auditory pathway of the rat: effects of survival time, duration, and intensity of stimulation. Brain Res Bull 39:75-82.

Zuschratter W, Gass P, Herdegen T, Scheich H (1995). Comparison of frequency-specific c-Fos expression and fluoro-2- deoxyglucose uptake in auditory cortex of gerbils (Meriones unguiculatus). Eur Journal of Neuroscience 7:1614-1626.

4

Immediate Early Genes and Sensory Maps of Olfactory and Gustatory Function

MONIQUE MONTAG-SALLAZ and DIRK MONTAG

Neurogenetics Research Group, Leibniz Institute for Neurobiology, Magdeburg, Germany

1. Introduction

Considerable progress has been achieved towards the understanding of molecular and physiological properties of sensory systems, transmission and encoding of sensory stimuli. However, it remains largely mysterious how, in molecular terms, perception of a sensory stimulus is associated and related to internal representations of previous experience and emotions resulting in the meaningful integration of sensory information. Furthermore, how sensory stimulus processing is finally transformed into long lasting cellular changes, and how these enable the formation of memories that affect behavior (e.g. a stimulus becomes familiar, associated with a cue and thus the identical stimulus later results in a different response) remains largely hypothetical. Investigations conducted in recent years have allowed researchers to gain insight into some of the molecular processes underlying synaptic plasticity and to develop concepts considering plasticity as fundamental to the formation of a memory. In this chapter, we will summarize some work using immediate early gene (IEG) expression in the olfactory and gustatory systems for functional mapping. These studies provide new insight about odor and taste coding, processing of novel versus familiar stimuli, and the neuronal networks involved and illustrate that IEGs are not simply markers of neuronal activity but rather indicate information processing and plasticity events. IEG expression identifies neurons undergoing long lasting molecular changes, thus, the study of IEG expression patterns reveals the activation of neuronal circuitry involved in the processing of information.

R. Pinaud, L.A. Tremere (Eds.), Immediate Early Genes in Sensory Processing, Cognitive Performance and Neurological Disorders, 57–72, ©2006 Springer Science + Business Media, LLC

2. Olfactory Information Processing, Functional Mapping, and Immediate Early Genes

Olfaction plays a prominent role in life. Indeed, the sense of smell is crucial for the survival of numerous species, as odors represent information about the environment, including the presence of prey and predators, location and identification of food, and sexual status of the mating partners. Odors in our daily world are usually complex blends of volatile compounds eliciting a singular percept (e.g. lavender, smoke, grass or soap), from which individual ingredients are difficult, or virtually impossible, to segment and identify. However, mammals are able to detect and discriminate a large variety of odor molecules. To achieve this capacity, animals possess odorant receptor proteins, each of which specific for the recognition of a chemical feature of a particular odor molecule. In humans, about 1000 odorant receptor genes have been identified (for review see Mombaerts, 2001); other mammals can have a significantly higher number of genes dedicated to this sensory modality, underlining the importance of olfaction.

Olfactory molecules bind to specific olfactory receptors expressed on the cilial membrane surface of olfactory sensory neurons, which are located in the nasal epithelium (Fig. 4.1). Olfactory sensory neurons carrying the same receptor send their axons to conserved locations in the main olfactory bulb (MOB), the olfactory glomeruli, where they form excitatory synaptic connections on dendrites of mitral and tufted cells, the output neurons of the MOB. Mitral and tufted cells, in turn, project to the higher olfactory centers via the olfactory tract. In mice for example,

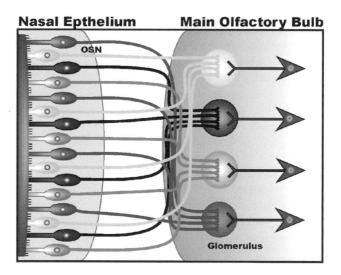

Figure 4.1. Schematic illustration of the olfactory system. Olfactory sensory neurons (ONS) located in the nasal epithelium send their axons to the glomeruli located in the main olfactory bulb. In general, each ONS expresses only one particular receptor. ONS expressing the same receptor (indicated by the shade) project to the same glomerulus in the main olfactory bulb innervating mitral and tufted cells. From the main olfactory bulb information is further transmitted via the olfactory tract to higher centers.

each of the 1800 glomeruli receives converging axonal inputs from several thousand olfactory sensory neurons innervating primary dendrites of $\sim$20 mitral cells (for review see Mori et al., 1999). Each glomerulus is only innervated by olfactory neurons expressing the same receptor; thus, the combination of activated glomeruli encodes the quality of the odor. In the MOB, olfactory information is modulated at different levels by inhibitory interneurons, the periglomerular cells in the glomerular layer and the granule cells in the granule cell layer. These neurons are responsible for lateral inhibition of the bulbar output neurons, which is a key for understanding the significance of odor maps, e.g. the spatial activity patterns within the olfactory bulb that are elicited by odor stimulation. Additionally, centrifugal fiber systems forming the olfactory peduncle and acting directly on inhibitory interneurons also modulate the activity of the mitral and tufted cells in the MOB. In particular, the cholinergic and noradrenergic afferents originating in the basal telencephalon and the locus coeruleus, respectively, are functionally involved in mechanisms of MOB plasticity associated with olfactory learning (Gervais et al., 1988; Ravel et al., 1992). Both local circuit interneurons and centrifugal systems take part in the initial events of odor information coding and central olfactory processing at the level of the MOB. As a consequence of this particular network organization and functioning, presentation of an individual odorant activates a specific set of glomeruli and the corresponding output cells and interneurons. This assembly is also known as the glomerular module and is viewed as a molecular feature-detecting unit (Mori et al., 1999).

Many techniques have been used to map the activity of the MOB after presentation of odorants. The first evidence of a spatial coding of odors in the MOB was obtained with electroencephalographic recordings (Freeman, 1982) and metabolic mapping using 14C-2-deoxyglucose (2-DG) uptake (Stewart et al., 1979; Jourdan et al., 1980; Royet et al., 1987; Astic et al., 1988). Subsequently, these principles were confirmed and extended by other methods such as voltage-sensitive dyes, Ca^{2+} imaging, intrinsic imaging, high-resolution functional magnetic resonance imaging (fMRI), local field potentials, and single unit electrophysiological recordings (for review see Xu et al., 2000). From these studies emerged a picture of the activity of the bulbar neuronal network after physiological stimulation, supporting the hypothesis of columnar processing of the olfactory information and spatial coding of odor quality in the MOB.

The analysis of IEG expression not only supported these data, but also— and more importantly—provided unique information about cellular and transcriptional events underlying odor processing, odor learning and familiarization with an odor. IEGs are characterized by their fast induction without the need of further protein synthesis, however, the nature and mechanisms of induction differ among IEGs (for review see Herdegen and Leah, 1998). The IEG *c-fos* encodes a transcription factor and is the most widely studied modulator IEG in the olfactory system. The activity-regulated gene 3.1 (arg 3.1; Link et al., 1995; also known as activity-regulated cytoskeleton-associated protein (arc; Lyford et al., 1995) is an IEG with the peculiarity that its mRNA selectively localizes in activated dendritic segments (Steward et al., 1998; Wallace et al., 1998), suggesting that the Arg 3.1

protein may be synthesized in dendrites at activated post-synaptic sites. These two IEGs will be the focus of the remainder of this chapter.

In 1993, we provided the first evidence that *c-fos* expression can be triggered by afferent olfactory input in the MOB in normally breathing awake rats (Sallaz and Jourdan, 1993). These results were supported later by other studies (Guthrie et al., 1995; Baba et al., 1997). By comparing the distribution of Fos protein and glomerular 2-DG uptake in adjacent sections of the same olfactory bulb, we were able to show that a 30-minute presentation of propionic acid vapor triggered a significant induction of Fos protein in the granule cells of the bulbar column defined by the foci of high 2-DG glomerular uptake. Interestingly, this study also provided the first evidence that *c-fos* expression could not be considered as a general marker of neuronal activity, in contrast to 2-DG metabolic mapping, because the bulbar relay neurons, mitral and tufted cells, in activated bulbar areas of odor-stimulated rats, did not exhibit Fos immunoreactivity despite being electrophysiologically active (Buonviso and Chaput, 1990). Likewise, electrical stimulation of either the olfactory nerve or the lateral olfactory tract, both able to elicit mitral cell bursting activity, failed to induce *c-fos* expression in neurons of the MOB (M. Sallaz, unpublished data). Furthermore, the coincidental patterns of glomerular 2-DG uptake and *c-fos* expression in the MOB of odor-stimulated rats demonstrated that *c-fos* expression in granule cells is triggered by the olfactory input. These experiments did not, however, rule out influences of other synaptic afferents to granule cells. Indeed, the odor-specific expression of Fos in the MOB of rats stimulated by propionic acid vapors also involved the activation of centrifugal afferents (Sallaz and Jourdan, 1996). In support of these findings, we found that unilateral olfactory peduncle section reduced the number of Fos-expressing neurons in ipsilateral and, to a lesser degree, in contralateral odor-selective areas of the olfactory bulb, without effects on the corresponding 2-DG foci (Sallaz and Jourdan, 1996). In addition, we provided evidence for an involvement of noradrenergic and cholinergic centrifugal systems in the modulation of *c-fos* expression in the MOB by use of a pharmacological approach (Sallaz and Jourdan, 1996). In summary, these data suggested that under certain conditions of olfactory stimulation involving the activation of centrifugal systems, Fos protein expression could be triggered resulting in further long-term modifications of the neuronal phenotype.

3. Coincidence as a Trigger of IEG Induction in the Olfactory System

Buonviso and co-workers showed that a single exposure of adult rats to an odorant, in the absence of any experimentally delivered reinforcement, led to a drastic decrease in mitral/tufted cell responsiveness (Buonviso et al., 1998; Buonviso and Chaput, 2000). Surprisingly, excitatory responses were decreased not only to familiar odors but also to other novel odors. In contrast, odor-induced expression of *c-fos* and arg 3.1/arc mRNA in specific olfactory bulb quadrants, as well as the previous familiarization of animals with the test odor, reduced the expression of both IEGs in these quadrants, thereby altering the odor-specific expression patterns (Montag-Sallaz and Buonviso, 2002). When different odors were used for familiarization and test, the odor-specific pattern was not affected. These data led us to hypothesize that IEG expression may occur only in granule cells receiving simultaneous inputs from the mitral/tufted cells and from the centrifugal fibers. According to our hypothesis, in naive rats, coincidence of both the odor-specific mitral/tufted cell and the diffuse centrifugal inputs would induce IEG expression in those granule cells receiving the dual input. As familiarization with the odor occurs, a decrease in attention and arousal occurs, thereby decreasing the release of acetylcholine and noradrenaline at the OB. In the absence (or during low strength) of centrifugal input, IEG expression would not be induced in the granule cells. We proposed that the spatio-temporal coincidence of both inputs would therefore be the limiting factor for IEG expression after presentation of an olfactory stimulus (Fig. 4.2).

The experiments monitoring *c-fos* and arg 3.1/arc expression discussed above highlight the importance of IEG expression analysis for the functional mapping of olfactory function. This approach provides a means for the characterization of fundamental mechanisms of olfactory information processing, quality encoding, memory formation, and familiarization with an odorant.

4. Gustatory Information Processing, Functional Mapping, and Immediate Early Genes

Taste perception is generally described in terms of four qualities: salty, sour, sweet, and bitter. In addition, it has been suggested that other taste categories exist as well—most notably umami, the sensation elicited by glutamate. These different qualities are detected by specific cellular mechanisms, for example, the interpretation of salty taste is mediated through sodium influx via Na^+ channels. Sour sensations are mediated by protons acting on Na^+ or K^+ channels, while umami is detected by glutamate receptors. Finally, bitter and sweet tastes are detected by specific receptors. From the taste cells, gustatory stimuli are transmitted by sensory fibers to the gustatory area of the nucleus of the solitary tract (Fig. 4.3) and from there to the gustatory cortex through a thalamic relay (for review see Sewards and Sewards, 2001; Katz et al., 2002). In contrast to olfactory neurons, each taste sensory fiber carries signals from different taste cells, which are activated by different taste stimuli. This activation is generally dominated by one stimulus and representation of gustatory stimuli is encoded by a unique combination of sensory fiber activity (for an overview on gustatory innervation and electrophysiological characterization of best stimuli in rats, see King et al., 1999).

A number of studies have used Fos expression to map activation patterns elicited by different tastes in the gustatory system (e.g., Travers and Hu, 2000; Chan et al., 2004). The principle findings of these works were that gustatory stimulation with tastants of different qualities [sweet (sucrose), bitter (quinine), or sour (citric acid), but not salty (NaCl)] induce the expression of Fos in the nucleus of the solitary tract. Within this nucleus, Fos expression was different across subnuclei, which lended support for a specific pattern of neuronal activation

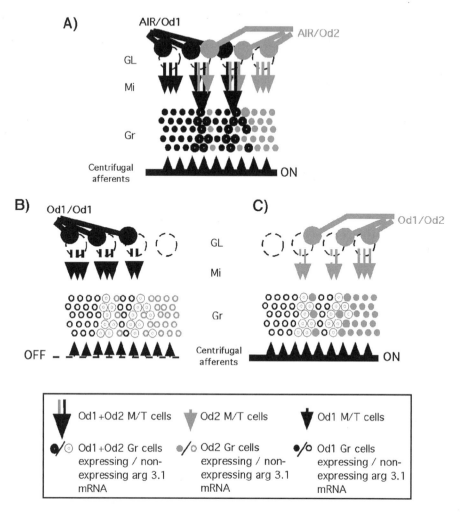

Figure 4.2. Familiarization and IEG expression in the olfactory bulb. (A) Presentation of odour 1 (Od1) to naive rats activates mitral and tufted (M/T) cells responding specifically to Od1 (Od1-M/T) and others responding to both Od1 and Odour 2 (Od2) (Od1+Od2-M/T). In addition, centrifugal systems are turned on because of the novelty of the stimulus, leading to expression of IEGs in granule cells (Gr) connected to OD1-M/T (Od1-Gr) and to the Od1+Od2-M/T (Od1+Od2-Gr) cells. Presentation of Od2 to naive rats results in the activation of a different but overlapping pattern of M/T cells (Od2-M/T and Od1+Od2-M/T) and expression of IEGs in the corresponding granule cells (Od2-Gr and Od1+Od2-Gr). GL indicates glomerulus. (B) Od1 is presented for the test after familiarization with the same odor (Od1/Od1) Centrifugal systems are turned off, and only a limited number of M/T cells (Od1-M/T and Od1+Od2-M/T) are responsive, resulting in a very limited IEG expression in the Od1-Gr and Od1+Od2-Gr and an alteration of the odor-specific pattern of IEG expression. (C) Od2 is presented for the test after familiarization with Od1 (Od1/Od2) Centrifugal systems are turned on, and the M/T cells processing both Od1 and Od2 features (Od1+Od2-M/T) remain silent while those processing the molecular features specific to Od2 (Od2-M/T) show a normal activity pattern. Consequently, IEG expression will be triggered selectively in the corresponding granule cells (Od2-Gr).

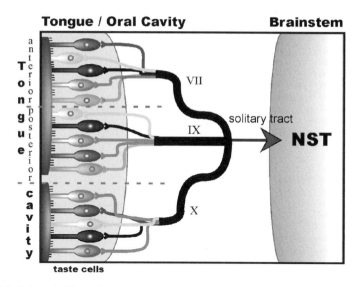

Figure 4.3. Schematic illustration of the gustatory system. Taste cells located in the tongue and oral cavity epithelium sense the different taste qualities. Branches of the cranial nerves VII, IX, and X connect the taste cells with the nucleus of the solitary tract in the brainstem. The anterior two thirds of the tongue transmit via cranial nerve VII (facial nerve, chorda tympani), the posterior third of the tongue via cranial nerve IX (glossopharyngeal nerve), and the back of the oral cavity via cranial nerve X (vagus).

for each tastant. However, the activity pattern revealed by Fos expression was identical for sucrose and quinine (Harrer and Travers, 1996; Travers and Hu, 2000). In contrast, in the C subnucleus, the region that receives the heaviest primary afferent gustatory projection, different Fos patterns were described for different taste qualities, even though different bitter tastes elicit a generalized and similar pattern (Chan et al., 2004).

For our studies investigating the processing of novel versus familiar stimuli, we analyzed the expression of the IEG *c-fos* and arg 3.1/arc in the gustatory system. These experiments required a new stimulation paradigm fulfilling particular requirements such as the precise control of the novel stimulus, lack of novelty-associated increases in locomotor activity, and the presentation of the stimulus, a solution of 0.5% saccharin in water, without significant disturbance. These features are of particular relevance to avoid non-specific expression of the IEGs. The exact timing of the novel stimulus is also particularly important considering the fast induction and short half life of IEG mRNAs. Furthermore, avoiding an increase in motor activity (e.g. due to the exploration of a novel environment) and stress (e.g., that induced by handling) are key requirements for this type of research. Indeed, both factors are likely to modify the liberation of neurotransmitters and the expression of IEGs in the nervous system. With these issues in mind, we developed a "taste novelty paradigm", which is detailed in Fig. 4.4, along with the brain areas investigated in our studies.

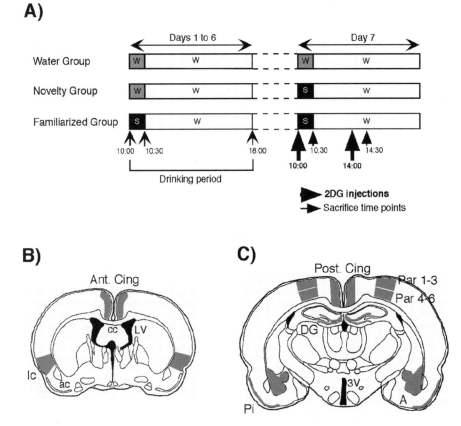

Figure 4.4. The "Taste Novelty Paradigm" and investigated brain areas. (A) Taste novelty paradigm. On each of the 7 consecutive days of the experiment, all mice were allowed to drink in their home cages between 10:00 and 18:00 and received tap water from 10:30 to 18:00. From 10:00 to 10:30, the water group received tap water (neutral taste), the familiarized group received saccharin (familiar taste), and the novelty group received water during the first 6 days but saccharin on day 7 (novel taste). On day 7, animals were sacrificed at different time points (0.5 h, 1 h, 4.5 h, and 6 h; small arrows) after the beginning of the drinking period. (^{14}C)-2-deoxyglucose was injected intraperitoneally 30 min before sacrifice (big arrows). (B, C) Brain areas investigated for expression of IEGs. Representation of two frontal brain sections showing the location of the areas where the density of labeled cells was evaluated (Paxinos and Franklin, 2001). B: for measurements in the insular cortex (granular part, Ic) and the anterior cingulate cortex (Ant. Cing), section located 0.34 mm caudally from Bregma. ac: anterior commissure; cc: corpus callosum; LV: lateral ventricle. C: for measurements in the following brain areas: Post. Cing, posterior cingulate cortex; CA1 and CA3, area CA1 and CA3 of the hippocampus; DG, dentate gyrus; Par 1-3 and Par 4-6, layers 1 to 3 and 4 to 6 of the posterior parietal associative area of the parietal cortex, respectively; A, amygdala (all nuclei); Pi, piriform cortex; 3v, 3rd ventricle, section located 1.82 mm caudally from Bregma.

Using this paradigm, we showed that the exposure to a taste, as long as it is novel to the animal, elicits the expression of IEGs in several brain areas. For example, we were able to document increased expression of *c-fos* and arg 3.1/arc mRNA in the cingulate cortex and deep layers of the parietal cortex (Montag-

Sallaz et al., 1999). In addition, *c-fos* mRNA expression was increased in the amygdala and arg 3.1/arc mRNA was increased in the dentate gyrus. Expression of *c-fos* and arg 3.1/arc was elevated 30 minutes after the exposure to novel stimuli. Interestingly, for arg 3.1/arc, we observed a second peak of expression 4.5 hours after onset of the novel stimulus. This biphasic expression had not been observed for any IEG before. Temporally, these two peaks of arg 3.1/arc expression, 30 minutes and 4.5 hours after the stimulus presentation, coincide with the 2 sensitive periods for memory formation that depend on protein synthesis. It is, thus, tempting to propose that arg 3.1/arc, in cooperation with other proteins generated during these two time windows, participates in the structural and functional synaptic reorganization underlying long-term memory formation.

5. IEGs: Indicators of Information Processing

5.1. Altered Processing of Gustatory Information in NCAM-Deficient Mice

The neural cell adhesion molecule (NCAM) has been implicated in the late phase of long-term memory formation because intracerebral injections of anti-bodies directed against NCAM resulted in amnesia only when injections were ad-ministered 5.5–8 hours post-training (Doyle et al., 1992; Scholey et al., 1993; for review see Rose, 1995). Furthermore, it has been reported that NCAM-deficient mice display a phenotype that includes abnormal exploration behavior, learning (Cremer et al., 1994), intermale aggression (Stork et al., 1997), and anxiety (Stork et al., 1999). Mice deficient for NCAM exhibit two major anatomical abnormalities: a size reduction of the olfactory bulb resulting from an altered cell migration process (Cremer et al., 1994), and strong alterations in mossy fiber growth and fasciculation (Cremer et al., 1997) associated with impaired long-term, but not short-term, plasticity at mossy fibers synapses (Cremer et al., 1998). Lisman and Otmakhova (2001) pointed out the crucial role of the mossy fiber/CA3 connections in novelty processing. Based on these points, we hypothesized that in the absence of NCAM, the expression patterns of *c-fos* and/or arg 3.1/arc after presentation of sensory stimuli may be altered. In NCAM-deficient mice submitted to the taste novelty paradigm described above, we observed higher *c-fos* mRNA expression in the amygdala after exposure to the novel taste (Montag-Sallaz et al., 2003b). In addition, arg 3.1/arc mRNA expression was higher 4.5 h after the neutral taste in the dentate gyrus and lower 4.5 h after the novel taste in the cingulate cortex in NCAM-deficient mice, when compared to wild-type controls. In conclusion, these results suggest that abnormal connectivity in NCAM-deficient mice appears to interfere with the appropriate processing of novel stimuli, leading to differences in neuronal activation patterns and IEG expression.

5.2. Altered Processing of Gustatory Information in CHL1-Deficient Mice

The Close Homolog of L1 (CHL1) is a member of the L1 family of cell adhe-sion molecules (Holm et al., 1996; Hillenbrand et al., 1999). Similar to NCAM mutants, mice deficient for CHL1 display aberrant connectivity of hippocampal mossy fibers and olfactory sensory axons (Montag-Sallaz et al., 2002), suggesting participation of CHL1 in the normal establishment of neuronal networks. Behavioral data showed that CHL1-deficient mice react differently towards novel environments (Montag-Sallaz et al., 2002), suggesting that processing of information, possibly novel versus familiar, may be altered in the absence of CHL1. In order to test this hypothesis, we analyzed the processing of gustatory stimuli in CHL1-deficient mice using the taste novelty paradigm described above (Montag-Sallaz et al., 2003a). 2-DG labeling revealed only small differences between CHL1-deficient and wild-type littermate mice. Furthermore, the specific increase in *c-fos* expression induced by the novel stimulus was similar in wild-type and mutant animals. In contrast, *c-fos* mRNA expression after familiarization with the taste was similarly elevated as after the novel taste in several brain areas of the CHL1-deficient and different from low basal levels in familiarized wild-type mice (Fig. 4.5). In addition, in these mutants, arg 3.1/arc expression was slightly reduced after novel taste exposure and increased after stimulation with the familiar taste, leading to a similar arg 3.1/arc mRNA expression after both stimuli. Our results indicate that CHL1-deficient mice may process novel and familiar information similarly and suggest that the altered neuronal connectivity in these mutants disturbs information processing at the molecular level.

These investigations on gustatory information processing in normal and mutant mice revealed changes at the genomic level that depend on the novelty/familiarity of the stimulus and on the appropriate wiring of the neuronal network involved. These data could only be obtained using IEGs as markers, as it will be discussed in the next paragraph.

6. Discussion, Perspectives

The early functional mapping techniques often relied on events occurring in the brain that were reflected in the local rates of cerebral glucose uptake, for example, axonal terminals that were labeled by 2-DG, as they represent the major sites of enhanced glucose utilization during increased neuronal activation (Dewar and McCulloch, 1992; Löwel, 2002). Analysis of IEG expression complemented and furthered our understanding of the brain mechanisms involved in sensory processing, learning and memory formation. As supported by numerous studies, IEG induction is not consistently coupled to simple metabolic or electrophysiological activity. Experiments using both 2-DG and IEGs as markers demonstrated that IEG activation can also occur when synaptic stimulation causes no change or even a decrease in metabolic activity (for review see Clayton, 2000). The IEG *c-fos*, for example, can certainly not be considered as a marker of neuronal activity *sensu stricto*, but rather induction of *c-fos* represents a read-out of intracellular second-messenger levels (Morgan and Curran, 1991). Furthermore, detection of IEG mRNAs by *in situ* hybridization usually reveals the somata of cells and immunohistochemical detection of the Fos protein labels the cell nuclei allowing for the precise identification of the cells undergoing changes at the transcriptional

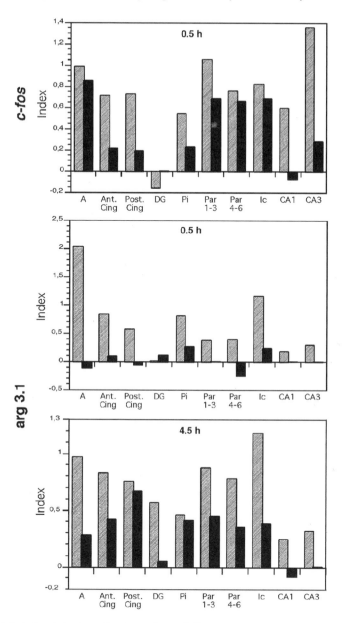

Figure 4.5. Relative induction of *c-fos* and arg 3.1/arc mRNA expression in wild-type and CHL1-deficient mice. For comparison of *c-fos* and arg 3.1/arc mRNA expression in CHL1-deficient (solid bars) and wild-type littermate (hatched bars) mice the index [(density of labeled cells after the novel taste divided by density of labeled cells after the familiar taste) − 1] was calculated; a value >0 indicates higher induction after the novel taste, whereas a value of = 0 indicates identical induction by novel and familiar stimulus. A, amygdala; Ant. Cing, anterior cingulate cortex; Post. Cing, posterior cingulate cortex; DG, dentate gyrus; Pi, piriform cortex; Par 1-3, layers 1 to 3 of the parietal cortex; Par 4-6, layers 4 to 6 of the parietal cortex; Ic, insular cortex; CA1 and CA3, area CA1 and CA3 of the hippocampus.

level. Comparison of IEG expression patterns with results collected using other activity mapping methods provides complementary information about neuronal and network activities, and is particularly useful to analyze the neuronal mechanisms underlying information processing, learning, and memory formation in the nervous system.

It is important to underline that the mRNA expression patterns of the transcription factor *c-fos* and the effector IEG arg 3.1/arc were always overlapping but distinct, indicating that they are induced differently and provide insights into distinct cellular activities (Montag-Sallaz et al., 1999; Clayton, 2000; Guzowski et al., 2001; Guzowski, 2002). Altered *c-fos* and/or arg 3.1/arc expression may reveal changes in neuronal activities shortly (30 min) after stimulation, while increased or decreased arg 3.1/arc expression 4.5 h after stimulation is likely to indicate changes in synaptic events associated with memory formation and consolidation (Montag-Sallaz and Montag, 2003). Interestingly, in the gustatory and olfactory systems, expression of the IEGs *c-fos* and arg3.1/arc was induced in wild-type animals when stimuli were new but not when animals were familiarized with the stimulus. Therefore, the induction of IEGs is not directly associated with the intrinsic molecular characteristics of a stimulus, such as a taste or an odor, but is rather related to the endogenous state of of the processing system, for example, the novelty or familiarity to a given sensory stimulus. Simply put, we propose that induction of IEGs depends on the convergence of information regarding stimulus reception and significance. Thus, although sensory stimuli are required, additional endogenous inputs are necessary for c-fos and arg3.1/arc expression. These inputs are independent of the nature of the stimulus (odor or taste molecule), but depend on the characteristic of the stimulus (e.g. novel or familiar). The spatio-temporal coincidence of both inputs would thus be the limiting factor for IEG expression after presentation of the stimulus, as demonstrated in the olfactory system. This hypothesis is strongly supported by the finding that learning itself, involving stimulus presentation and activation of central afferents, is sufficient to induce IEG expression (Montag-Sallaz and Montag, 2003).

Comparison of IEG expression in sensory systems of wild-type and mutant mice may shed light in understanding cognitive alterations linked to particular brain diseases. For example, the role of centrifugal afferents in olfactory information processing that we discussed above may be related to the finding that in early stages of several neurological diseases e.g. Parkinson's and Alzheimer's disease and schizophrenia, patients suffer from olfactory sensitivity loss (for review see Murphy, 1999; Hawkes, 2003; Kovacs, 2004). The data obtained with NCAM-deficient and CHL1-deficient mice, support a crucial role of the mossy fiber organization and connectivity for the processing of novelty and familiarity of stimuli. Furthermore these data support the connection between molecular processes with information processing and brain diseases. Connections involving hippocampal mossy fibers to CA3 appear to play a crucial role in novelty processing (Lisman and Otmakhova, 2001), while the CA3 area itself may be regarded as a "comparator device" (Vinogradova, 2001). By computing the match-mismatch of signal weight (presence and level of a signal) in its two inputs (medial septal nucleus/nucleus of the diagonal band and dentate gyrus), CA3 may detect the novelty of a stimulus. Abnormal connections between mossy fibers and CA3 pyramidal cells are, therefore, likely to interfere with the correct assignment of novelty or familiarity to incoming information.

In NCAM-deficient mice, IEG expression was predominantly altered after the novel stimulus but normal after the familiar taste, whereas in CHL1-deficient mice, the expression of *c-fos* and arg 3.1/arc in CA3 was similar after both novel and familiar tastes. As a consequence, other brain structures in the mutant mice may receive information incorrectly "marked" as novel or familiar, leading to an inappropriate processing of the stimulus and possibly to differences in neuronal activation and IEG expression. Alterations in "marking" an information as new or known may underlie some symptoms of schizophrenia, because familiar or internal informations receive the quality novel instead of familiar in schizophrenic patients (Arnold, 1999). In addition, numerous studies have shown that subtle changes affecting hippocampal neuronal circuitry are involved in schizophrenia (for review see Greene, 2001). Indeed, mice lacking the 180 kDa isoform of NCAM display increased lateral ventricle size and reduced pre-pulse inhibition of the startle response (Wood, 1998). In addition a mis-sense polymorphism in the *CHL1/CALL* gene has been shown to be associated with schizophrenia (Sakurai et al., 2002). Together these findings provide parallels between aberrancies in neuronal connectivity in mouse models with schizophrenic patients.

In conclusion, the studies on functional mapping in the olfactory and gustatory systems using IEGs as markers revealed new insights about sensory information processing and memory formation. In particular, these accomplishment derive from the fact that IEG expression does not simply reflect electrical activity elicited by stimulation, but it is more indicative of information processing. Monitoring IEG expression using animal models allows researchers to characterize pathways that involve genetic and molecular events to shed light into network and system properties that can be correlated with behavior, all of which further enhancing our understanding of the complex interplay in an ever changing nervous system. In addition to revealing sites where information is processed by different sensory systems, analysis of the expression patterns and conditions underlying the induction of IEGs can help us to understand the basis of certain neuropathologies and their consequences at the cellular and cognitive level.

References

Arnold O (1999). Schizophrenia—A disturbance of signal interaction between the entorhinal cortex and the dentate gyrus? The contribution of experimental dibenamine psychosis to the pathogenesis of schizophrenia: A hypothesis. Neuropsychobiology 40:21-32.

Astic L, Saucier D, Jourdan F, Holley A (1988). Functional activity of the rat olfactory bulb related to different duration of odor exposure. Chem Senses 13:333-343.

Baba K, Ikeda M, Houtani T, Nakagawa H, Ueyama T, Sato K, Sakuma S, Yamashita T, Tsukahara Y, Sugimoto T (1997). Odour exposure reveals non-uniform expression profiles of c-Jun protein in rat olfactory bulb neurons. Brain Res 774:142-148.

Barbeau D, Liand JJ, Quirion R, Robataille Y, Srivastava LK (1994). Decreased expression of the embryonic form of the neural cell adhesion molecule in schizophrenic brains. Proc Natl Acad Sci USA 92:2785-2789.

Buonviso N, Gervais R, Chalansonnet M, Chaput M (1998). Short-lasting exposure to one odour decreases general reactivity in the olfactory bulb of adult rats. Eur J Neurosci 10:2472-2475.

Buonviso N, Chaput M (2000). Olfactory experience decreases responsiveness of the olfactory bulb in the adult rat. Neuroscience 95:325-332.

Chan CY, Yoo JE, Travers SP (2004). Diverse bitter stimuli elicit highly similar patterns of Fos-like immunoreactivity in the nucleus of the solitary tract. Chem Senses 29:573-581.

Clayton DF (2000). The genomic action potential. Neurobiol Learn Mem 74:185-216.

Cremer H, Lange R, Christoph A, Plomann M, Vopper G, Roes J, Brown R, Baldwin S, Kraemer P, Scheff S, Barthels D, Rajewsky K, Wille W (1994). Inactivation of the N-CAM gene in mice results in size reduction of the olfactory bulb and deficits in spatial learning. Nature 367:455-459.

Cremer H, Chazal G, Goridis C, Represa A (1997). N-CAM is essential for axonal growth and fasciculation in the hippocampus. Mol Cell Neurosci 8:323-335.

Cremer H, Chazal G, Carleton A, Goridis C, Vincent JD, Lledo PM (1998). Long-term but not short-term plasticity at mossy fiber synapses is impaired in neural cell adhesion molecule-deficient mice. Proc Natl Acad Sci USA 95:13242-13247.

Dewar D, McCulloch J (1992). Mapping functional events in the CNS with 2-deoxyclucose autoradiography. In: Quantitative Methods in Neuroanatomy (Stewart M, ed.), pp. 57-84. John Wiley & Sons.

Doyle E, Nolan PM, Bell R, Regan CM (1992). Intraventricular infusions of anti-neural cell adhesion molecules in a discrete post training period impair consolidation of a passive avoidance response in the rat. J Neurochem 59:1570-1573.

Freeman WJ, Schneider W (1982). Changes in spatial patterns of rabbit olfactory EEG with conditioning to odors. Psychophysiol 19:44-56.

Gervais R, Holley A, Keverne B (1988). The importance of central noradrenergic influences on the olfactory bulb in the processing of learned olfactory cues. Chemical Senses 13:3-12.

Greene R (2001). Circuit analysis of NMDAR hypofunction in the hippocampus, in vitro, and psychosis of schizophrenia. Hippocampus 11:569-577.

Guthrie KM, Nguyen T, Gall CM (1995). Insulin-like growth factor-1 mRNA is increased in deafferented hippocampus: spatiotemporal correspondence of a trophic event with axon sprouting. J Comp Neurol 352:147-160.

Guzowski J, Setlow B, Wagner E, McGaugh J (2001). Experience-dependent gene expression in the rat hippocampus after spatial learning: a comparison of the immediate-early genes Arc, c-fos, and zif268. J Neurosci 21:5089-5098.

Guzowski J (2002). Insights into immediate-early gene function in hippocampal memory consolidation using antisense oligonucleotide and fluorescent imaging approaches. Hippocampus 12:86-104.

Harrer MI, Travers SP (1996). Topographic organization of Fos-like immunoreactivity in the rostral nucleus of the solitary tract evoked by gustatory stimulation with sucrose and quinine. Brain Res 711:125-137.

Hawkes C (2003). Olfaction in neurodegenerative disorder. Mov Disord 18:364-372.

Hillenbrand R, Molthagen M, Montag D, Schachner M (1999). The close homologue of the neural adhesion molecule L1 (CHL1): patterns of expression and promotion of neurite outgrowth by heterophilic interactions. Eur J Neurosci 11:813-826.

Holm J, Hillenbrand R, Steuber V, Bartsch U, Moos M, Lübbert H, Montag D, Schachner M (1996). Structural features of a close homologue of L1 (CHL1) in the mouse: a new member of the L1 family of neural recognition molecules. Eur J Neurosci 8:1613-1629.

Jourdan FJ, Duveau A, Astic L, Holley A (1980). Spatial distribution of (14C)-2-deoxyglucose uptake in the olfactory bulbs of rats stimulated with two different odours. Brain Res 188:139-154.

Katz DB, Nicolelis MAL, Simon SA (2002). Gustatory processing is dynamic and distributed. Curr Opin Neurobiol 12:448-454.

King CT, Travers SP, Rowland NE, Garcea M, Spector AC (1999). Glossopharyngeal nerve transection eliminates quinine-stimulated fos-like immunoreactivity in the nucleus of the solitary tract: implications for a functional topography of gustatory nerve input in rats. J Neurosci 19:3107-3121.

Kovacs T (2004). Mechanisms of olfactory dysfunction in aging and neurodegenerative disorders. Ageing Res Rev 3:215-232.

Link W, Konietzko U, Kauselmann G, Krug M, Schwanke B, Frey U, Kuhl D (1995). Somatodendritic expression of an immediate early gene is regulated by synaptic activity. Proc Natl Acad Sci USA 92:5734-5738.

Lisman J, Otmakhova N (2001). Storage, recall, and novelty detection of sequences by the hippocampus: elaborating on the SOCRATIC model to account for normal and aberrant effects of dopamine. Hippocampus 11:551-568.

Löwel S (2002). 2-Deoxyglucose architecture of cat primary visual cortex. In: The Cat Primary Visual Cortex (Payne B, Peters A, eds), pp. 167-193. Academic Press.

Lyford G, Yamagata K, Kaufmann W, Barnes C, Sanders L, Copeland N, Gilbert D, Jenkins N, Lanahan A, Worley P (1995). Arc, a growth factor and activity-regulated gene, encodes a novel cytoskeleton-associated protein that is enriched in neuronal dendrites. Neuron 14:433-445.

Mombaerts, P. (2001). The human repertoire of odorant receptor genes and pseudogenes. Annu Rev Genomics Hum Genet 2:493-510.

Montag-Sallaz M, Welzl H, Kuhl D, Montag D, Schachner M (1999). Novelty-induced increased expression of the immediate early genes c-fos and arg 3.1/arc in the mouse brain. J Neurobiol 38:234-246.

Montag-Sallaz M, Buonviso N (2002). Altered odor-induced expression of c-fos and arg 3.1/arc immediate early genes in the olfactory system after familiarization with an odor. J Neurobiol 52:61-72.

Montag-Sallaz M, Schachner M, Montag D (2002). Misguided axonal projections, NCAM180 mRNA upregulation, and altered behavior in mice deficient for the Close Homologue of L1 (CHL1) Mol Cell Biol 22:7967-7981.

Montag-Sallaz M, Montag D (2003). Learning-induced arg 3.1/arc mRNA expression in the mouse brain. Learn Mem 10:99-107.

Montag-Sallaz M, Baarke A, Montag D (2003a). Altered information processing in CHL1-deficient mice: analysis of c-fos and arg 3.1 expression and 2DG uptake. J Neurobiol 57:67-80.

Montag-Sallaz M, Montag D, Schachner M (2003b). Altered processing of novel information in N-CAM-deficient mice. Neuroreport 14:1343-1346.

Morgan J, Curran T (1991). Stimulus-transcription coupling in the nervous system: involvement of the inducible proto-oncogenes fos and jun. Annu Rev Neurosci 14:421-451.

Mori K, Nagao H, Yoshihara Y (1999). The olfactory bulb: coding and processing of odor molecule information. Science 286:711-715.

Murphy C (1999). Loss of olfactory function in dementing disease. Physiol Behav 66:177-182.

Paxinos G, Franklin K (2001). The Mouse Brain in Stereotaxic Coordinates. Second Edition. Academic Press.

Ravel N, Vigouroux M, Elaagouby A, Gervais R (1992). Scopolamine impairs delayed matching in an olfactory task in rats. Psychopharmacology 109:439-443.

Rose SPR (1995). Glycoproteins and memory formation. Behav Brain Res 66:73-78.

Royet JP, Sicard G, Souchier C, Jourdan F (1987). Specificity of spatial patterns of glomerular activation in the mouse olfactory bulb: computer-assisted image analysis of 2-deoxyglucose autoradiograms. Brain Res 417:1-11.

Sakurai K, Migita O, Toru M, Arinami T (2002). An association between a missense polymorphism in the close homologue of L1 (CHL1, CALL) gene and schizophrenia. Mol Psychiatry 7:412-415.

Sallaz M, Jourdan F (1993). C-fos expression and 2-deoxyglucose uptake in the olfactory bulb of odour-stimulated awake rats. NeuroReport 4:55-58.

Sallaz M, Jourdan F (1996). Odour-induced c-fos expression in the rat olfactory bulb: involvement of centrifugal afferents. Brain Res 721:66-75.

Scholey AB, Rose SPR, Zamani MR, Bock E, Schachner M (1993). A role for the neural cell adhesion molecule in a late, consolidating phase of glycoprotein synthesis six hours following passive avoidance training of the young chick. Neurosci 55:499-509.

Sewards TV, Sewards, MA (2001). Cortical association areas in the gustatory system. Neurosci Biobeh Rev 25:395-407.

Steward O, Wallace CS, Lyford G, Worley PF (1998). Synaptic activation causes the mRNA for the IEG Arc to localize selectively near activated postsynaptic sites on dendrites. Neuron 21:741-751.

Stewart WB, Kauer JS, Shepherd GM (1979). Functional organization of rat olfactory bulb analyzed by the 2-deoxyglucose method. J Comp Neurol 185:715-734.

Stork O, Welzl H, Cremer H, Schachner M (1997). Increased internal aggression and neuroendocrine response in mice deficient for the neural cell adhesion molecule (N-CAM). Eur J Neurosci 9:1117-1125.

Stork O, Welzl H, Wotjak CT, Hoyer D, Delling M, Cremer H, Schachner M (1999). Anxiety and increased 5-HT$_{1A}$ receptor response in N-CAM null mutant mice. J Neurobiol 40:343-355.

Travers SP, Hu H (2000). Extranuclear projections of rNST neurons expressing gustatory-elicited Fos. J Comp Neurol 427:124-138.

Vicente AM, Macciardi F, Verga M, Bassett AS, Honer WG, Bean G, Kennedy JL (1997). NCAM and schizophrenia: genetic studies. Mol Psychiatry 2:65-69.

Vinogradova O (2001). Hippocampus as comparator: role of the two input and two output systems of the hippocampus in selection and registration of information. Hippocampus 11:578-598.

Wallace CS, Lyford G, Worley PF, Steward O (1998). Differential intracellular sorting of IEG mRNAs depends on signals in the mRNA sequence. J Neurosci 18:26-35.

Wood GK, Tomasiewicz H, Rutishauser U, Magnuson T, Quirion R, Rochford J, Srivastava LK (1998). NCAM-180 knockout mice display increased lateral ventricle size and reduced prepulse inhibition of startle. NeuroReport 9:461-466.

Xu F, Kida I, Hyder F, Shulman RG (2000). Assessment and discrimination of odor stimuli in rat olfactory bulb by dynamic functional MRI. Proc Natl Acad Sci USA 97:10601-10606.

5

Immediate Early Gene Expression in the Primary Somatosensory Cortex: Focus on the Barrel Cortex

RAPHAEL PINAUD [a], ROBERT K. FILIPKOWSKI [b], ANTONIO F. FORTES [c] and LIISA A. TREMERE [a]

[a] Department of Neurobiology, Duke University Medical Center, Durham, NC, USA
[b] Nencki Institute, Warsaw, Poland
[c] Department of Neuroscience, University of Minnesota, Minneapolis, MN, USA

1. Introduction

The rodent somatosensory system has been extensively and successfully used as a model to understand the neural basis of sensory processing and experience-dependent plasticity in the central nervous system (CNS). The primary areas of investigation in the somatosensory system encompass the neural mechanisms of tactile information processing and activity-dependent plasticity, as well as the neural substrates of pain. The goal of this chapter is to review how immediate early gene (IEG) expression is regulated by tactile experience in the rodent primary somatosensory cortex (S1). We will focus our discussions on the rodent brain representation of the mechanoreceptors associated with long facial whiskers, a region commonly referred to as the barrel cortex (discussed below). The regulation of IEG expression by pain is discussed in Chapter 6 of this volume.

The rodent barrel cortex has become one of the most widely used experimental models in the study of how peripheral experience is represented across central cortical circuits, and how these networks are changed as a function of activity. This is in part due to the remarkable organization in which each peripheral whisker is represented, throughout the ascending somatosensory pathway, onto discrete columnar modules in the rodent S1 (Fig. 5.1). Information regarding the deflection of individual whiskers is conveyed ipsilaterally from its follicles to the trigeminal nuclei within the brainstem. These second order neurons project contralaterally to cells of the ventrobasal complex within the thalamus which, in turn, send fibers to the thalamorecipient layer IV cells within the S1. The terminals

R. Pinaud, L.A. Tremere (Eds.), Immediate Early Genes in Sensory Processing, Cognitive Performance and Neurological Disorders, 73–92, ©2006 Springer Science + Business Media, LLC

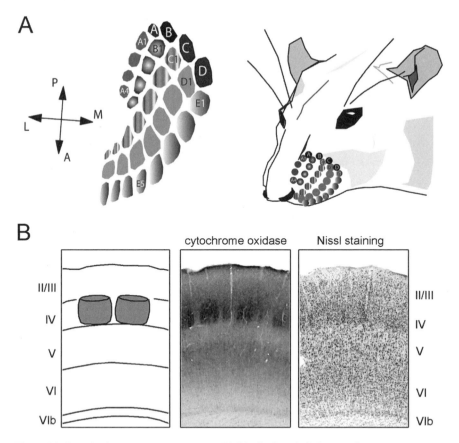

Figure 5.1. Barrels of rat somatosensory cortex. (A) Distribution of vibrissae on the snout corresponds with the distribution of the barrels in layer IV of the cortex in horizontal plane, which can be visualized with histological stainings (B, coronal sections). (Modified from Filipkowski, 2000).

of cells that comprise the thalamo-cortical projections terminate in discrete drum-like anatomical modules, each of which composed of a large density of neurons. These modules are referred to as "barrels" and are exclusively observed in layer IV. However, the barrel organization is part of a larger module, or column, that encompasses all cortical layers. Thus, the representation of whiskers within the S1 is commonly referred to as the "barrel cortex" (reviewed in Glazewski, 1998; Kossut, 1998; Fox, 2002) (Fig. 5.1).

Information from the whiskers that is processed in the barrel cortex, and possibly higher-order cortical areas, is critical for a number of behaviors in rodents that range from spatial orientation and sensory learning to reproductive behavior and texture discrimination (Schiffman et al., 1970; Gustafson and Felbain-Keramidas, 1977; Ahl, 1986; Guic-Robles et al., 1989, 1992; Bialy and Beck, 1993). Thus, the study of the somatic representation and plasticity in the rodent somatosensory cortex, especially as it relates to the barrel cortex, is associated with high behavioral relevance.

2. Immediate Early Gene Expression in the Barrel Cortex

As discussed in previous chapters, IEG expression has been successfully used to map the activity patterns of neural circuits in the vertebrate brain, associated with sensory input (also reviewed in Hughes and Dragunow, 1995; Kaczmarek and Chaudhuri, 1997; Herdegen and Leah, 1998; Mello et al., 2004; Pinaud, 2004). The expression of most IEGs is highly associated with neuronal depolarization and occurs shortly after cell activation in most sensory systems, without *de novo* protein synthesis. Thus, the mapping of the specific products that result from IEG expression (i.e., specific mRNAs and proteins encoded by each IEG) has been used to probe the functional organization of various regions of the vertebrate brain, including sensory systems (see Part I of this volume). In particular, the anatomical and functional organization of the rodent barrel cortex has been studied using IEG expression as a mapping tool following a number of experimental protocols (Filipkowski, 2000) that range from mechanical brushing of whiskers to exposure of animals to an enriched environment (EE). Moreover, the patterns of IEG expression that arise following tactile stimulation have been probed with various histological techniques. On the forefront of these methodologies, *in-situ* hybridization and immunocytochemistry (ICC) have been the most extensively used to map specific IEG mRNA and protein distribution, respectively (see Chapter 1). In the following paragraphs we will review key findings on the IEG expression in the S1, focusing primarily in barrel cortex, in the context of somatosensory processing and plasticity. Our discussions will be centered on the expression of the IEGs *c-fos*, *fosB*, *junB*, *c-jun*, ICER, NGFI-A, NGFI-B, NGFI-C and *cpg15* although the large majority of the research conducted in this field has been focused upon *c-fos* and NGFI-A expression.

3. The Fos and Jun Families: AP-1

Proteins encoded by elements of the Fos family of transcription factors, which include c-Fos, FosB, Fra-1 and Fra-2, have been shown to specifically dimerize with members of the Jun family, that include c-Jun, JunB and JunD, through leucine-zipper interactions, to form a transcription factor complex known as the activator protein-1 (AP-1) (Morgan and Curran, 1991; Herdegen and Leah, 1998). AP-1 recognizes and binds to a specific DNA motif encountered in a relatively large number of promoters of genes expressed in CNS, thus regulating their expression. The specific composition of AP-1, which is based on the formation of Fos and Jun homo- and hetero-dimers, appears to be key in the directionality of target gene expression; for example, Fos/JunD dimers have been shown to positively regulate, while FosB/JunB heterodimers negatively regulate target gene expression (Angel and Karin, 1991; Kobierski et al., 1991; Hughes and Dragunow, 1995). Extensive research on the regulation of members of the AP-1 complex has implicated this transcription factor in a large array of cell biological processes that range from stimulus-transcription coupling following sensory stimulation, to apoptosis and cell proliferation (reviewed in Herdegen and Leah, 1998; Kaminska et al., 2000; Eferl and Wagner, 2003).

3.1. c-fos

Pioneering work conducted by Mack and Mack (1992) investigated the effects of tactile stimulation on c-Fos expression in the rat barrel cortex. In these experiments, adult animals were mildly anesthetized with urethane and had their large whiskers stimulated unilaterally with a paintbrush for 15 minutes. c-Fos expression was assessed with immunocytochemical approaches 2 hours after stimulation onset. This early work demonstrated that although c-Fos was found at low basal levels in the rat barrel cortex, unilateral stimulation of the vibrissal system triggered a marked upregulation of this transcription factor in the contralateral barrel cortex, with the highest induction detected in the thalamo-recipient layer IV (Mack and Mack, 1992). These findings were the first to suggest that physiological stimulation of the peripheral whiskers led to the regulation of IEG expression in the rodent barrel cortex.

In subsequent work, Filipkowski and colleagues (2000) provided a more detailed description of the patterns of c-Fos induction following manual whisker stimulation in rats. Importantly, these efforts were conducted in non-anesthetized animals, given that anesthesia had been shown to interfere with IEG expression. In these experiments, rats were habituated to sit on top of a copper cylinder—as to avoid direct animal manipulation and restraint—where they underwent manual brushing of their whiskers for 20 minutes (Filipkowski et al., 2000) (Fig. 5.2). After stimulation, animals were returned to their home cages for 2 hours and c-Fos levels were assessed through immunocytochemical methods. c-Fos immunoreactive nuclei were detected across all cortical layers, with the exception of layer I. A significant upregulation following brushing was detected in the contralateral cortex for all layers studied: the highest increase in c-Fos immunoreactivity was described for cortical layer IV (4.6-fold), followed by significant increases observed in both supragranular (II/III) and infragranular layers (V/VI)—(4.1 and 2.5-fold, respectively) (Filipkowski et al., 2000) (Fig. 5.2). Despite the presence of c-Fos immunoreactive neurons in layer VIb, no significant changes were detected following sensory stimulation. In an attempt to characterize the neuro-chemical identity of these c-Fos-positive neurons, Filipkowski and co-workers conducted double-labeling immunocytochemical experiments between c-Fos and parvalbumin, a calcium-binding protein that reveals a sub-population of inhibitory neurons. It was reported that approximately 90% of c-Fos positive neurons in the stimulated rat barrel cortex did not co-localize with parvalbumin, indicating that the stimulation-induced c-Fos expression involved predominantly excitatory neurons. However, the remaining 10% of neurons exhibited co-localization and presumably reflected inhibitory neurons that were activated by the brushing paradigm. No evidence for an asymmetric distribution of these double-labeled neurons across layers was reported (Filipkowski et al., 2000).

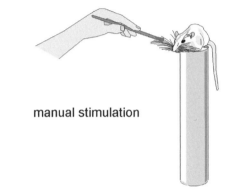

manual stimulation

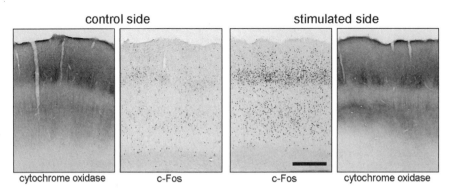

control side		stimulated side	
cytochrome oxidase	c-Fos	c-Fos	cytochrome oxidase

Figure 5.2. c-Fos protein induction in the barrel cortex (barrels visualized by cytochrome oxidase staining) following 20 min of manual stimulation of vibrissae on one side of the snout while rats were sitting on the top of a copper cylinder. The control (non-stimulated) side is compared with the stimulated one. Scale bar = 0.5 mm. (From Filipkowski et al., 2000).

In a second paradigm, authors removed rats from their home cages and, instead of placing them in the copper cylinder for manual stimulation, animals were transferred to a new environment (wired cage), where they remained for 20 minutes. This step was followed by their return to the original home cages where they remained for an additional 2 hours (Fig. 5.3). During the animals' experience in the wired cage, authors reported significant sniffing and whisking behavior, leading to a more "natural" stimulation of their vibrissal system. Consequently, significant c-Fos expression was found in the barrel cortex of animals that had been exposed to the wired cage, as compared to the low protein levels detected in controls. c-Fos induction following this paradigm yielded similar laminar expression patterns as compared to those observed after manual brushing, with highest expression detected in layer IV, followed by significant increases in protein levels detected in the infra- and supra-granular layers (Fig. 5.3). However, the extent of c-Fos induction was numerically different across experimental paradigms, with cortical layer IV, infragranular and supragranular layers exhibiting 6.4, 3.5 and 5-fold induction, respectively (Filipkowski et al., 2000). Likewise, no

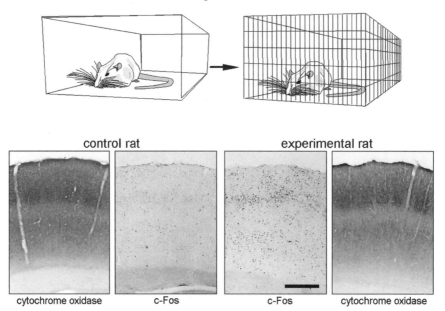

Figure 5.3. c-Fos protein induction in the barrel cortex (barrels visualized by cytochrome oxidase staining) following 20 min of exposition to a new environment, a wired cage. The control (naive) rats are compared with the rats placed in the new environment. Scale bar = 0.5 mm. (From Filipkowski et al., 2000).

effects of experience in the wired cage were detected for layer VIb. Although the specific reasons for the differences in c-Fos expression between manual brushing and exposure to a wired cage have not been systematically addressed, it has been proposed that neuronal activity (and perhaps c-Fos expression) associated with whisking behavior is highest following detection of holes, rather than the discrimination of specific textures, in the environment (Lipp and Van der Loos, 1991; Filipkowski et al., 2000).

Other methodologies have also been used to investigate IEG expression in the rodent barrel cortex with minimal behavioral interference. A commonly used approach involves the treatment of animals with apomorphine (APO), a dopamine receptor agonist that has been shown to increase motor activity, sniffing and whisking behavior. In 1994, Steiner and Gerfen coupled APO treatment with the unilateral clipping of whiskers in the rat and investigated the effects of these manipulations on *c-fos* mRNA expression using *in-situ* hybridization with specific ^{35}S-labeled probes. Animals had all whiskers clipped on the left side of the snout, followed by a wait period of 4 hours before sacrifice. An additional group of animals underwent these same conditions, however, before sacrifice, they were treated with APO followed by a 1-hour waiting period. These authors reported that whereas uninjected rats exhibited low activity levels, APO-treated rats displayed whisking, sniffing and robust snout contact with environmental surfaces (Steiner and Gerfen, 1994). In these experiments, while *c-fos* mRNA was expressed at very low basal levels in control animals, APO treatment triggered a robust induction of this molecule in the barrel cortex. Consistent with the protein data discussed above, highest *c-fos* expression was observed in layers IV and V/VI. However, in this preparation, a significant induction was also detected in deep layer VI. Clipping of the whiskers on the left side of the snout led to a significant decrease in *c-fos* mRNA levels in the contralateral barrel cortex in both control and APO-treated animals, an effect that was also observed when comparing the deprived against the spared cortex (Steiner and Gerfen, 1994). Overall, APO-treated rats exhibited higher expression of *c-fos* mRNA in the barrel cortex, as compared to non-injected controls, with the exception of the sensory deprived area.

Similar experiments have been conducted by other research groups, using ICC directed at c-Fos protein. For example, Filipkowski and colleagues (2001) treated rats with APO and evaluated c-Fos expression in the rat somatosensory cortex. In these authors' paradigm, APO injections were also used to increase whisking behavior and were coupled to unilateral whisker clipping (Filipkowski et al., 2001) (Fig. 5.4). It was found that APO triggered c-Fos level increases in both deprived and non-deprived cortex. Highest significance was detected in the supragranular layers (2.2-fold) of the deprived side, while significant increases were detected for all cortical layers studied in the non-deprived cortex (II/III, 3.5-fold; IV, 4.6-fold; V/VI, 5.6-fold; layer VIb, 12.3-fold). In the APO group, c-Fos protein levels were significantly decreased on the deprived side for all layers (II/III, 2-fold; IV, 6-fold; V/VI, 2-fold), although changes in layer VIb were not detected when comparing both hemispheres (Filipkowski et al., 2001) (Fig. 5.4). These authors also tested the contributions of NMDA receptor activation to APO-induced c-Fos expression, by co-treating animals with MK-801, an NMDA-receptor antagonist. It was found that, at low doses, MK-801 did not impact APO-induced c-Fos protein expression while, at high doses, the behavior of animals was significantly impaired, which prevented, to a large extent, the comparisons of c-Fos levels at high and low doses (Filipkowski et al., 2001). These results suggested that increased whisking behavior is paralleled by a significant increase in *c-fos* expression levels in the rodent barrel cortex, and that the expression of this IEG is not likely to be regulated by NMDA receptor activation in this system.

Melzer and Steiner (1997) investigated *c-fos* mRNA expression in animals that had metal filaments glued to specific whiskers on the left snout and were exposed to a pulsating magnetic field that moved whiskers in the rostro-caudal direction, mimicking the direction and frequency of natural sniffing behavior. For these experiments, left C2 whisker or left C1, C2 and C3 whiskers were stimulated by the magnetic field for 5–15 minutes. In some animals, specific whiskers were clipped. In addition, these authors studied the impact of the intensity of the stimulus on the patterns of *c-fos* expression in the rodent barrel cortex (Melzer and Steiner, 1997). mRNA levels were assessed by *in-situ* hybridization 5–30 minutes after stimulation offset. Autoradiogram analysis revealed that C2 stimulation led

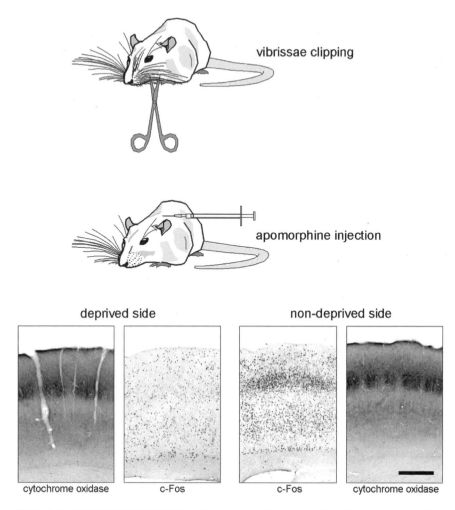

Figure 5.4. c-Fos protein induction in the rat barrel cortex (barrels visualized by cytochrome oxidase staining) following apomorphine injection and whisking behavior induced by the drug. Rats had their vibrissae clipped from one side of the snout so that the deprived side can be compared with non-deprived one. Scale bar = 0.5 mm. (From Filipkowski et al., 2001).

to a significant increase in *c-fos* mRNA levels in the C2 representation in the contralateral barrel cortex. In addition, increased signal was detected in the full column associated with the stimulated whisker. Consistent with findings obtained using other methodologies, highest hybridization levels were identified in the thalamo-recipient layer IV, although smaller, but still significant, increases were detected for both supra- and infra-granular layers, especially in deep layer V and layer VI (Melzer and Steiner, 1997). Emulsion-dipped sections that were counter-stained with Nissl yielded similar results, where highest *c-fos* mRNA levels were found in the thalamo-recipient layer of the stimulated column. Morphological analyses suggested that most *c-fos*-expressing neurons in this preparation were stellate neurons. Surprisingly, these authors reported that no changes in the overall *c-fos* expression levels were detected across the various stimulating time periods (5, 10 and 15 minutes), as well as for the different survival time points (5, 15 and 30 minutes) (but see below). Finally, Melzer and Steiner reported that the strength of the stimulation directly impacted the mRNA levels of *c-fos* in layer IV, with high levels being detected in animals exposed to strong magnetic fields, and low levels observed in rats exposed to weak fields (Melzer and Steiner, 1997). Interestingly, there is considerable overlap in the activity profiles revealed using either histological techniques for IEG expression and activity maps generated by the 2-deoxyglucose method. In these cases there is evidence to connect the increases in IEG expression or activity with putative locations of thalamic terminations (Bernardo and Woolsey, 1987; Jensen and Killackey, 1987).

c-Fos protein levels were also assessed and shown to be regulated in the barrel cortex of rats following EE exposure (Montero, 1997; Staiger et al., 2000). For instance, Staiger and colleagues used ICC to detect c-Fos in animals that underwent unilateral clipping of whiskers, with the sparing of 1 or 2 whiskers, followed by overnight exposure to an EE. All whiskers on the other side of the snout were left intact (Staiger et al., 2000). In accordance with previous reports, these authors detected very low basal levels of c-Fos protein in the barrel cortex, which were similar to those levels of immunoreactivity detected in the deprived cortex. EE exposure triggered a robust c-Fos induction in the contralateral cortex, associated with the unclipped snout. Conversely, EE exposure in animals that had intact C1 and/or C2 whiskers elicited a marked increase in c-Fos immunoreactivity (as quantified by densitometric analysis) in the associated barrel, while virtually no increases were detected in the deprived cortex. Similarly to previous reports, c-Fos induction was more robust in the thalamo-recipient layer and lower layer III, followed by the interface of layers V and VI. Moreover increased c-Fos was also detected in layer II and upper layer III (Staiger et al., 2000).

Subsequent work from the same group investigated the neurochemical identity of c-Fos positive neurons following the EE paradigm briefly described above. Using a combination of morphological criteria, as evidenced by Lucifer Yellow filling of c-Fos positive neurons, as well as double-immunocytochemical staining for excitatory and inhibitory markers, these authors were able to document that both inhibitory and excitatory neurons participate in the response of the barrel cortex to EE exposure. For the detection of inhibitory neurons, Staiger and colleagues used an antibody directed at the 65 KDa glutamic acid decarboxylase (GAD) protein, one of the synthetic enzymes for GABA, as well as the calcium-binding proteins parvalbumin, calbindin and calretinin, which are thought to be expressed in different, largely non-overlapping, populations of inhibitory neurons. It was found that after EE exposure, the proportion of GAD-positive neurons that co-localized with c-Fos underwent a significant increase for both supragranular layers and layer IV. This effect was not detected in the infragranular layers (Staiger et al., 2002). Likewise, this outcome was detected for the population of parvalbumin-positive neurons that co-localized with c-Fos, for all cortical layers.

Co-localization studies conducted with calbindin revealed that this phenomenon was significant for both granular and infragranular layers, but not supragranular layers. Finally, calretinin encompassed the only population of inhibitory neurons where no changes were detected as a function of experience in the EE (Staiger et al., 2002). In the barrel cortex, highest numbers of c-Fos positive neurons were detected in layer IV, while the lowest numbers were consistently detected in the supragranular layers following EE exposure, with supragranular, granular and infragranular layers exhibiting 2.6-, 4.4- and 1.8-fold induction, respectively (Staiger et al., 2002). These findings suggested that not only excitatory, but also inhibitory neurons in the rodent S1 are directly activated by tactile experience that results from EE exposure.

Bisler and co-workers (2002), using the same paradigm of whisker clipping and EE exposure, set out to systematically characterize the time-course of the effects of EE on c-Fos expression levels. In order to achieve this goal, after appropriate clipping, authors exposed rats to the EE setting for various time-points that ranged from 10 minutes to 5 days. ICC directed at c-Fos was conducted and analyzed in the somatosensory cortex of rats (Bisler et al., 2002). It was reported that c-Fos expression was detected at significant levels after 10 minutes of exposure in the EE. Highest c-Fos protein levels were detected in layer IV and lower layer III at the 1 hour time-point, where the columns associated with the stimulated whiskers became highly evident. c-Fos levels markedly decreased after that time-point, but were still enhanced, as compared to basal levels, following 6 and 14 hours of EE exposure. Immunoreactivity levels for this IEG were not significantly different from basal levels after 2 and 5 days of EE experience (Bisler et al., 2002).

3.2. fosB

FosB expression has been studied through immunocytochemical methods in the paradigm of APO-induced whisking behavior following complete unilateral whisker clipping (Filipkowski et al., 2001). Whisker clipping alone or in combination with APO treatment did not yield any measurable alterations in the pattern of FosB immunoreactivity, for any cortical layer in the rodent S1. Furthermore, co-treatment of APO and MK-801 did not alter the number of FosB positive cells in the barrel cortex (Filipkowski et al., 2001).

3.3. junB

The regulation of *junB* expression has been studied in the rat barrel cortex following different experimental paradigms. Filipkowski and colleagues have investigated the expression of JunB, using ICC, following whisking induced by APO-treatment, with and without clipping of certain sets of whiskers (Filipkowski et al., 2001). Sensory deprivation induced by whisker clipping significantly decreased JunB in all cortical layers, with the exception of layer VIb. Conversely, APO treatment increased JunB levels in layer VIb on both hemispheres. Increases in JunB levels following APO injection were measured to be 4.3-fold for the deprived, and 6.4-fold for the spared cortex. In addition, co-treatment of animals with a low dose of MK-801 and APO elicited an increase in JunB levels in layers IV and VI, suggesting that JunB regulation by somatosensory experience in the barrel cortex does not depend, to a large extent, on NMDA receptor activation (Filipkowski et al., 2001).

Staiger and colleagues (2000) investigated JunB expression in rats that had intact whiskers on one side of the snout while, on the other side, only one or two whiskers were spared, and were exposed overnight to an EE. Using ICC, these authors were able to document that JunB is found at very low basal levels in the rat barrel cortex. A similar profile was also detected for the deprived cortex, where low levels of immunoreactivity for JunB were detected using densitometric analysis (Staiger et al., 2000). In contrast, a marked increase in JunB immunoreactivity was detected in stimulated columns within the S1 of animals exposed to the EE. Highest increases in JunB levels were detected in the thalamo-recipient layer and lower layer III, followed by the transition regions between layers V and VI. Immunoreactive profiles were also reported in layer II and upper layer III (Staiger et al., 2000). The time-course of EE-induced JunB regulation was addressed by Bisler and co-workers (2002). In these experiments, authors were able to reproduce previous findings indicating that JunB was expressed at very low basal levels in the rat barrel cortex. After 10 minutes of exposure to an EE, JunB levels were significantly increased, as compared to basal levels. Peak protein levels were detected after 1 hour of experience in the EE, when a clear columnar pattern of immunoreactivity could be detected (Bisler et al., 2002). JunB immunoreactivity levels decreased significantly after that time point and, after 14 hours of EE exposure, were not significantly different from those levels found in control, baseline animals.

3.4. c-jun

c-Jun expression in the somatosensory cortex has been studied in the context of overnight exposure of animals to an EE following clipping of specific whiskers (Staiger et al., 2000). In contrast to the low basal JunB immunoreactive levels, c-Jun was observed at high basal levels in the rat barrel cortex. Interestingly, c-Jun levels in the barrel cortex were neither affected by sensory deprivation, induced by whisker clipping, nor by somatosensory stimulation, that occurred during EE exposure (Staiger et al., 2000). These data suggest that c-Jun is not regulated by experience in the rodent barrel cortex and is unlikely to participate in the cascade of molecular events that take place during cortical reorganization induced by whisker clipping and/or exposure of animals to an EE.

4. The Immediate Early Genes NGFI-A, NGFI-B and NGFI-C

The IEGs NGFI-A (a.k.a., *zif268*, *egr-1*, *krox-24* and *zenk*) and NGFI-C (a.k.a., *egr-4* and *pAT133*) are also rapidly and transiently induced following cell stimulation and encode zinc-finger transcriptional regulators that recognize a highly conserved DNA motif. Upon binding to this consensus, these transcription factors regulate the expression of target genes which are thought to exert more stable, long-lasting changes in cellular physiology, ranging from neuronal plasticity and neurotransmitter release to cell growth and differentiation (reviewed in Beckmann and Wilce, 1997; O'Donovan et al., 1999; Pinaud, 2004, 2005). Conversely, the IEG NGFI-B (a.k.a., Nur77 and Nr4a1) encodes an orphan nuclear steroid receptor that has been shown to exert powerful transcriptional regulatory activities and is, therefore, a member of a different family of genes. Prominent roles have been described for the protein encoded by this IEG in hormone biosynthesis and development (Parker and Schimmer, 1994; Maruyama et al., 1998).

The expression of these three IEGs has been studied in the rodent somatosensory cortex. The regulation of NGFI-A has been, by far, the most studied in the barrel cortex in a number of experimental paradigms. On the other hand, few studies have tackled the expression patterns of NGFI-B and NGFI-C in this system.

4.1. NGFI-A

The effects of tactile stimulation on the regulation of NGFI-A expression in the rodent barrel cortex were first studied by Mack and Mack (1992). Their paradigm of choice was the unilateral stimulation of whiskers (15 minutes) with a paintbrush in urethane anesthetized rats. NGFI-A protein levels were assessed with ICC 2 hours after stimulation. This work reported that NGFI-A was encountered in the rat S1 at high basal levels. Stimulation triggered the highest increases in NGFI-A immunoreactivity in the thalamorecipient layer. Moreover, significant increases were also detected in layers II, III and VI (Mack and Mack, 1992).

Steiner and Gerfen (1994) investigated the regulation of NGFI-A (zif268) mRNA in the rat barrel cortex following APO-induced whisking/tactile behavior in animals that underwent clipping of all whiskers in the left side of the snout. In accordance with previous observations at the protein level, it was found that NGFI-A mRNA is expressed at high basal levels in the S1; highest in-situ hybridization signal in the baseline condition was detected in layers IV and V/VI. Significant increases in mRNA levels, following the same laminar profile, were evidenced after APO injection. In addition, this treatment triggered the upregulation of this IEG in deep layer VI. As for other IEGs such as c-fos, whisker clipping in one side of the snout significantly downregulated NGFI-A mRNA levels in the contralateral cortex in both control and APO-treated animals (Steiner and Gerfen, 1994). Interestingly, however, the effects of clipping on NGFI-A levels were disparate from basal levels: whereas the strongest downregulation of mRNA levels were detected in layers V/VI, followed by cortical layer IV in control animals, APO-injected rats exhibited the peak suppression in the thalamorecipient layer, followed by layers V/VI.

Using the same paradigm of APO-induced whisking, associated with unilateral whisker clipping, Filipkowski and colleagues (2001) investigated NGFI-A protein regulation in the barrel cortex, using immunocytochemical approaches.

As previously observed at the mRNA level, clipping significantly decreased basal NGFI-A immunoreactivity in the deprived cortex. Conversely, APO-treatment did not elicit increases in NGFI-A levels in the barrel cortex, with the exception of a 7.5-fold increase in layer VIb of the deprived cortex. Sensory deprivation induced by clipping in this condition decreased the number of NGFI-A positive neurons in cortical layers II/III, IV and V/VI. Interestingly, blockade of NMDA receptor activation by a low dose of MK-801 did not interfere with any of the effects observed, irrespective of APO administration (Filipkowski et al., 2001).

In-situ hybridization was also used to study NGFI-A regulation in the barrel cortex by tactile experience in animals that had metal filaments glued to specific sets of whiskers and exposed to a magnetic field (methodology discussed in more detail in the *c-fos* section) (Melzer and Steiner, 1997). Single- or paired-whisker stimulation led to a significant upregulation of NGFI-A mRNA in the column specifically associated with the stimulus. Although significant increases were detected in supra- and infra-granular layers, highest upregulation was detected in the thalamorecipient layer. In conditions where whiskers were clipped surrounding the stimulated vibrissa, there was an increase in the signal-to-noise ratio of the hybridization signal; this phenomenon was attributed to a decrease in the basal levels in the columns associated with the clipped whiskers, along with the upregulation related to the stimulated vibrissa. Finally, similarly to findings obtained with *c-fos* expression, NGFI-A mRNA levels varied as a function of the strength of the stimulus in cortical layer IV (Melzer and Steiner, 1997).

Overnight exposure of rats to an EE following unilateral clipping of different sets of whiskers (details in *c-fos* section) also provided evidence for the regulation of NGFI-A by tactile experience (Staiger et al., 2000). In this paradigm, highest basal levels were detected in cortical layers IV and VI; this laminar organization was preserved in both deprived and spared cortex. Similar findings were obtained with mRNA analysis (Melzer and Steiner, 1997). The signal-to-noise ratio of NGFI-A protein was higher in animals where whiskers surrounding the stimulated vibrissa were clipped, given that EE-induced upregulation associated with the stimulated column occurred in parallel with the downregulation of protein levels in the deprived modules (Staiger et al., 2000). In a systematic time-course analysis using this same experimental paradigm, Bisler and colleagues (2002) found that NGFI-A protein levels, as measured by a grey-level index, were significantly increased after 10 minutes of EE exposure. Peak protein levels were evidenced at the 1 hour time point, being greatly reduced thereafter, although still significant at the 6-hour time-point, and reaching basal levels after 14 hours of exposure (Bisler et al., 2002). In these experiments, highest immunoreactive levels were consistently detected in lower layer III and IV.

4.2. NGFI-B

In 1992, Mack and Mack investigated the effects of unilateral whisker paint-brush stimulation on NGFI-B protein levels in the barrel cortex of anesthetized animals. In these experiments it was found that NGFI-B was expressed at moderate basal levels in the S1. In addition, sensory stimulation triggered an increase in NGFI-B protein levels in the supragranular layers and cortical layer IV (Mack and Mack, 1992).

4.3. NGFI-C

The regulation of NGFI-C in the barrel cortex following tactile (and seizure) experience was studied in anesthetized rats by Mack and colleagues (1995). Using RT-PCR from S1 mRNA extracts these authors were able to document that NGFI-C is significantly upregulated in the cortex that is contralateral to the stimulation side. NGFI-C levels peaked by 30–60 minutes, and returned to basal levels 4–5 hours, after stimulation (Mack et al., 1995). Peak NGFI-C levels were quantified to 1.8-fold of the basal levels, while no changes were detected for both ipsilateral cortex and unstimulated controls. *In-situ* hybridization with a specific NGFI-C probe revealed that although this mRNA is expressed in most cortical layers, peak signal was detected in layers II and IV of the contralateral cortex (Mack et al., 1995).

5. Other Genes

5.1. ICER

ICER (inducible cAMP early repressor) encodes a set of leucine-zipper transcriptional regulators that are originated from the use of an alternative promoter located in an intron embedded in the CREM (cAMP responsive element modulator) gene (reviewed in Sassone-Corsi, 1998; Mioduszewska et al., 2003). ICER negatively regulates the expression of genes impacted by the cAMP pathway, via CRE (cAMP responsive element) binding, and has been proposed to constitute an antagonist of the transcription factor CREB, which generally positively regulates cAMP sensitive genes through CRE (Sassone-Corsi, 1998; Mioduszewska et al., 2003). Finally, ICER may actively participate in activity-dependent neuronal plasticity and recent evidence suggests that these proteins may also contribute to the regulation of apoptosis (Jaworski et al., 2003; Mioduszewska et al., 2003).

ICER expression has been investigated in the barrel cortex after overnight exposure of rats to an EE following unilateral clipping of C1 and/or C1 and C2 whiskers (Staiger et al., 2000). Immunoreactivity for ICER was detected at moderately high basal levels. Exposure of animals to the EE triggered a significant induction of ICER and revealed cortical columns associated with the stimulated whiskers. Importantly, authors suggested that, based on a qualitative assessment, increased ICER immunoreactivity may have been associated with increased immunolabeling signal per neuron, rather than a general increase in the ICER-expressing neuronal population (quantification in this experiment was performed through optical density analysis, rather than cell counts) (Staiger et al., 2000). Laminar analysis revealed that EE-induced ICER upregulation affected more strongly cortical layers IV and lower layer III in the stimulus-associated columns. The time-course of EE-induced ICER regulation was addressed by

Bisler and colleagues. Unlike most IEGs discussed in this chapter, ICER expression levels were very low after a 1 hour experience in the EE setting and peaked after 6 hours (Bisler et al., 2002). At peak levels, ICER immunoreactive neurons anatomically revealed stimulated columns. Interestingly, although ICER levels decreased after 6 hours of exposure to the EE, immunolabeling detected at 2 and 5 days of continuous exposure was significantly above basal levels, suggesting that the expression of this repressor protein is not only delayed in comparison to other IEGs, but also sustained.

5.2. cpg15

cpg15 is an IEG induced by Ca^{2+} influx through NMDA receptors, and L-type voltage-sensitive calcium channels, and has been repeatedly implicated in neuronal plasticity by promoting dendritic and axonal expansion in an activity-dependent fashion (Nedivi et al., 1998; Fujino et al., 2003). It has been shown that following trimming of all whiskers, with the exception of D1, *cpg15* expression is significantly depressed in the deprived barrels and enhanced in the barrel column associated with the spared whisker, in 4 week-old mouse barrel cortex (Harwell et al., 2005).

6. Plasticity in the Somatosensory Cortex

The various protocols of ablation and stimulation of whiskers, including the ones discussed in this review, have been successfully used to characterize the molecular, cellular, biophysical and physiological foundations of experience-dependent plasticity within sensory systems. A series of elegant experiments conducted in the rodent barrel cortex demonstrated that a significant depression of electrophysiological responses occurs in neurons recorded within the deprived column after unilateral whisker ablation (Glazewski et al., 1998; Fox, 2002). This effect has been shown to be stronger in cells located in the supragranular layers of the deprived cortex. Interestingly, partial whisker ablation led to a much more substantial depression in the deprived columns, as compared to the complete unilateral removal of the vibrissae, especially in layers II/III, suggesting that activity suppression induced by whisker removal occurs through a synergistic mechanism involving the depression of activity in the deprived cortex, as well as the stimulus-driven activity in the spared cortex (Glazewski et al., 1998).

In experiments where a single or multiple whiskers were spared, it was also observed that cells in layers II/III and the thalamo-recipient layer of the deprived cortex undergo potentiation of electrophysiological responses over time. This outcome occurs significantly faster when multiple whiskers are spared, as compared to paradigms where only one whisker is left intact, again corroborating for hypothesis for a cooperative effect between deprived and spared cortices in mediating plastic changes in the S1 (Wallace and Fox, 1999).

The experience-dependent electrophysiological changes that are associated with these types of plasticity-inducing paradigms have been repeatedly shown, as discussed above, to markedly impact the regulation of IEGs in the barrel cortex. For most IEGs discussed in this chapter, the most robust changes in gene expression patterns were detected in layer IV, as well as layers II/III. These findings on the laminar profiles of IEG expression are in close topographical accordance with the electrophysiological findings discussed here and implicate layers II/III and IV in experience-dependent plasticity. Coincidently, the highest density of GABAergic neurons in the barrel cortex and other primary sensory areas is also found the supragranular layers and the thalamo-recipient layer of S1 (Gabbott and Somogyi, 1986; Keller and White, 1987; Jones, 1993), suggesting that GABAergic neurons may play a prominent role in the regulation of plasticity in sensory systems, including within the barrel cortex. In fact, GABAergic transmission and cell distribution have been shown to be associated with both experience-, as well as injury-induced plasticity, in the somatosensory cortex of a number of species (Welker et al., 1989a, b; Micheva and Beaulieu, 1995; Tremere et al., 2001a, b; Li et al., 2002). Likewise, in these paradigms, the roles of GABA in activity-regulated plastic changes are also more robust in supragranular layers and layer IV (reviewed in Tremere et al., 2003; Tremere and Pinaud, 2005). In fact, roles for lateral inhibition, especially in layer IV, in mediating plastic changes in the barrel cortex associated with whisker removal have been previously proposed (for review see Fox, 2002). These findings are in accordance with data discussed above indicating that EE exposure triggers c-Fos expression in a significant population of inhibitory neurons in the rat barrel cortex (Filipkowski et al., 2000; Staiger et al., 2002) (Fig. 5.5).

Although experience-dependent changes in GABAergic tone tend to occur on a relatively fast time scale, long-term effects (e.g., rewiring of GABAergic synapses associated with lateral inhibition or regulation of GAD activity) may be orchestrated in part by IEG expression within GABAergic neurons. In fact, as discussed above, some IEGs play a role in the regulation of a number of proteins involved in neurotransmitter release and other vital processes associated with the plasticity response (Pospelov et al., 1994; Thiel et al., 1994; Petersohn et al., 1995; Petersohn and Thiel, 1996; Pinaud, 2004, 2005). Even though, to the best of our knowledge, only the expression of c-Fos has been investigated in inhibitory neurons of the barrel cortex, it is highly likely that other IEGs associated with activity-dependent plasticity may also be expressed in GABAergic neurons in the rodent S1. Such possibility has been addressed in other sensory systems (Pinaud et al., 2004) and implicates IEG expression in GABAergic neurons as a likely major contributor to long-term plasticity.

Overall, these experience-dependent IEG expression programs that occur both in excitatory and inhibitory neurons may be aimed, via late-response genes, at rearranging the connectivity of lateral connections, especially within the granular and supragranular layers of the whisker columns, so that long-term circuit reorganization ensues. According to this hypothesis, previous work showed that alpha-CaMKII and some CREB isoforms are essential for the experience-dependent changes observed in the rodent barrel cortex (Glazewski et al., 1996, 1999; Fox, 2002). These two molecules have been shown to be heavily associated with the induction of a number of IEGs (reviewed in Pinaud, 2005), which

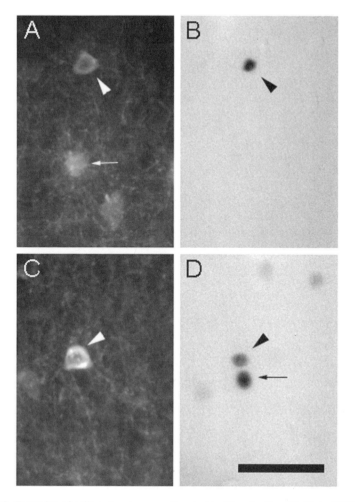

Figure 5.5. Examples of cells expressing parvalbumin (A, C) and c-Fos (B, D) in II/III layer of the barrel cortex following manual stimulation (Fig. 5.2). Cells containing one protein are marked with small arrows; cells with both proteins are marked with arrowheads. Scale bar = 50 μm. (From Filipkowski et al., 2000).

further implicates the expression of this class of genes in the molecular pathways that promote experience-dependent changes within the vertebrate somatosensory cortex, and potentially throughout the vertebrate sensory systems.

References

Ahl AS (1986). The role of vibrissae in behavior: a status review. Vet Res Commun 10:245-268.

Angel P, Karin M (1991). The role of Jun, Fos and the AP-1 complex in cell-proliferation and transformation. Biochim Biophys Acta 1072:129-157.

Beckmann AM, Wilce PA (1997). Egr transcription factors in the nervous system. Neurochem Int 31:477-510; discussion 517-476.

Bernardo KL, Woolsey TA (1987). Axonal trajectories between mouse somatosensory thalamus and cortex. J Comp Neurol 258:542-564.

Bialy M, Beck J (1993). The influence of vibrissae removal on copulatory behaviour in male rats. Acta Neurobiol Exp (Wars) 53:415-419.

Bisler S, Schleicher A, Gass P, Stehle JH, Zilles K, Staiger JF (2002). Expression of c-Fos, ICER, Krox-24 and JunB in the whisker-to-barrel pathway of rats: time course of induction upon whisker stimulation by tactile exploration of an enriched environment. J Chem Neuroanat 23:187-198.

Eferl R, Wagner EF (2003). AP-1: a double-edged sword in tumorigenesis. Nat Rev Cancer 3:859-868.

Filipkowski RK (2000). Inducing gene expression in barrel cortex—focus on immediate early genes. Acta Neurobiol Exp (Wars) 60:411-418.

Filipkowski RK, Rydz M, Kaczmarek L (2001). Expression of c-Fos, Fos B, Jun B, and Zif268 transcription factor proteins in rat barrel cortex following apomorphine-evoked whisking behavior. Neuroscience 106:679-688.

Filipkowski RK, Rydz M, Berdel B, Morys J, Kaczmarek L (2000). Tactile experience induces c-fos expression in rat barrel cortex. Learn Mem 7:116-122.

Fox K (2002). Anatomical pathways and molecular mechanisms for plasticity in the barrel cortex. Neuroscience 111:799-814.

Fujino T, Lee WC, Nedivi E (2003). Regulation of cpg15 by signaling pathways that mediate synaptic plasticity. Mol Cell Neurosci 24:538-554.

Gabbott PL, Somogyi P (1986). Quantitative distribution of GABA-immunoreactive neurons in the visual cortex (area 17) of the cat. Exp Brain Res 61:323-331.

Glazewski S (1998). Experience-dependent changes in vibrissae evoked responses in the rodent barrel cortex. Acta Neurobiol Exp (Wars) 58:309-320.

Glazewski S, Chen CM, Silva A, Fox K (1996). Requirement for alpha-CaMKII in experience-dependent plasticity of the barrel cortex. Science 272:421-423.

Glazewski S, McKenna M, Jacquin M, Fox K (1998). Experience-dependent depression of vibrissae responses in adolescent rat barrel cortex. Eur J Neurosci 10:2107-2116.

Glazewski S, Barth AL, Wallace H, McKenna M, Silva A, Fox K (1999). Impaired experience-dependent plasticity in barrel cortex of mice lacking the alpha and delta isoforms of CREB. Cereb Cortex 9:249-256.

Guic-Robles E, Valdivieso C, Guajardo G (1989). Rats can learn a roughness discrimination using only their vibrissal system. Behav Brain Res 31:285-289.

Guic-Robles E, Jenkins WM, Bravo H (1992). Vibrissal roughness discrimination is barrelcortex-dependent. Behav Brain Res 48:145-152.

Gustafson JW, Felbain-Keramidas SL (1977). Behavioral and neural approaches to the function of the mystacial vibrissae. Psychol Bull 84:477-488.

Harwell C, Burbach B, Svoboda K, Nedivi E (2005). Regulation of cpg15 expression during single whisker experience in the barrel cortex of adult mice. J Neurobiol 65:85-96.

Herdegen T, Leah JD (1998). Inducible and constitutive transcription factors in the mammalian nervous system: control of gene expression by Jun, Fos and Krox, and CREB/ATF proteins. Brain Res Brain Res Rev 28:370-490.

Hughes P, Dragunow M (1995). Induction of immediate-early genes and the control of neurotransmitter-regulated gene expression within the nervous system. Pharmacol Rev 47:133-178.

Jaworski J, Mioduszewska B, Sanchez-Capelo A, Figiel I, Habas A, Gozdz A, Proszynski T, Hetman M, Mallet J, Kaczmarek L (2003). Inducible cAMP early repressor, an endogenous antagonist of cAMP responsive element-binding protein, evokes neuronal apoptosis in vitro. J Neurosci 23:4519-4526.

Jensen KF, Killackey HP (1987). Terminal arbors of axons projecting to the somatosensory cortex of the adult rat. I. The normal morphology of specific thalamocortical afferents. J Neurosci 7:3529-3543.

Jones EG (1993). GABAergic neurons and their role in cortical plasticity in primates. Cereb Cortex 3:361-372.

Kaczmarek L, Chaudhuri A (1997). Sensory regulation of immediate-early gene expression in mammalian visual cortex: implications for functional mapping and neural plasticity. Brain Res Brain Res Rev 23:237-256.

Kaminska B, Pyrzynska B, Ciechomska I, Wisniewska M (2000). Modulation of the composition of AP-1 complex and its impact on transcriptional activity. Acta Neurobiol Exp (Wars) 60:395-402.

Keller A, White EL (1987). Synaptic organization of GABAergic neurons in the mouse SmI cortex. J Comp Neurol 262:1-12.

Kobierski LA, Chu HM, Tan Y, Comb MJ (1991). cAMP-dependent regulation of proenkephalin by JunD and JunB: positive and negative effects of AP-1 proteins. Proc Natl Acad Sci USA 88:10222-10226.

Kossut M (1998). Experience-dependent changes in function and anatomy of adult barrel cortex. Exp Brain Res 123:110-116.

Li CX, Callaway JC, Waters RS (2002). Removal of GABAergic inhibition alters subthreshold input in neurons in forepaw barrel subfield (FBS) in rat first somatosensory cortex (SI) after digit stimulation. Exp Brain Res 145:411-428.

Lipp HP, Van der Loos H (1991). A computer-controlled Y-maze for the analysis of vibrissotactile discrimination learning in mice. Behav Brain Res 45:135-145.

Mack KJ, Mack PA (1992). Induction of transcription factors in somatosensory cortex after tactile stimulation. Brain Res Mol Brain Res 12:141-147.

Mack KJ, Yi SD, Chang S, Millan N, Mack P (1995). NGFI-C expression is affected by physiological stimulation and seizures in the somatosensory cortex. Brain Res Mol Brain Res 29:140-146.

Maruyama K, Tsukada T, Ohkura N, Bandoh S, Hosono T, Yamaguchi K (1998). The NGFI-B subfamily of the nuclear receptor superfamily (review). Int J Oncol 12:1237-1243.

Mello CV, Velho TA, Pinaud R (2004). Song-induced gene expression: a window on song auditory processing and perception. Ann NY Acad Sci 1016:263-281.

Melzer P, Steiner H (1997). Stimulus-dependent expression of immediate-early genes in rat somatosensory cortex. J Comp Neurol 380:145-153.

Micheva KD, Beaulieu C (1995). An anatomical substrate for experience-dependent plasticity of the rat barrel field cortex. Proc Natl Acad Sci USA 92:11834-11838.

Mioduszewska B, Jaworski J, Kaczmarek L (2003). Inducible cAMP early repressor (ICER) in the nervous system—a transcriptional regulator of neuronal plasticity and programmed cell death. J Neurochem 87:1313-1320.

Montero VM (1997). c-fos induction in sensory pathways of rats exploring a novel complex environment: shifts of active thalamic reticular sectors by predominant sensory cues. Neuroscience 76:1069-1081.

Morgan JI, Curran T (1991). Stimulus-transcription coupling in the nervous system: involvement of the inducible proto-oncogenes fos and jun. Annu Rev Neurosci 14:421-451.

Nedivi E, Wu GY, Cline HT (1998). Promotion of dendritic growth by CPG15, an activity-induced signaling molecule. Science 281:1863-1866.

O'Donovan KJ, Tourtellotte WG, Millbrandt J, Baraban JM (1999). The EGR family of transcription-regulatory factors: progress at the interface of molecular and systems neuroscience. Trends Neurosci 22:167-173.

Parker KL, Schimmer BP (1994). The role of nuclear receptors in steroid hormone production. Semin Cancer Biol 5:317-325.

Petersohn D, Thiel G (1996). Role of zinc-finger proteins Sp1 and zif268/egr-1 in transcriptional regulation of the human synaptobrevin II gene. Eur J Biochem 239:827-834.

Petersohn D, Schoch S, Brinkmann DR, Thiel G (1995). The human synapsin II gene promoter. Possible role for the transcription factor zif268/egr-1, polyoma enhancer activator 3, and AP2. J Biol Chem 270:24361-24369.

Pinaud R (2004). Experience-dependent immediate early gene expression in the adult central nervous system: evidence from enriched-environment studies. Int J Neurosci 114:321-333.

Pinaud R (2005). Critical calcium-regulated biochemical and gene expression programs involved in experience-dependent plasticity. In: Plasticity in the Visual System: From Genes to Circuits (Pinaud R, Tremere LA, De Weerd P, eds), pp. 153-180. New York: Springer-Verlag.

Pinaud R, Velho TA, Jeong JK, Tremere LA, Leao RM, von Gersdorff H, Mello CV (2004). GABAergic neurons participate in the brain's response to birdsong auditory stimulation. Eur J Neurosci 20:1318-1330.

Pospelov VA, Pospelova TV, Julien JP (1994). AP-1 and Krox-24 transcription factors activate the neurofilament light gene promoter in P19 embryonal carcinoma cells. Cell Growth Differ 5:187-196.

Sassone-Corsi P (1998). Coupling gene expression to cAMP signalling: role of CREB and CREM. Int J Biochem Cell Biol 30:27-38.

Schiffman HR, Lore R, Passafiume J, Neeb R (1970). Role of vibrissae for depth perception in the rat (Rattus norvegicus). Anim Behav 18:290-292.

Staiger JF, Bisler S, Schleicher A, Gass P, Stehle JH, Zilles K (2000). Exploration of a novel environment leads to the expression of inducible transcription factors in barrel-related columns. Neuroscience 99:7-16.

Staiger JF, Masanneck C, Bisler S, Schleicher A, Zuschratter W, Zilles K (2002). Excitatory and inhibitory neurons express c-Fos in barrel-related columns after exploration of a novel environment. Neuroscience 109:687-699.

Steiner H, Gerfen CR (1994). Tactile sensory input regulates basal and apomorphine-induced immediate-early gene expression in rat barrel cortex. J Comp Neurol 344:297-304.

Thiel G, Schoch S, Petersohn D (1994). Regulation of synapsin I gene expression by the zinc finger transcription factor zif268/egr-1. J Biol Chem 269:15294-15301.

Tremere L, Hicks TP, Rasmusson DD (2001a). Expansion of receptive fields in raccoon somatosensory cortex in vivo by GABA(A) receptor antagonism: implications for cortical reorganization. Exp Brain Res 136:447-455.

Tremere L, Hicks TP, Rasmusson DD (2001b). Role of inhibition in cortical reorganization of the adult raccoon revealed by microiontophoretic blockade of GABA(A) receptors. J Neurophysiol 86:94-103.

Tremere LA, Pinaud R (2005). Intra-cortical inhibition in the regulation of receptive field properties and neural plasticity in the primary visual cortex. In: Plasticity in the Visual System: From Genes to Circuits (Pinaud R, Tremere LA, De Weerd P, eds), pp. 229-243. New York: Spinger-Verlag.

Tremere LA, Pinaud R, De Weerd P (2003). Contributions of inhibitory mechanisms to perceptual completion and cortical reorganization. In: Filling-in: From Perceptual Completion to Cortical Reorganization (Pessoa L, De Weerd P, eds), pp. 295-322. New York: Oxford University Press.

Wallace H, Fox K (1999). Local cortical interactions determine the form of cortical plasticity. J Neurobiol 41:58-63.

Welker E, Soriano E, Van der Loos H (1989a). Plasticity in the barrel cortex of the adult mouse: effects of peripheral deprivation on GAD-immunoreactivity. Exp Brain Res 74:441-452.

Welker E, Soriano E, Dorfl J, Van der Loos H (1989b). Plasticity in the barrel cortex of the adult mouse: transient increase of GAD-immunoreactivity following sensory stimulation. Exp Brain Res 78:659-664.

6

Immediate Early Genes Induced in Models of Acute and Chronic Pain

SHANELLE W. KO

Department of Physiology, University of Toronto, Toronto, Canada

1. Introduction

Since the initial observation that peripheral noxious stimulation could induce expression of c-Fos in the spinal cord (Hunt et al., 1987), a myriad of studies have reported the activation of immediate early genes (IEGs) in response to nociceptive stimulation. Members of this class of genes are rapidly and transiently induced after cell stimulation, without the requirement of new protein synthesis. Given that IEG expression heavily depends on neuronal depolarization, the probing of both mRNA and protein products encoded by this class of genes have been extensively used to map neural circuits associated with a variety of experimental paradigms, including acute and chronic pain (Zimmermann and Herdegen, 1994; Herdegen and Zimmermann, 1995; Herdegen and Leah, 1998; Bester and Hunt, 2002). In fact, a variety of noxious stimuli, differing in type, duration and site of application have been shown to induce the expression of IEGs in both the spinal cord and brain (Zimmermann and Herdegen, 1994; Herdegen and Zimmermann, 1995; Herdegen and Leah, 1998; Bester and Hunt, 2002). These changes in gene expression have been proposed to mediate a cascade of events that ultimately lead to long-term alterations in neuronal functioning (Kaczmarek and Chaudhuri, 1997; Herdegen and Leah, 1998; Pinaud, 2004, 2005). Studying the induction of IEGs not only allows for the mapping of nociceptive pathways but also provides information about how abnormal activity within the central nervous system contributes to pathological pain. Increased pain sensitivity may arise from long-term changes in the excitability of neurons so that they are more sensitive to subsequent stimuli, and these changes may be mediated in part through IEG expression after noxious insult (Woolf and Salter, 2000; Scholz and Woolf, 2002).

The following chapter seeks to introduce the reader to a broad array of literature covering the induction of IEGs after a range of noxious stimuli.

R. Pinaud, L.A. Tremere (Eds.), Immediate Early Genes in Sensory Processing, Cognitive Performance and Neurological Disorders, 93–110, ©2006 Springer Science + Business Media, LLC

Consideration is given to the character of the stimulation (for example, acutely applied mechanical stimuli versus the more persistent nature of inflammatory pain) and different spatial and temporal aspects of application. This review seeks to provide a comprehensive introduction to the different pain models used to map nociceptive pathways and map neuronal activation with IEG expression in the brain.

While IEG activation associated with noxious stimuli has been reported throughout brain and spinal cord regions, the body of literature is unequally weighted in favor of the study of c-Fos activation in the spinal cord. Given that several complete reviews have been published in recent years on IEG regulation in the spinal cord (Zimmermann and Herdegen, 1994; Herdegen and Leah, 1998; Zimmermann, 2001; Bester and Hunt, 2002; Coggeshall, 2005), I will discuss the activation of IEGs during pain processing in supraspinal structures, focusing on the transcriptional regulators *c-fos*, *egr-1* (also known as *zif268*, NGFI-A, *krox-24* and *zenk*) and *c-jun*. I will begin by discussing IEG expression after acutely applied thermal, mechanical and electrical stimuli, and then discuss models of pathological pain, including inflammatory, phantom and visceral pain, as well as migraines.

2. Immediate Early Gene Expression After Acute Noxious Thermal, Mechanical or Electrical Stimulation

Thermal, mechanical and electrical stimuli are commonly applied to the tail or hind paw of rodents, although distinct patterns of neuronal activation can be induced by varying the site of application, for example, mechanical stimulation of the vibrissal pad (Sugimoto et al., 1994) or thermal stimulation of the neck skin (Keay and Bandler, 1993). Thermal stimulation can arise from the immersion of the test site in a hot water bath or the application of an intense beam of light, while pinching, crushing or brushing the skin can be used to deliver mechanical stimulation. Electrical stimulation can be delivered to the hind paw, tail, or directly to the sciatic nerve (Herdegen et al., 1990; Le Bars et al., 2001). Compared to electrical stimulation, thermal and mechanical stimulation may represent more naturally and commonly occurring stimuli. Importantly, the spatial and temporal nature of the noxious stimulus chosen has been shown to affect the expression patterns of different IEGs. For example, Hunt and colleagues (1987) reported that both noxious and non-noxious peripheral stimulation evoke c-Fos expression in the spinal cord; however, the specific laminar distribution of this protein is associated with the nature of the stimulus. It is therefore important to keep in mind that patterns of IEG expression may depend on the type, duration, intensity and site of the stimulation, and that specific noxious stimulus-induced changes in IEG expression are potentially influenced by characteristics of the experimental design.

2.1. Repeated or Coincident Stimulation Induces a Distinct Pattern of IEG Expression in Subcortical Structures

Noxious mechanical stimulation delivered by pinching the hind paw increased c-Fos in the central lateral, paracentral, central medial, parafascicular, reuniens and paraventricular nuclei of the thalamus, in both hemispheres (Bullitt, 1989). A unilateral increase in c-Fos immunoreactivity was detected in the contralateral ventral posterior lateral nucleus (VPL) of the thalamus. This thalamic nucleus contains third order neurons of the anterolateral system that target the primary somatosensory cortex (S1). A separate study reported the absence of c-Fos upregulation in the VPL after noxious mechanical stimulation (Pearse et al., 2001). The discrepancy between the results obtained in these two studies may be due to the duration of the mechanical stimulus used in each study, since one study administered a single stimulus (Pearse et al., 2001) and the other repeatedly applied mechanical stimulation over a long period of time (Bullitt, 1989).

The expression of Egr-1 is regulated by noxious stimuli in subcortical stations of the rodent brain. Herdegen and colleagues (1990) demonstrated that Egr-1 (Krox-24) immunoreactivity was significantly increased in the dorsal hypothalamus, amygdala and the lateral habenula of animals that underwent electric stimulation of the sciatic nerve. In addition, a less pronounced but still significant upregulation of Egr-1 was detected in the arcuate nucleus of the hypothalamus, the lateral reticular nucleus and the periaqueductal gray matter (PAG) (Herdegen et al., 1990). Subsequent studies using repeated and coincident stimulation highlighted the spatial and temporal aspects of IEG expression. Although noxious mechanical hind paw stimulation was reported to not affect c-Fos or Egr-1 expression in the medial thalamus (including the mediodorsal, centrolateral, centromedial and intermediodorsal thamalic nuclei) and lateral thalamus (including VPL, ventral posterior medial (VPM) and the posterior group) and to reduce baseline levels of c-Jun protein, a combination of repeated mechanical and electrical stimulation of the sciatic nerve triggered a significant increase in Egr-1 expression in these stations (Pearse et al., 2001). Electrical stimulation of the sciatic nerve again failed to induce c-Fos and Egr-1 but increased c-Jun expression in the medial thalamus. In this single-stimulus paradigm, expression of all three IEGs was not changed in the lateral thalamus. Conversely, when mechanical stimulation was applied to the contralateral hind paw after electrical stimulation of the sciatic nerve (repeated stimulation) there was a significant increase in the number of c-Jun and Egr-1 immunolabeled neurons but no induction of c-Fos in both the medial and lateral thalamus. With coincident stimulation (i.e. hind paws were stimulated simultaneously), there was a decrease in the number of c-Jun-positive cells and an absence of c-Fos and Egr-1 expression in the lateral thalamus. Together, these findings demonstrate that IEGs are regulated by noxious stimuli in subcortical regions previously implicated in pain processing, as evidenced by a combination of anatomical and electrophysiological approaches (Gauriau and Bernard, 2002; Vanegas, 2004; Vanegas and Schaible, 2004). In addition, IEG expression revealed the activation of certain brain areas only after combined or sequential pain stimulation, suggesting that multiple brain regions get recruited as a function of the complexity of the pain experience. Finally, these studies indicate that the regions of interest, stimulation paradigms and IEGs being studied should be taken into account when using IEG activation to map neuronal activity after noxious stimulation.

2.2. IEG Expression Depends on the Duration and Intensity of the Stimulus

Stimulus intensity and duration required to elicit c-Fos expression in the thalamus is different when compared to the spinal cord. Whereas fifteen minutes of peripheral mechanical stimulation elicited c-Fos expression in the spinal cord (Hunt et al., 1987), continuous stimulation for four hours was required to induce consistent labeling of this IEG in various thalamic stations including the VPL (Bullitt, 1989). Electrical stimulation of the sciatic nerve at an intensity insufficient to activate A-delta and C fibers did not elicit Egr-1 expression, however, when the stimulus intensity was increased, the number of Egr-1 positive neurons was significantly enhanced in the spinal dorsal horn and several brain regions, including the PAG, the lateral habenula and the amygdala, as well as the arcuate nucleus (Herdegen et al., 1990). Pinching the tail induced c-Fos in the anterior medial preoptic area, paraventricular nucleus of the hypothalamus, paraventricular nucleus of the thalamus, medial amygdala, basolateral amygdala, lateral habenula and ventral tegmental area and, importantly, the degree of IEG expression corresponded to the number of tail pinches applied (Smith et al., 1997). These studies suggest that stimulus intensity plays a role in the regulation of IEGs and should be carefully considered when studying and standardizing IEG expression following noxious stimulation.

2.3. The Site of Stimulus Application Affects the Expression Patterns of IEGs

Mechanical stimulation delivered to different anatomical locations induces c-Fos in distinct topographical locations within the same brain area. For example, mechanical stimulation of the oral mucous membrane increased c-Fos immunoreactivity in the subnucleus caudalis, bilaterally, while stimulation of the vibrissal pad induced expression of this IEG in a different laminar sub-domain of this nucleus (Sugimoto et al., 1994). Different patterns of c-Fos expression were also induced in the PAG following deep versus superficial stimulation. For example, thermal stimuli delivered to the skin of the neck elicited different expression profiles of c-Fos in the PAG compared to rats that received deep noxious stimuli through intramuscular formalin injections; whereas the skin stimulation paradigm increased the expression of c-Fos in the caudal lateral PAG, formalin-injected animals showed a selective increase of this IEG within the caudal ventrolateral PAG (vlPAG) (Keay and Bandler, 1993). Further demonstrating the importance of the stimulus type, a later study reported that noxious cutaneous stimulation induced by clipping the skin of the neck induced c-Fos expression in the caudal vlPAG but not in the lateral PAG. This difference may be attributable to the nature of the stimulus (thermal versus mechanical) and the more persistent nature of the skin clip (Keay et al., 2001). Overall these findings suggest that the topographic organization of IEG expression patterns may reveal different anatomical and functional properties of areas involved in the processing of noxious stimulation.

3. Immediate Early Gene Expression in Models of Persistent Inflammatory Pain

Several models have been developed to mimic persistent inflammatory pain. Compared to acute noxious stimuli (mechanical/electrical/thermal) the longer duration and inescapable nature of inflammatory pain may reflect a more clinically-relevant pain model. The most commonly used chemical irritants are formalin, capsaicin, carrageenan and Complete Freund's adjuvant (CFA). Behavioral responses to a subcutaneous hind paw injection of formalin consist of licking and biting the injected hind paw and is characterized by distinct phases (Dubuisson and Dennis, 1977; Wheeler-Aceto et al., 1990; Crawley, 2000). Phase 1, representing acute pain fiber activity, lasts about ten minutes and is followed by a second phase that may represent responses to tissue damage and inflammation (Crawley, 2000). A third phase of the formalin licking response has been reported and is proceeded by a brief period of inactivity (5–10 min) separating the second and third phases (Kim et al., 1999). Injection of CFA into a hind paw leads to an increase in behavioral responses to a previously non-noxious stimulus, a phenomenon referred to as allodynia, in both the ipsalateral and contralateral hind paw. Allodynia, or pain from a stimulus which is not normally painful, is measured by the application of finely calibrated hairs called Von Frey filaments (Crawley, 2000). The filaments are applied to the hind paw and a response (raising, licking or flinching) is recorded (Pitcher et al., 1999).

IEG induction by noxious chemical injection is widely used to map nociceptive pathways activated by persistent pain. While most studies have focused on the activation of c-Fos, a number of more recent studies highlight the role of Egr-1 in both the behavioral and neuronal responses to noxious stimuli.

3.1. Unilateral Hind Paw Inflammation Induces IEG Expression Bilaterally in the Brain

In contrast to the mostly ipsilateral increase in IEG expression seen in the spinal cord following peripheral inflammation (Leah et al., 1996; Wei et al., 1998), unilateral hind paw injections of chemical irritants result in bilateral increases in IEG expression in a number of brain areas associated with pain processing. For example, unilateral hind paw injection of formalin results in a bilateral increase of c-Fos immunoreactivity throughout the hippocampal formation (Aloisi et al., 1997; Aloisi et al., 2000). Interestingly, this response is sex-dependent, since c-Fos activation was stronger in female rats (Aloisi et al., 1997) and attenuated in gonadectomized animals (Aloisi et al., 2000). Intracerebroventricular injections of neurokinin 1 and 2 receptor antagonists can prevent formalin-induced upregulation of c-Fos in the prefrontal cortex, locus coeruleus (LC), PAG and in the paraventricular, dorsomedial and ventromedial nuclei, and are also effective at reducing formalin-induced pain behaviors (Baulmann et al., 2000). A role for the pontine parabrachial nucleus in nociceptive processes was demonstrated by a bilateral increase in c-Fos immunoreactivity six and twenty-four hours after unilateral hind paw injection of CFA in rats (Bellavance and Beitz, 1996). Carrageenan inflammation increased c-Fos bilaterally in the LC and the nucleus subcoeruleus (SC). Importantly, lesion of the LC/SC ipsilateral or contralateral to the site of hind paw injection did not affect pain behavior suggesting that unilateral noradrenergic projections via LC/SC activation is sufficient for modulating nociceptive processing (Tsuruoka et al., 2003). The expression of the transcriptional regulators c-Fos, Fos B, Jun B, Jun D, c-Jun and Egr-1 was measured after the induction of monoarthritis or peripheral inflammation in rats, however, of these six transcription factors, only c-Fos was robustly activated bilaterally throughout the hindbrain. The expression of c-Fos was increased in the caudal medulla oblongata, more specifically the caudal intermediate reticular nucleus, subnucleus reticularis dorsalis, ventrolateral reticular formation and the lateral paragigantocellular nucleus, beginning four hours after inflammation or the induction of monoarthritis, and returned to baseline levels within two weeks. Expression in caudal reticular areas was only present four (inflammation) or twenty-four (monoarthritis) hours after induction, however, in the parabrachial area, c-Fos expression was still significantly increased over controls two weeks after the induction of monoarthirits (Lanteri-Minet et al., 1994). This study demonstrates the spatial and temporal differences in IEG expression elicited by different models of persistent pain.

3.2. Stimulus-Dependent Activation of c-Fos in the PAG

Deep noxious stimulation resulting from formalin injection into the neck muscles stimulates c-Fos expression throughout the PAG, but predominantly in the vlPAG (Keay and Bandler, 1993; Keay et al., 2000). Nociceptive connections between the vlPAG and the ventrolateral medulla (RVLM) were demonstrated by combining the use of c-Fos as a neuronal activity marker and rhodamine beads as a retrograde tracer. After injection of rhodamine into the RVLM of formalin-injected rats, double labeled neurons (c-Fos and rhodamine) were observed in the PAG that were not observed in control animals (Keay et al., 2000). Subsequently it was shown that distinct patterns of c-Fos activation could be evoked in the vlPAG by noxious activation of spinal versus vagal afferents. Activation of spinal afferents by a hind paw injection of formalin or carrageenan induced an increase in c-Fos expression within the caudal vlPAG, while c-Fos expression after stimulation of vagal afferents was restricted to the roastral vlPAG (Keay et al., 2002). These findings suggest that the descending modulatory roles of the PAG in pain processing are mediated through distinct neuronal subdomains embedded in this region, as revealed by IEG expression.

3.3. IEGs Induced by Noxious Chemical Injections in Oral/Facial Regions

Noxious facial stimulation produced by formalin injection into the rodent vibrissal pad increased c-Fos in the caudal medullary formation (Mineta et al., 1995). Noxious stimuli delivered to the tongue or nasal cavity induced c-Fos in the trigeminal nucleus caudalis (TNC) and laminae I and II of the subnucleus caudalis (Anton et al., 1991; Carstens et al., 1995). Both Egr-1 and c-Fos were induced in the TNC of rats after formalin was injected in the whisker pad. The increase in Egr-1 and c-Fos expression was concentrated in the superficial layers of the TNC and was reversible by pretreatment with the NMDA receptor antagonist MK801; this pharmacological agent had little effect on the expression of both IEGs within the deeper layers (Otahara et al., 2003). Disorders of the temporo-mandibular joint (TMJ) can cause facial pain and headaches. To study the areas that are activated by noxious input from the TMJ, mustard oil was unilaterally injected into the TMJ region of rats; it was found that this procedure significantly increases c-Fos in the subnucleus caudalis (Hathaway et al., 1995).

3.4. Persistent Pain Models and Genetically Modified Mice

The increasing availability of knockout and transgenic mice creates new opportunities for the study of IEG activation as it relates to nociception and allows for directly studying the contributions of specific genes to pain processing. Mice overexpressing the NMDA receptor subunit NR2B in forebrain regions have increased behavioral responses to formalin compared to control mice. In this study, a hind paw injection of formalin induced c-Fos in the anterior cingulate cortex (ACC), somatosensory cortex, insular cortex, CA1 region of the hippocampus, PAG and spinal cord of control mice. However, this formalin-induced increase in c-Fos expression was significantly greater in the ACC, insular cortex and CA1 region in NR2B over expressing mice (Wei et al., 2001). Thus, it seems that increased sensitivity to persistent inflammatory pain corresponded to an increase in the strength of c-Fos activation in the ACC, a region known to play a role in pain processing (Devinsky et al., 1995; Hathaway et al., 1995; Davis et al., 1997; Wu et al., 2005).

Although many studies report the activation of IEGs in response to noxious stimuli, few studies have addressed the role of specific IEGs in the behavioral responses to acute and chronic pain. Mice with a deletion of egr-1 exhibit normal responses to acute thermal and mechanical stimuli but show deficits in models of more persistent pain (Ko et al., 2005). While there was no difference in early responses to formalin, egr-1 knockout mice showed attenuated late phase licking behaviors. In addition, these knockout animals displayed reduced responses to mechanical stimulation after CFA injection compared to control mice (Figure 6.1).

These results suggest that egr-1 plays a selective role in the behavioral responses to persistent pain but not acute noxious stimuli. In addition, egr-1 expression in the ACC of wild-type mice was similar to saline injected animals one hour after formalin injection but was significantly increased by three hours

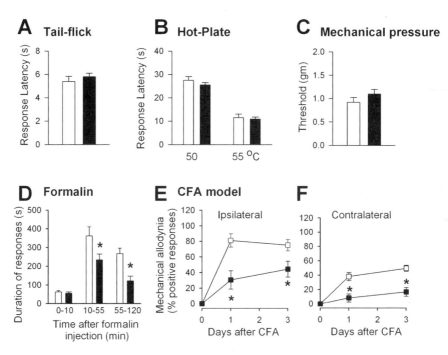

Figure 6.1. Behavioral nociceptive responses to acute and inflammatory pain in *egr-1* knockout and wild-type mice. Behavioral responses to (A) tail-flick, (B) hot-plate, and (C) mechanical stimuli were similar between knockout (black bars) and wild-type mice (white bars). (D) Late-phase formalin-induced nociceptive behaviors were significantly reduced in *egr-1* knockout mice (black bars) compared to wild-type mice (white bars). (E-F) Mechanical allodynia, induced by CFA, was greater in the ipsilateral (E) or contralateral (F) hind paw of wild-type (open squares) mice compared to *egr-1* knockout mice (closed squares).

(Ko et al., 2005) (Figure 6.2). While other studies report Egr-1 upregulation as early as one hour after noxious stimulation (Herdegen et al., 1990; Wei et al., 2000), Egr-1 expression has been shown to vary over time and in relation to the stimulus type and intensity (Lanteri-Minet et al., 1993b; Wei et al., 1999; Pearse et al., 2001; Leah and Wilce, 2002). These studies highlight the importance of including multiple time points when studying IEG expression, as mRNA or protein levels may vary over time and depend on the pain model used.

4. Rapid Induction of IEGs in the Forebrain Following Tail and Digit Amputation

A large number of human amputees experience phantom limb sensation or phantom pain (Sherman et al., 1980; Katz and Melzack, 1990; Jenson and P., 1994; Ramachandran and Rogers-Ramachandran, 2000). Cortical reorganization occurs after limb or digit amputation (Wall, 1977; Pons et al., 1991; Ramachandran et al., 1992; Kaas and Florence, 1997; Kaas, 1998; Merzenich, 1998; Tremere et al., 2001; Kaas and Collins, 2003) and the degree of this reorganization correlates with the amount of phantom pain (Flor et al., 1995; Birbaumer et

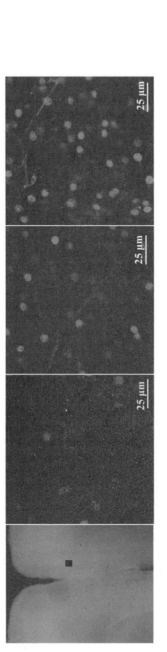

Figure 6.2. Increased Egr1 expression in the ACC after formalin injection. Left to right, coronal section of the brain at the level of the ACC, the region marked with the black box is shown on the right; Egr1 expression 3 hours after saline injection; 1 and 3 hours after formalin injection.

al., 1997). IEG-mediated long-term changes in gene expression and neuronal functioning in the ACC, a region with a well-defined role in pain, may underlie the development of phantom pain (Talbot et al., 1991; Coghill et al., 1994; Devinsky et al., 1995; Davis et al., 1997). To mimic phantom limb pain found in humans, a single digit of one hind paw or the distal tip of the tail is amputated. Unilateral digit amputation of the rat hind paw resulted in a bilateral upregulation of c-Fos and Egr-1 in the ACC. Expression of both IEGs appeared fifteen minutes following amputation reaching a maximum at forty-five minutes and returning to baseline levels by two weeks. Interestingly, this expression was accompanied by a loss of long-term depression of electrophysiological responses in the ACC (Wei et al., 1999). Egr-1 expression was also increased in the rat hippocampus following digit amputation and was selective to the CA1 region since only minor changes were detected in the CA3 region or dentate gyrus (DG). Amputation of the distal tip of a mouse tail also resulted in a significant increase in the expression of Egr-1 and, similar to results found in rats, this increase was most dramatic in the CA1 region of the hippocampus. Egr-1 expression induced by amputation could be attenuated by MK-801 or morphine pre-treatment. Long-term potentiation was enhanced in the hippocampus of mice with tail amputation. Egr-1 may play a role in this injury-induced alteration in synaptic plasticity given that *egr-1* knockout mice did not show a similar enhancement in long-term potentiation (Wei et al., 2000). These studies suggest that the upregulation of Egr-1 may play a role in the synaptic changes that occur with the development of pathological pain.

5. Induction of IEGs in Models for Visceral Pain

Pain of visceral origin is difficult to localize and is often referred to cutaneous structures. Common models for visceral pain include colorectal distension (CRD), intraperitoneal injection of acetic acid and chemical stimulation of the colon (Lanteri-Minet et al., 1993a; Traub et al., 1996). Much like other pain models discussed in this chapter, IEG expression induced by visceral pain can vary with the type of stimulus, site of application and intensity.

Visceral pain induced by intraperitoneal injections of acetic acid induced IEG expression in the hindbrain in a pattern distinct from that evoked by CRD. In addition, of the six IEGs studied (Egr-1, c-Fos, Fos B, Jun D, c-Jun), only c-Fos and Egr-1 showed robust activation after noxious visceral stimulation. Expression was increased in the caudal intermediate reticular nucleus as part of the caudalmost ventrolateral medulla and the superior lateral nucleus of the rostrolateral parabrachial area. CRD produced more robust expression in the caudal intermediate reticular nucleus compared to peritoneal inflammation, which lead to a stronger increase in the superior lateral nucleus (Lanteri-Minet et al., 1993a). Rats receiving CRD with light restraint or restraint alone showed significant increases in c-Fos expression throughout cortical and subcortical regions although expression was more robust in mice receiving CRD (but see Chapter 11). CRD, but not restraint, activated the infralimbic and prelimbic cortices, mediodorsal thalamic nucleus and the central amygdaloid nucleus (Traub et al., 1996). CRD at different intensities also produced distinct patterns of c-Fos expression. An intensity-dependent increase in c-Fos activity was found in the nucleus of the solitary tract, RVLM, nucleus cuneiformis, PAG and amygdala, while expression in the dorsomedial and ventromedial nucleus of the hypothalamus and thalamus was independent of the stimulus intensity. Expression in the nucleus of the solitary tract was selective for noxious CRD in a manner reversible by perivagal treatment with capsaicin and peripheral injection of a serotonin receptor antagonist (Monnikes et al., 2003). Studies like these further demonstrate the selective nature of IEG activation for the site and modality of the stimulus and suggest that multiple IEGs and stimulus variations should be studied to account for these differences.

6. Models for Migraine Induce IEGs in the Brainstem

Migraines, characterized by head pain, nausea, vomiting, sensitivity to light, sound and movement, are a neurovascular disorder whose underlying cause is still being intensely scrutinized (Mitsikostas and Sanchez del Rio, 2001; Goadsby et al., 2002). Functional imaging studies suggest that dysfunction in the brainstem, specifically the PAG, underlies the pathophysiology of migraines (Weiller et al., 1995; Bahra et al., 2001; Welch et al., 2001). A number of models (mechanical/electrical/chemical), all of which increase the expression of c-Fos in brainstem regions, including the PAG, are used to mimic migraine pain (Mitsikostas and Sanchez del Rio, 2001). Migraine models allow for the testing of novel therapeutic drugs and in many cases the attenuation of c-Fos activation is used as a direct measure of the drugs potential efficacy.

6.1. Two Types of Electrical Stimulation Induce c-Fos Expression in the Trigeminal Nucleus Caudalis

Nociceptive information arising from the trigeminal nucleus caudalis (TNC) is transmitted to other brainstem nuclei and on to higher cortical structures for further processing of migraine pain. One technique used to activate the trigeminal vascular system and mimic migraine pain is the direct electrical stimulation of the trigeminal ganglion (TG) (Uhl et al., 1991; Walther et al., 1993; Mitsikostas and Sanchez del Rio, 2001). For example, electrical stimulation of the TG induced expression of c-Fos ipsilaterally in the superficial layers of the TNC (Uhl et al., 1991; Walther et al., 1993). In addition, both low and high intensity stimulation of the TG induced c-Fos in the rat trigeminal sensory nucleus (Takemura et al., 2000).

A migraine model that is thought to mimic those associated with auras is cortical spreading depression (CSD), which is characterized by the suppression of cortical neuronal activity following an initial burst of seizure discharge (Leao, 1944; de Oliveira Castro et al., 1985; Leao, 1986). CSD can be induced by potassium-induced depolarization, as well as mechanical or electrical stimulation of the cortex (Moskowitz et al., 1993; Ingvardsen et al., 1997; Plumier et al., 1997) and induces c-Fos expression in the TNC ipsilaterally (Moskowitz et al., 1993). Expression of c-Fos in the ipsilateral TNC and cerebral cortex could be reduced by chronic surgical transection of the meningeal afferents or by pretreatment with a serotonin receptor agonist (Moskowitz et al., 1993). The pattern of c-Fos, Egr-1 and c-Jun activation after CSD varied between the different layers of the piriform, perirhinal, entorhinal, and insular cortex in that c-Jun was seen mostly in layers II, III and VI while Egr1 expression was concentrated in layers II, IV and VI (Herdegen et al., 1993). CSD induced c-Fos expression in the magnocellular region of the hypothalamic paraventricular nucleus (PVN) and throughout the ipsilateral cortex, but decreased c-Fos in both the parvocellular region of the PVN and in the cortex contralateral to the site of CSD (Iqbal Chowdhury et al., 2003).

6.2. IEG Expression Following Meningeal Stimulation

Noxious chemical stimuli can be delivered to meningeal tissue through a small catheter placed in the cisterna magna. A chemical irritant, for example, blood (Nozaki et al., 1992b), capsaicin (Nozaki et al., 1992a; Cutrer et al., 1995b) or carrageenan (Nozaki et al., 1992a) is often injected intracisternally to induce c-Fos expression in the TNC. This c-Fos activation can be blocked or reduced by prior administration of serotonin receptor agonists (Nozaki et al., 1992a; Cutrer et al., 1995c, 1999b; Mitsikostas et al., 1999a), NK1 receptor antagonists (Cutrer et al., 1995a; Clayton et al., 1997), $GABA_A$ antagonists (Cutrer et al., 1995b, 1999a; Cutrer and Moskowitz, 1996), NMDA and AMPA receptor antagonists (Mitsikostas et al., 1998, 1999b) and morphine (Nozaki et al., 1992a). The degree of capsaicin induced c-Fos activation in layers I and II of the TNC, parabrachial nucleus, vlPAG, amygdala and somatosensory cortex and nociceptive behaviors (immobility, scratching and grooming of the head) followed a linear-like relationship with the dose of capsaicin (Ter Horst et al., 2001). In this study, behavioral responses corresponded to IEG induction and the strength of the nociceptive stimulus, supporting a role for noxious stimulus induced IEG expression in the long-term changes following injury (persistent pain).

6.3. Stimulation of the Superior Sagittal Sinus Induces IEG Expression

Stimulation of the superior sagittal sinus (SSS) in humans evokes head pain and may reproduce the neurovascular activation of the pain sensitive structures observed during a migraine attack (Mitsikostas and Sanchez del Rio, 2001). Mechanical and electrical stimulation of the SSS induces c-Fos expression in the TNC of cats, monkeys and rats (Kaube et al., 1993b; Strassman et al., 1994; Goadsby and Hoskin, 1997). This upregulation can be attenuated by serotonin agonists (Kaube et al., 1993a; Goadsby and Hoskin, 1996; Hoskin et al., 1996) and NK1 receptor antagonists (Goadsby et al., 1998). SSS stimulation in cats increases c-Fos expression in the vlPAG, trigeminal nucleus and superior salivatory nucleus (Keay and Bandler, 1998; Knight et al., 2005). When SSS stimulation was combined with PAG stimulation, a strong increase was seen in the nucleus raphe magnus ipsilateral to the site of PAG stimulation. This study identifies the nucleus raphe magnus as a structure involved in the modulation of trigeminal nociceptive information in the PAG and maps neuronal activity in structures that may be involved in the processing/transfer of this information (Knight et al., 2005). Dissecting these pathways will lead to insight into the underlying mechanisms of migraine pain.

7. Discussion

The expression of IEGs after noxious stimulation leads to long-term changes in neuronal functioning that may underlie the development of pathological pain. Mapping neuronal activity by studying IEG activation provides a useful tool for investigating the processing of nociceptive information in the brain. Advances in genetic engineering allows for the direct study of a particular IEG in behavioral nociceptive responses and also provides models in which to study how IEG activation is affected in animals with altered pain sensitivity. In addition, the use of double labeling techniques and retrograde tracers provides information on the type of cells expressing a particular IEG and the projections they maintain.

Throughout this chapter, an emphasis was placed on the importance of the stimulus type, location, duration and intensity in the IEG expressed and what brain regions are activated. It is therefore important to consider each of these variables when using IEG activation to map neuronal activity induced by noxious stimulation. Sampling of IEG expression at different time points following noxious insult is important since IEG expression may be different at early time points compared to expression at later times (reflecting long-term changes in gene expression). This consideration may be especially important when examining changes in neuronal activity in models of persistent, inflammatory pain, when behavioral alterations can last for long periods of time. Varying the duration and intensity of a noxious stimulus affects IEG expression (Herdegen et al., 1990; Smith et al., 1997; Ter Horst et al., 2001; Monnikes et al., 2003), therefore, incorporation of multiple stimulus strengths and durations may provide more meaningful information about the particular brain region and IEG being considered.

Other important variables to be considered when using IEG expression to map nociceptive pathways include the basal expression of IEGs, stress from handling, and the use of anesthetics, since basal expression of different IEGs can vary (Herdegen et al., 1995), handling stress can induce IEG activity (Traub et al., 1996), and anesthesia alone can induce IEG expression (Takayama et al., 1994). A number of studies that have specifically addressed these questions reinforce their importance. For example, Egr-1 activation induced in the hippocampus after tail amputation in mice was blocked when the animal was kept under constant anesthesia (Wei et al., 2000). On the other hand, anesthesia-induced increases in IEG expression have been reported (Lanteri-Minet et al., 1993a; Clement et al., 1996). An additional benefit to studying unanesthetized animals is the ability to correlate changes in nociceptive behavior with changing IEG expression, strengthening the association between IEGs and the neuronal activity underlying the transmission of nociceptive information.

References

Aloisi AM, Zimmermann M, Herdegen T (1997). Sex-dependent effects of formalin and restraint on c-Fos expression in the septum and hippocampus of the rat. Neuroscience 81:951-958.

Aloisi AM, Ceccarelli I, Herdegen T (2000). Gonadectomy and persistent pain differently affect hippocampal c-Fos expression in male and female rats. Neurosci Lett 281:29-32.

Anton F, Herdegen T, Peppel P, Leah JD (1991). c-FOS-like immunoreactivity in rat brainstem neurons following noxious chemical stimulation of the nasal mucosa. Neuroscience 41:629-641.

Bahra A, Matharu MS, Buchel C, Frackowiak RS, Goadsby PJ (2001). Brainstem activation specific to migraine headache. Lancet 357:1016-1017.

Baulmann J, Spitznagel H, Herdegen T, Unger T, Culman J (2000). Tachykinin receptor inhibition and c-Fos expression in the rat brain following formalin-induced pain. Neuroscience 95:813-820.

Bellavance LL, Beitz AJ (1996). Altered c-fos expression in the parabrachial nucleus in a rodent model of CFA-induced peripheral inflammation. J Comp Neurol 366:431-447.

Bester H, Hunt SP (2002). The expression of c-Fos in the spinal cord: Mapping of nociceptive pathways. In: Immediate Early Genes and Inducible Transcription Factors in Mapping of the Central Nervous System Function and Dysfunction (Kaczmarek L, Robertson HA, eds), pp. 171-184. Elsevier.

Birbaumer N, Lutzenberger W, Montoya P, Larbig W, Unertl K, Topfner S, Grodd W, Taub E, Flor H (1997). Effects of regional anesthesia on phantom limb pain are mirrored in changes in cortical reorganization. J Neurosci 17:5503-5508.

Bullitt E (1989). Induction of c-fos-like protein within the lumbar spinal cord and thalamus of the rat following peripheral stimulation. Brain Res 493:391-397.

Carstens E, Saxe I, Ralph R (1995). Brainstem neurons expressing c-Fos immunoreactivity following irritant chemical stimulation of the rat's tongue. Neuroscience 69:939-953.

Clayton JS, Gaskin PJ, Beattie DT (1997). Attenuation of Fos-like immunoreactivity in the trigeminal nucleus caudalis following trigeminovascular activation in the anaesthetised guinea-pig. Brain Res 775:74-80.

Clement CI, Keay KA, Owler BK, Bandler R (1996). Common patterns of increased and decreased fos expression in midbrain and pons evoked by noxious deep somatic and noxious visceral manipulations in the rat. J Comp Neurol 366:495-515.

Coggeshall RE (2005). Fos, nociception and the dorsal horn. Prog Neurobiol 77:299-352.

Coghill RC, Talbot JD, Evans AC, Meyer E, Gjedde A, Bushnell MC, Duncan GH (1994). Distributed processing of pain and vibration by the human brain. J Neurosci 14:4095-4108.

Crawley JN (2000). What's Wrong with my Mouse?: Behavioral Phenotyping of Transgenic and Knockout Mice. New York: Wiley-Liss.

Cutrer FM, Moskowitz MA (1996). Wolff Award 1996. The actions of valproate and neurosteroids in a model of trigeminal pain. Headache 36:579-585.

Cutrer FM, Moussaoui S, Garret C, Moskowitz MA (1995a). The non-peptide neurokinin-1 antagonist, RPR 100893, decreases c-fos expression in trigeminal nucleus caudalis following noxious chemical meningeal stimulation. Neuroscience 64:741-750.

Cutrer FM, Limmroth V, Ayata G, Moskowitz MA (1995b). Attenuation by valproate of c-fos immunoreactivity in trigeminal nucleus caudalis induced by intracisternal capsaicin. Br J Pharmacol 116:3199-3204.

Cutrer FM, Mitsikostas DD, Ayata G, Sanchez del Rio M (1999a). Attenuation by butalbital of capsaicin-induced c-fos-like immunoreactivity in trigeminal nucleus caudalis. Headache 39:697-704.

Cutrer FM, Schoenfeld D, Limmroth V, Panahian N, Moskowitz MA (1995c). Suppression by the sumatriptan analogue, CP-122,288 of c-fos immunoreactivity in trigeminal nucleus caudalis induced by intracisternal capsaicin. Br J Pharmacol 114:987-992.

Cutrer FM, Yu XJ, Ayata G, Moskowitz MA, Waeber C (1999b). Effects of PNU-109,291, a selective 5-HT1D receptor agonist, on electrically induced dural plasma extravasation and capsaicin-evoked c-fos immunoreactivity within trigeminal nucleus caudalis. Neuropharmacology 38:1043-1053.

Davis KD, Taylor SJ, Crawley AP, Wood ML, Mikulis DJ (1997). Functional MRI of pain- and attention-related activations in the human cingulate cortex. J Neurophysiol 77:3370-3380.

de Oliveira Castro G, Martins-Ferreira H, Gardino PF (1985). Dual nature of the peaks of light scattered during spreading depression in chick retina. An Acad Bras Cienc 57:95-103.

Devinsky O, Morrell MJ, Vogt BA (1995). Contributions of anterior cingulate cortex to behaviour. Brain 118(Pt 1):279-306.

Dubuisson D, Dennis SG (1977). The formalin test: a quantitative study of the analgesic effects of morphine, meperidine, and brain stem stimulation in rats and cats. Pain 4:161-174.

Flor H, Elbert T, Knecht S, Wienbruch C, Pantev C, Birbaumer N, Larbig W, Taub E (1995). Phantom-limb pain as a perceptual correlate of cortical reorganization following arm amputation. Nature 375:482-484.

Gauriau C, Bernard JF (2002). Pain pathways and parabrachial circuits in the rat. Exp Physiol 87:251-258.

Goadsby PJ, Hoskin KL (1996). Inhibition of trigeminal neurons by intravenous administration of the serotonin (5HT)1B/D receptor agonist zolmitriptan (311C90): are brain stem sites therapeutic target in migraine? Pain 67:355-359.

Goadsby PJ, Hoskin KL (1997). The distribution of trigeminovascular afferents in the nonhuman primate brain Macaca nemestrina: a c-fos immunocytochemical study. J Anat 190(Pt 3):367-375.

Goadsby PJ, Hoskin KL, Knight YE (1998). Substance P blockade with the potent and centrally acting antagonist GR205171 does not effect central trigeminal activity with superior sagittal sinus stimulation. Neuroscience 86:337-343.

Goadsby PJ, Lipton RB, Ferrari MD (2002). Migraine—current understanding and treatment. N Engl J Med 346:257-270.

Hathaway CB, Hu JW, Bereiter DA (1995). Distribution of Fos-like immunoreactivity in the caudal brainstem of the rat following noxious chemical stimulation of the temporomandibular joint. J Comp Neurol 356:444-456.

Herdegen T, Zimmermann M (1995). Immediate early genes (IEGs) encoding for inducible transcription factors (ITFs) and neuropeptides in the nervous system: functional network for long-term plasticity and pain. Prog Brain Res 104:299-321.

Herdegen T, Leah JD (1998). Inducible and constitutive transcription factors in the mammalian nervous system: control of gene expression by Jun, Fos and Krox, and CREB/ATF proteins. Brain Res Brain Res Rev 28:370-490.

Herdegen T, Walker T, Leah JD, Bravo R, Zimmermann M (1990). The KROX-24 protein, a new transcription regulating factor: expression in the rat central nervous system following afferent somatosensory stimulation. Neurosci Lett 120:21-24.

Herdegen T, Sandkuhler J, Gass P, Kiessling M, Bravo R, Zimmermann M (1993). JUN, FOS, KROX, and CREB transcription factor proteins in the rat cortex: basal expression and induction by spreading depression and epileptic seizures. J Comp Neurol 333:271-288.

Herdegen T, Kovary K, Buhl A, Bravo R, Zimmermann M, Gass P (1995). Basal expression of the inducible transcription factors c-Jun, JunB, JunD, c-Fos, FosB, and Krox-24 in the adult rat brain. J Comp Neurol 354:39-56.

Hoskin KL, Kaube H, Goadsby PJ (1996). Sumatriptan can inhibit trigeminal afferents by an exclusively neural mechanism. Brain 119(Pt 5):1419-1428.

Hunt SP, Pini A, Evan G (1987). Induction of c-fos-like protein in spinal cord neurons following sensory stimulation. Nature 328:632-634.

Ingvardsen BK, Laursen H, Olsen UB, Hansen AJ (1997). Possible mechanism of c-fos expression in trigeminal nucleus caudalis following cortical spreading depression. Pain 72:407-415.

Iqbal Chowdhury GM, Liu Y, Tanaka M, Fujioka T, Ishikawa A, Nakamura S (2003). Cortical spreading depression affects Fos expression in the hypothalamic paraventricular nucleus and the cerebral cortex of both hemispheres. Neurosci Res 45:149-155.

Jenson TS, P. R (1994). Phantom pain and other phenomena after amputation, 3rd Edition. Edinburgh; New York: Churchill Livingstone.

Kaas JH (1998). Phantoms of the brain. Nature 391:331, 333.

Kaas JH, Florence SL (1997). Mechanisms of reorganization in sensory systems of primates after peripheral nerve injury. Adv Neurol 73:147-158.

Kaas JH, Collins CE (2003). Anatomic and functional reorganization of somatosensory cortex in mature primates after peripheral nerve and spinal cord injury. Adv Neurol 93:87-95.

Kaczmarek L, Chaudhuri A (1997). Sensory regulation of immediate-early gene expression in mammalian visual cortex: implications for functional mapping and neural plasticity. Brain Res Brain Res Rev 23:237-256.

Katz J, Melzack R (1990). Pain 'memories' in phantom limbs: review and clinical observations. Pain 43:319-336.

Kaube H, Hoskin KL, Goadsby PJ (1993a). Inhibition by sumatriptan of central trigeminal neurones only after blood-brain barrier disruption. Br J Pharmacol 109:788-792.

Kaube H, Keay KA, Hoskin KL, Bandler R, Goadsby PJ (1993b). Expression of c-Fos-like immunore-activity in the caudal medulla and upper cervical spinal cord following stimulation of the superior sagittal sinus in the cat. Brain Res 629:95-102.

Keay KA, Bandler R (1993). Deep and superficial noxious stimulation increases Fos-like immunoreac-tivity in different regions of the midbrain periaqueductal grey of the rat. Neurosci Lett 154:23-26.

Keay KA, Bandler R (1998). Vascular head pain selectively activates ventrolateral periaqueductal gray in the cat. Neurosci Lett 245:58-60.

Keay KA, Li QF, Bandler R (2000). Muscle pain activates a direct projection from ventrolateral periaqueductal gray to rostral ventrolateral medulla in rats. Neurosci Lett 290:157-160.

Keay KA, Clement CI, Depaulis A, Bandler R (2001). Different representations of inescapable noxious stimuli in the periaqueductal gray and upper cervical spinal cord of freely moving rats. Neurosci Lett 313:17-20.

Keay KA, Clement CI, Matar WM, Heslop DJ, Henderson LA, Bandler R (2002). Noxious activation of spinal or vagal afferents evokes distinct patterns of fos-like immunoreactivity in the ventrolateral periaqueductal gray of unanaesthetised rats. Brain Res 948:122-130.

Kim SJ, Calejesan AA, Li P, Wei F, Zhuo M (1999). Sex differences in late behavioral response to subcutaneous formalin injection in mice. Brain Res 829:185-189.

Knight YE, Classey JD, Lasalandra MP, Akerman S, Kowacs F, Hoskin KL, Goadsby PJ (2005). Patterns of fos expression in the rostral medulla and caudal pons evoked by noxious craniovascular stimulation and periaqueductal gray stimulation in the cat. Brain Res 1045:1-11.

Ko SW, Vadakkan KI, Ao H, Gallitano-Mendel A, Wei F, Milbrandt J, Zhuo M (2005). Selective contribution of Egr1 (zif/268) to persistent inflammatory pain. J Pain 6:12-20.

Lanteri-Minet M, Isnardon P, de Pommery J, Menetrey D (1993a). Spinal and hindbrain structures involved in visceroception and visceronociception as revealed by the expression of Fos, Jun and Krox-24 proteins. Neuroscience 55:737-753.

Lanteri-Minet M, Weil-Fugazza J, de Pommery J, Menetrey D (1994). Hindbrain structures involved in pain processing as revealed by the expression of c-Fos and other immediate early gene proteins. Neuroscience 58:287-298.

Lanteri-Minet M, de Pommery J, Herdegen T, Weil-Fugazza J, Bravo R, Menetrey D (1993b). Differential time course and spatial expression of Fos, Jun, and Krox-24 proteins in spinal cord of rats undergoing subacute or chronic somatic inflammation. J Comp Neurol 333:223-235.

Le Bars D, Gozariu M, Cadden SW (2001). Animal models of nociception. Pharmacol Rev 53:597-652.

Leah J, Wilce PA (2002). The Egr1 transcription factors and their utility in mapping brain functioning. In: Immediate Early Genes and Inducible Transcription Factors in Mapping of the Central Nervous System Function and Dysfunction (Kaczmarek L, Robertson HA, eds), pp. 309-323. Elsevier.

Leah JD, Porter J, de-Pommery J, Menetrey D, Weil-Fuguzza J (1996). Effect of acute stimulation on Fos expression in spinal neurons in the presence of persisting C-fiber activity. Brain Res 719:104-111.

Leao AA (1944). Spreading depression of activity in the cerebral cortex. J Neurophysiol 7:379-390.

Leao AA (1986). Spreading depression. Funct Neurol 1:363-366.

Merzenich M (1998). Long-term change of mind. Science 282:1062-1063.

Mineta Y, Eisenberg E, Strassman AM (1995). Distribution of Fos-like immunoreactivity in the caudal medullary reticular formation following noxious facial stimulation in the rat. Exp Brain Res 107:34-38.

Mitsikostas DD, Sanchez del Rio M (2001). Receptor systems mediating c-fos expression within trigeminal nucleus caudalis in animal models of migraine. Brain Res Brain Res Rev 35:20-35.

Mitsikostas DD, Sanchez del Rio M, Moskowitz MA, Waeber C (1999a). Both 5-HT1B and 5-HT1F receptors modulate c-fos expression within rat trigeminal nucleus caudalis. Eur J Pharmacol 369:271-277.

Mitsikostas DD, Sanchez del Rio M, Waeber C, Moskowitz MA, Cutrer FM (1998). The NMDA receptor antagonist MK-801 reduces capsaicin-induced c-fos expression within rat trigeminal nucleus caudalis. Pain 76:239-248.

Mitsikostas DD, Sanchez del Rio M, Waeber C, Huang Z, Cutrer FM, Moskowitz MA (1999b). Non-NMDA glutamate receptors modulate capsaicin induced c-fos expression within trigeminal nucleus caudalis. Br J Pharmacol 127:623-630.

Monnikes H, Ruter J, Konig M, Grote C, Kobelt P, Klapp BF, Arnold R, Wiedenmann B, Tebbe JJ (2003). Differential induction of c-fos expression in brain nuclei by noxious and non-noxious colonic distension: role of afferent C-fibers and 5-HT3 receptors. Brain Res 966:253-264.

Moskowitz MA, Nozaki K, Kraig RP (1993). Neocortical spreading depression provokes the expression of c-fos protein-like immunoreactivity within trigeminal nucleus caudalis via trigeminovascular mechanisms. J Neurosci 13:1167-1177.

Nozaki K, Moskowitz MA, Boccalini P (1992a). CP-93,129, sumatriptan, dihydroergotamine block c-fos expression within rat trigeminal nucleus caudalis caused by chemical stimulation of the meninges. Br J Pharmacol 106:409-415.

Nozaki K, Boccalini P, Moskowitz MA (1992b). Expression of c-fos-like immunoreactivity in brainstem after meningeal irritation by blood in the subarachnoid space. Neuroscience 49:669-680.

Otahara N, Ikeda T, Sakoda S, Shiba R, Nishimori T (2003). Involvement of NMDA receptors in Zif/268 expression in the trigeminal nucleus caudalis following formalin injection into the rat whisker pad. Brain Res Bull 62:63-70.

Pearse DD, Bushell G, Leah JD (2001). Jun, Fos and Krox in the thalamus after C-fiber stimulation: coincident-input-dependent expression, expression across somatotopic boundaries, and nucleolar translocation. Neuroscience 107:143-159.

Pinaud R (2004). Experience-dependent immediate early gene expression in the adult central nervous system: evidence from enriched-environment studies. Int J Neurosci 114:321-333.

Pinaud R (2005). Critical calcium-regulated biochemical and gene expression programs in experience-dependent plasticity. In: Plasticity in the Visual System: From Genes to Circuits (Pinaud R, Tremere L, Weerd PD, eds), pp. 153-180. New York: Springer-Verlag.

Pitcher GM, Ritchie J, Henry JL (1999). Paw withdrawal threshold in the von Frey hair test is influenced by the surface on which the rat stands. J Neurosci Methods 87:185-193.

Plumier JC, David JC, Robertson HA, Currie RW (1997). Cortical application of potassium chloride induces the low-molecular weight heat shock protein (Hsp27) in astrocytes. J Cereb Blood Flow Metab 17:781-790.

Pons TP, Garraghty PE, Ommaya AK, Kaas JH, Taub E, Mishkin M (1991). Massive cortical reorganization after sensory deafferentation in adult macaques. Science 252:1857-1860.

Ramachandran VS, Rogers-Ramachandran D (2000). Phantom limbs and neural plasticity. Arch Neurol 57:317-320.

Ramachandran VS, Stewart M, Rogers-Ramachandran DC (1992). Perceptual correlates of massive cortical reorganization. Neuroreport 3:583-586.

Scholz J, Woolf CJ (2002). Can we conquer pain? Nat Neurosci 5 Suppl:1062-1067.

Sherman RA, Sherman CJ, Gall NG (1980). A survey of current phantom limb pain treatment in the United States. Pain 8:85-99.

Smith WJ, Stewart J, Pfaus JG (1997). Tail pinch induces fos immunoreactivity within several regions of the male rat brain: effects of age. Physiol Behav 61:717-723.

Strassman AM, Mineta Y, Vos BP (1994). Distribution of fos-like immunoreactivity in the medullary and upper cervical dorsal horn produced by stimulation of dural blood vessels in the rat. J Neurosci 14:3725-3735.

Sugimoto T, Hara T, Shirai H, Abe T, Ichikawa H, Sato T (1994). c-fos induction in the subnucleus caudalis following noxious mechanical stimulation of the oral mucous membrane. Exp Neurol 129:251-256.

Takayama K, Suzuki T, Miura M (1994). The comparison of effects of various anesthetics on expression of Fos protein in the rat brain. Neurosci Lett 176:59-62.

Takemura M, Shimada T, Sugiyo S, Nokubi T, Shigenaga Y (2000). Mapping of c-Fos in the trigeminal sensory nucleus following high- and low-intensity afferent stimulation in the rat. Exp Brain Res 130:113-123.

Talbot JD, Marrett S, Evans AC, Meyer E, Bushnell MC, Duncan GH (1991). Multiple representations of pain in human cerebral cortex. Science 251:1355-1358.

Ter Horst GJ, Meijler WJ, Korf J, Kemper RH (2001). Trigeminal nociception-induced cerebral Fos expression in the conscious rat. Cephalalgia 21:963-975.

Traub RJ, Silva E, Gebhart GF, Solodkin A (1996). Noxious colorectal distention induced-c-Fos protein in limbic brain structures in the rat. Neurosci Lett 215:165-168.

Tremere L, Hicks TP, Rasmusson DD (2001). Role of inhibition in cortical reorganization of the adult raccoon revealed by microiontophoretic blockade of GABA(A) receptors. J Neurophysiol 86:94-103.

Tsuruoka M, Arai YC, Nomura H, Matsutani K, Willis WD (2003). Unilateral hindpaw inflammation induces bilateral activation of the locus coeruleus and the nucleus subcoeruleus in the rat. Brain Res Bull 61:117-123.

Uhl GR, Walther D, Nishimori T, Buzzi MG, Moskowitz MA (1991). Jun B, c-jun, jun D and c-fos mRNAs in nucleus caudalis neurons: rapid selective enhancement by afferent stimulation. Brain Res Mol Brain Res 11:133-141.

Vanegas H (2004). To the descending pain-control system in rats, inflammation-induced primary and secondary hyperalgesia are two different things. Neurosci Lett 361:225-228.

Vanegas H, Schaible HG (2004). Descending control of persistent pain: inhibitory or facilitatory? Brain Res Brain Res Rev 46:295-309.

Wall PD (1977). The presence of ineffective synapses and the circumstances which unmask them. Philos Trans R Soc Lond B Biol Sci 278:361-372.

Walther D, Takemura M, Uhl G (1993). Fos family member changes in nucleus caudalis neurons after primary afferent stimulation: enhancement of fos B and c-fos. Brain Res Mol Brain Res 17:155-159.

Wei F, Ren K, Dubner R (1998). Inflammation-induced Fos protein expression in the rat spinal cord is enhanced following dorsolateral or ventrolateral funiculus lesions. Brain Res 782:136-141.

Wei F, Li P, Zhuo M (1999). Loss of synaptic depression in mammalian anterior cingulate cortex after amputation. J Neurosci 19:9346-9354.

Wei F, Xu ZC, Qu Z, Milbrandt J, Zhuo M (2000). Role of EGR1 in hippocampal synaptic enhancement induced by tetanic stimulation and amputation. J Cell Biol 149:1325-1334.

Wei F, Wang GD, Kerchner GA, Kim SJ, Xu HM, Chen ZF, Zhuo M (2001). Genetic enhancement of inflammatory pain by forebrain NR2B overexpression. Nat Neurosci 4:164-169.

Weiller C, May A, Limmroth V, Juptner M, Kaube H, Schayck RV, Coenen HH, Diener HC (1995). Brain stem activation in spontaneous human migraine attacks. Nat Med 1:658-660.

Welch KM, Nagesh V, Aurora SK, Gelman N (2001). Periaqueductal gray matter dysfunction in migraine: cause or the burden of illness? Headache 41:629-637.

Wheeler-Aceto H, Porreca F, Cowan A (1990). The rat paw formalin test: comparison of noxious agents. Pain 40:229-238.

Woolf CJ, Salter MW (2000). Neuronal plasticity: increasing the gain in pain. Science 288:1765-1769.

Wu LJ, Toyoda H, Zhao MG, Lee YS, Tang J, Ko SW, Jia YH, Shum FW, Zerbinatti CV, Bu G, Wei F, Xu TL, Muglia LJ, Chen ZF, Auberson YP, Kaang BK, Zhuo M (2005). Upregulation of forebrain NMDA NR2B receptors contributes to behavioral sensitization after inflammation. J Neurosci 25:11107-11116.

Zimmermann M (2001). Pathobiology of neuropathic pain. Eur J Pharmacol 429:23-37.

Zimmermann M, Herdegen T (1994). Control of Gene Transcription by Jun and Fos Proteins in the Nervous System. APS Journal 3(1):33-48.

PART II

Immediate Early Gene Expression in Complex Systems and Higher Order Cognitive Function

7

Mapping Sleep-Wake Control with the Transcription Factor c-Fos

SAMUEL DEURVEILHER and KAZUE SEMBA

Department of Anatomy and Neurobiology, Dalhousie University, Halifax, NS, Canada

1. Introduction

States of wakefulness and sleep are remarkably different not only in behavior, polygraphic signs, and neuronal firing patterns, but also in the expression of genes, such as immediate-early genes (IEGs; reviewed in Bentivoglio and Grassi-Zucconi, 1999; Cirelli and Tononi, 2000b). IEGs are a class of genes that are rapidly and transiently expressed in neurons (and other cells) in response to various stimuli, such as sensory stimuli, trophic factors, neurotransmitters, and drugs (reviewed in Morgan and Curran, 1991; Hughes and Dragunow, 1995; Herdegen and Leah, 1998). This responsiveness to stimuli forms the basis for using IEGs as markers of neuronal activation. Of the commonly studied IEGs in contemporary neuroscience, *c-fos* has been the most extensively studied and its protein product, c-Fos, is the focus of this review. We will first discuss the advantages and limitations of using c-Fos as an anatomical marker of neuronal activity for studying sleep-wake control, and how it has been used successfully to identify the location, neurochemical phenotype, and connectivity of sleep- and wake-active neurons. We will then discuss recent advances in understanding the role of c-Fos in the transcriptional regulation of sleep and wake states.

2. c-Fos as an Anatomical Marker for Identifying Neuronal Activation Patterns in the Sleep-Wake Control System

2.1. Advantages and Limitations of c-Fos Expression Mapping

Immunohistochemistry for c-Fos protein and *in situ* hybridization for *c-fos* mRNA have provided an opportunity to map neuronal activation patterns during the sleep-wake cycle and after sleep deprivation. The advantages are the following: (1) The mapping can be done at the cellular level, and it is possible to examine concurrently all neurons within the neuronal network that controls behavioral

R. Pinaud, L.A. Tremere (Eds.), Immediate Early Genes in Sensory Processing, Cognitive Performance and Neurological Disorders, 113–136, ©2006 Springer Science + Business Media, LLC

states, including the basal forebrain, hypothalamus, thalamus and brainstem (reviewed in Lydic and Baghdoyan, 1999; Pace-Schott and Hobson, 2002; Jones, 2005). This is a clear advantage compared to such alternatives as single-unit electrophysiological recordings which are much more time-consuming; single-unit recording data are also anatomically less precise although they do offer more functional detail. (2) Simultaneous visualization of c-Fos and neurotransmitter markers allows for identification of the transmitter phenotype of c-Fos-expressing, sleep- and wake-active neurons. This is a useful feature because the sleep-wake regulatory system uses various neurotransmitters, and neurons with different neurochemical phenotypes are often intermixed in the same brain region (Lydic and Baghdoyan, 1999; Pace-Schott and Hobson, 2002; Jones, 2005). The same goal can be achieved by combining extracellular recording with juxtacellular labeling, but this is a highly demanding technique. (3) c-Fos immunohistochemistry can be combined with retrograde axonal tracing to identify the connectivity of sleep- and wake-active neurons. Electrophysiologically, this could be accomplished by antidromic activation of recorded neurons, which, however, is considerably less efficient for characterizing a large population of neurons. (4) Finally, the initiation and maintenance of sleep and wake states are influenced by environmental and physiological signals (e.g., stress, temperature and metabolic signals), as well as drugs (e.g., stimulants and sedatives/hypnotics), and the examination of regional patterns of c-Fos expression could offer insights into the possible neuronal populations and pathways involved in these modulations.

Despite these major advantages of using c-Fos as a marker of neuronal activation, this technique is not without limitations (Dragunow and Faull, 1989; Hoffman and Lyo, 2002). (1) A conundrum is that some neurons do not express c-Fos during a behavioral state in which they are known to be active based on other techniques. It is therefore important to keep in mind that the absence of c-Fos in some neurons does not indicate the absence of activity, and other information, such as neuronal firing and transmitter release, needs to be considered. (2) The temporal resolution of c-Fos expression is relatively low because the synthesis of c-Fos protein occurs with a peak at 30 to 90 minutes after stimulation, and the half-life of the protein is a few hours (Morgan and Curran, 1991; Herdegen and Leah, 1998). Thus, c-Fos imaging is not as effective as electrophysiological methods for identifying neuronal activity during a short-lasting behavioral state, e.g., rapid eye movement (REM) sleep (also known as paradoxical or active sleep). Attempts have been made to circumvent this problem by using pharmacological and other methods to induce enhanced REM sleep periods. As an alternative to using c-Fos, early events in *c-fos* transcription can be examined by using probes designed to detect preprocessed RNA still containing introns (heteronuclear RNA). (3) c-Fos is also of limited value when the stimulus lasts for more than several hours, because c-Fos induction is transient and subject to auto-inhibition (Sassone-Corsi et al., 1988). Thus, the interpretation of c-Fos expression after long-term or chronic sleep deprivation becomes challenging; the duration of sleep deprivation could influence c-Fos expression in a complex and non-linear manner (Cirelli et al., 1995a). (4) c-Fos data can be collected only at a single time point in each animal, and multiple groups of animals are required to study a time course. Recently, however, *in vivo* c-Fos imaging techniques that allow repetitive scanning of the same animal have been developed using transgenic mice with the firefly luciferase gene under the control of the *c-fos* promoter (Collaco and Geusz, 2003). (5) c-Fos immunolabeling does not allow identification of neurons whose activity is inhibited, because c-Fos expression usually requires an increase in intracellular calcium concentration in response to excitatory synaptic input (Finkbeiner and Greenberg, 1998; Herdegen and Leah, 1998). (6) The c-Fos technique cannot distinguish whether the presence of c-Fos in a neuron is a direct or an indirect effect of a stimulus. In spite of these limitations, the c-Fos technique has been used successfully with appropriate controls in sleep research, and these data have provided important insights into the mechanisms underlying sleep and wakefulness, as discussed below.

2.2. Identification of Sleep-Active Nuclei in the Preoptic Area

One of the most successful examples of the use of c-Fos in sleep research was the identification of two preoptic nuclei that contain neurons that are active during sleep: the ventrolateral (VLPO; Sherin et al., 1996; Fig. 7.1) and median preoptic nuclei (MnPO; Gong et al., 2000). The increased c-Fos activity during sleep in neurons of these nuclei is remarkable because c-Fos levels in the brain are generally low during sleep (Bentivoglio and Grassi-Zucconi, 1999; Cirelli and Tononi, 2000b). The number of c-Fos-positive cells in both the VLPO and MnPO was positively correlated with the amount of sleep during the preceding 1 or 2 hours (Sherin et al., 1996; Gong et al., 2000). Both VLPO and MnPO neurons increased c-Fos immunoreactivity not only during spontaneous sleep but also during "recovery" sleep subsequent to sleep deprivation (Sherin et al., 1996; Wagner et al., 2000; Gong et al., 2004). The presence of sleep-active neurons in the VLPO and MnPO was subsequently confirmed by single-unit recordings (Szymusiak et al., 1998; Suntsova et al., 2002). The majority of VLPO and MnPO neurons that expressed c-Fos during sleep contained glutamic acid decarboxylase, a marker for GABAergic neurons (Gong et al., 2004; Modirrousta et al., 2004), and galanin, a neuropeptide primarily with an inhibitory action (Gaus et al., 2002). The number of GABAergic neurons in the VLPO and MnPO that were c-Fos immunoreactive during sleep was positively correlated with the amount of sleep in the preceding 2 hours (Fig. 7.2; Gong et al., 2004). Congruent patterns of sleep-associated c-Fos expression in galanin-containing neurons in the VLPO area were demonstrated in a number of mammalian species (Gaus et al., 2002). Many sleep-responsive neurons in the VLPO send projections to the tuberomammillary nucleus, which contains histaminergic neurons that become active during wakefulness (Sherin et al., 1996, 1998). Furthermore, many VLPO and MnPO GABAergic neurons that expressed c-Fos during sleep bore the inhibitory α_{2A} adrenergic receptor (Modirrousta et al., 2004), which is consistent with the inhibitory action of noradrenaline on most GABAergic neurons in the VLPO area *in vitro* (Gallopin et al., 2000). These findings suggest that sleep-active VLPO neurons probably inhibit tuberomammillary histaminergic neurons

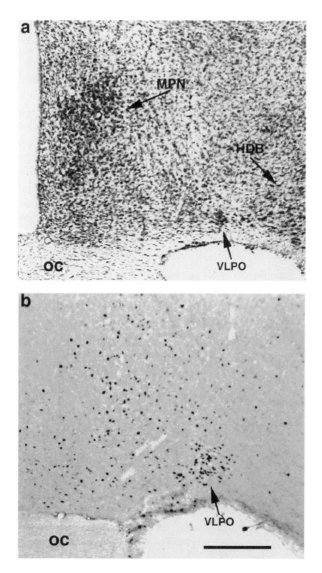

Figure 7.1. A cluster of c-Fos-immunoreactive neurons identifies the ventrolateral preoptic nucleus (VLPO, *b*) in a rat that slept most of the preceding hour (Sherin et al., 1998), as originally reported by Sherin et al. (1996). The VLPO is also visible as a cluster of neurons along the ventral surface of the brain in a Giemsa-stained section (*a*, Sherin et al., 1998). Giemsa dye stains cell bodies; although not discernible at this magnification, c-Fos-immunoreactivity is localized to cellular nuclei. This original VLPO (*b*) has since been redefined as the VLPO core or cluster in order to differentiate it from the "extended" VLPO (Lu et al., 2000) representing the adjacent medial and dorsal areas that contain primarily REM sleep-active neurons (Lu et al., 2002). Abbreviations: HDB, horizontal limb of the diagonal band nucleus; MPN, medial preoptic nucleus; oc, optic chiasm. Scale bar (shown in *b*): *a*, 500 μm; *b*, 300 μm. Reproduced from J. Neurosci., Vol. 18, Sherin JE, Elmquist JK, Torrealba F, Saper CB, Innervation of histaminergic tuberomammillary neurons by GABAergic and galaninergic neurons in the ventrolateral preoptic nucleus of the rat, pp. 4705-4721, Copyright (1998), by courtesy of the original authors and with permission from The Society for Neuroscience.

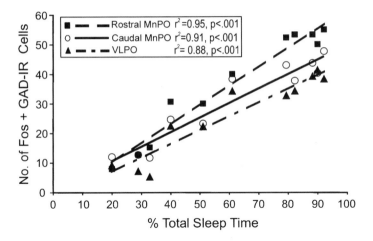

Figure 7.2. Dual immunolabeling techniques are used to demonstrate that the numbers of glutamic acid decarboxylase (GAD)-immunoreactive or GABAergic neurons expressing c-Fos in the rostral and caudal parts of the median preoptic nucleus (MnPO) and in the ventrolateral preoptic nucleus (VLPO) are positively correlated with the total spontaneous sleep time in the preceding 2 hours in rats (Gong et al., 2004). Anatomical abbreviations in the legend box have been changed from the original for consistency. Adapted from J. Physiol., Vol. 556, Gong H, McGinty D, Guzman-Marin R, Chew KT, Stewart D, Szymusiak R, Activation of *c-fos* in GABAergic neurones in the preoptic area during sleep and in response to sleep deprivation, pp. 935-946, Copyright (2004), by courtesy of the original authors and with permission from Blackwell Publishing.

by releasing GABA and galanin during sleep. In turn, VLPO neurons may be inhibited during waking by noradrenaline released from the terminals of the arousal-promoting locus coeruleus neurons.

It is important to recall that the preoptic area has long been known to contain sleep-active neurons on the basis of single-unit recording data (see above). The factor that played a key role in the identification especially of the VLPO, which does not have clear nuclear boundaries, was the ability of c-Fos mapping to reveal a cluster of active neurons with a high anatomical precision (Sherin et al., 1996). It would be much more difficult to identify such a cluster with single-unit recordings. Thus, the VLPO was identified at least initially on a functional basis as a cluster of sleep-active neurons located in the ventrolateral region of the preoptic area, and its identification was uniquely facilitated by the c-Fos mapping technique.

In addition to sleep state-dependent c-Fos expression, VLPO and MnPO neurons respond, with c-Fos expression, to various stimuli that are known to affect the amount of sleep. VLPO neurons increased c-Fos immunoreactivity in response to sleep-promoting factors, such as prostaglandin D_2 (Scammell et al., 1998) and an adenosine A_{2A} receptor agonist (Scammell et al., 2001), as well general anesthetics, such as the GABAergic agents muscimol, propofol, pentobarbital, and the α_2 adrenergic receptor agonist dexmedetomidine (Nelson et al., 2002, 2003). They also expressed c-Fos following systemic estrogen treatment (Peterfi et al., 2004). MnPO neurons increased c-Fos immunoreactivity in response to elevated ambient temperatures (Gong et al., 2000), cytokine interleukin 1β-induced fever (Baker et al., 2004), and osmotic challenges (Gvilia et al., 2005). Interestingly, MnPO neurons that increased c-Fos immunoreactivity during sleep were GABAergic, whereas those responding to osmotic challenges were non-GABAergic (Gvilia et al., 2005). However, the degree of convergence, or divergence, of other stimulus inputs onto single VLPO and MnPO neurons remains to be determined. Dual-activity mapping techniques that allow concurrent identification of separate populations of neurons responsive to two different stimuli may be valuable for addressing this question (Farivar et al., 2004).

In addition to the VLPO and MnPO, two other brain areas are known to contain neurons that increase firing during sleep: the basal forebrain, located lateral to the preoptic area (Szymusiak and McGinty, 1986; Koyama and Hayaishi, 1994; Lee et al., 2004), and the dorsal medulla, in particular the nucleus of the solitary tract (Eguchi and Satoh, 1980a, b). Neurons in the basal forebrain increased c-Fos expression during recovery sleep, and many of them were GABAergic (Modirrousta et al., 2004). Previous results of behavioral state-dependent c-Fos expression in the nucleus of the solitary tract are inconclusive (Yamuy et al., 1993; Pompeiano et al., 1994; Merchant-Nancy et al., 1995; Verret et al., 2005).

2.3. Identification of REM Sleep-Active Neurons in the Forebrain

As previously mentioned, one limitation of c-Fos mapping is low temporal resolution, on the order of tens of minutes, which does not permit determination of neuronal activity selective to short-lasting sleep stages, such as REM sleep. REM sleep episodes in commonly studied species are too short (e.g., an average of 1.5 minutes per episode in rat) for c-Fos protein synthesis. To overcome this problem, a number of studies have used pharmacological treatments, dark exposure during the light phase, auditory stimulation during REM sleep, or prior REM sleep deprivation to enhance REM sleep and they subsequently examined c-Fos after periods of such enhanced REM sleep (reviewed in Cirelli and Tononi, 2000b).

Two groups of REM sleep-active neurons in the forebrain have been identified recently using c-Fos immunohistochemistry. One group is located in the "extended" VLPO, which includes areas both medial and dorsal to the original VLPO (subsequently called the VLPO core), and these extended-VLPO neurons expressed c-Fos after REM sleep enhancement following dark exposure at the beginning of the light phase (Lu et al., 2000, 2002). Like the sleep-active neurons in the VLPO core, the majority of extended-VLPO neurons that increased c-Fos immunoreactivity during REM sleep contained GABA and galanin, and many of them sent projections to known wake/REM sleep-regulatory regions in the brainstem (i.e., dorsal raphe nucleus, laterodorsal tegmental nucleus and locus coeruleus; Lu et al., 2002). Electrophysiological recordings have confirmed the presence of REM sleep-active neurons in the VLPO region (Szymusiak et al., 1998).

The second group of REM sleep-active neurons identified with the c-Fos method is a population of neurons containing melanin-concentrating hormone

(MCH) located in the lateral hypothalamus and zona incerta. These neurons showed increased c-Fos immunoreactivity during REM sleep "rebound" after REM sleep deprivation (Verret et al., 2003), or during recovery sleep after total sleep deprivation (Modirrousta et al., 2005), in rats. Interestingly, like many VLPO and MnPO GABAergic neurons (see above), many MCH neurons that expressed c-Fos during recovery sleep also bore the α_{2A} adrenergic receptor (Modirrousta et al., 2005). The presence of this receptor on MCH neurons is consistent with the inhibitory action of noradrenaline on these neurons *in vitro* (van den Pol et al., 2004; Bayer et al., 2005), suggesting that, like the preoptic neurons, MCH neurons may be inhibited during wake by noradrenaline released by locus coeruleus and other brainstem neurons. It should be reminded that the basic mechanisms for generating REM sleep are located within the brainstem (Jouvet, 1962), and, therefore, the role of MCH-containing hypothalamic neurons and extended-VLPO neurons in the regulation of REM sleep is likely to be modulatory.

2.4. Identification of REM Sleep-Active GABAergic Neurons in the Brainstem

The c-Fos imaging technique has also been successful in identifying several populations of REM sleep-active GABAergic neurons in the brainstem. GABA concentrations were elevated during REM sleep in several brainstem regions, suggesting that certain GABAergic neurons were active during REM sleep (Nitz and Siegel, 1997a, b). Using c-Fos and double immunolabeling, it was possible to show that GABAergic neurons that co-distributed with cholinergic and monoaminergic neurons in the mesopontine tegmentum selectively increased c-Fos expression during REM sleep rebound in rats (Maloney et al., 1999), and during a REM sleep-like state after microinjection of carbachol (a cholinergic receptor agonist) into the pontine reticular formation in cats (Torterolo et al., 2000, 2001b). Note that the adjacent monoaminergic neurons were without c-Fos, consistent with their virtual silence during REM sleep (McGinty and Harper, 1976; Aston-Jones and Bloom, 1981). Some of the GABAergic neurons may be inhibitory interneurons and target these monoaminergic neurons.

Another population of GABAergic neurons that increase c-Fos immunoreactivity during REM sleep rebound is located in the reticular nuclei of the caudal pons and medulla, and the medullary raphe nuclei. The number of GABAergic, c-Fos-expressing neurons in these regions was negatively correlated with the amplitude of muscle tone, which is the lowest during REM sleep, in rats (Maloney et al., 2000). Interestingly, many neurons in the ventral medullary reticular formation that expressed c-Fos during carbachol-induced REM sleep sent projections to the motor trigeminal nucleus in cats (Morales et al., 1999). These REM sleep-active, pontine and medullary GABAergic neurons may therefore inhibit somatic motor neurons, thereby initiating the loss of muscle tone (atonia) characteristic of REM sleep.

Other GABAergic populations that have been identified as REM sleep-active using c-Fos are located in the dorsal tegmental nucleus of Gudden (Torterolo et al., 2002), substantia nigra pars compacta, and ventral tegmental area (Maloney et al., 2002). These findings corroborate electrophysiological studies showing that fast-spiking ventral tegmental neurons, which are thought to be GABAergic, discharged at their highest rate during REM sleep (Lee et al., 2001). The role of these neurons in REM sleep remains to be investigated.

2.5. Identification of Other REM Sleep-Active Neurons in the Brainstem

c-Fos immunohistochemistry has been used to identify several other REM sleep-active neuronal groups in the brainstem. One of them is a group of medullary reticular formation neurons containing the inhibitory amino acid, glycine (Boissard et al., 2002). These neurons may participate, along with the GABAergic neurons described above, in the inhibition of somatic motor neurons during REM sleep. A second cell group is a population of nitric oxide-producing neurons in the cuneiform nucleus (Pose et al., 2000), whose function in REM sleep is currently unknown. A third group consists of dopaminergic neurons of the ventral tegmental area (Maloney et al., 2002); these neurons may participate in the activation of limbic forebrain regions during REM sleep. A fourth group of neurons identified as REM sleep-active is a population of neurotrophin tyrosine-kinase receptor-containing neurons in the dorsal pontine tegmentum; neurotrophins may act on pontine neurons to facilitate REM sleep (Yamuy et al., 2005).

2.6. Orexin/Hypocretin Neurons in the Perifornical Lateral Hypothalamus

Single-unit recordings of immunohistochemically identified orexin (also known as hypocretin) neurons have shown that these neurons increase firing rates during wakefulness, particularly with motor activity, compared to during slow-wave sleep or REM sleep (Lee et al., 2005a; Mileykovskiy et al., 2005). Consistently, orexin neurons increased c-Fos immunoreactivity during spontaneous wakefulness or during 2–3 hours of sleep deprivation compared to spontaneous sleep (Estabrooke et al., 2001; Martínez et al., 2002; España et al., 2003; Modirrousta et al., 2005). The proportion of orexin neurons that were c-Fos-positive during wakefulness in rats varied considerably between studies, ranging from 10% (Martínez et al., 2002; España et al., 2003; Modirrousta et al., 2005) to 50% (Estabrooke et al., 2001). Technical factors, including the specificity and sensitivity of antibodies, the procedures for cell sampling, and the criteria for identifying a cell as c-Fos-positive, may be responsible in part for these differences. An important determinant of c-Fos expression in orexin neurons appears to be motor activity. For example, up to 79% of orexin neurons expressed c-Fos during active wake, while less than 3% were c-Fos-positive during quiet wakefulness in cats (Torterolo et al., 2003). These results suggest that the activity of orexin neurons is more closely associated with motor activity particularly during high arousal states, than with wake state *per se.*

Although the above studies were focused on the c-Fos immunoreactivity of orexin neurons as a function of behavioral states, c-Fos mapping with other

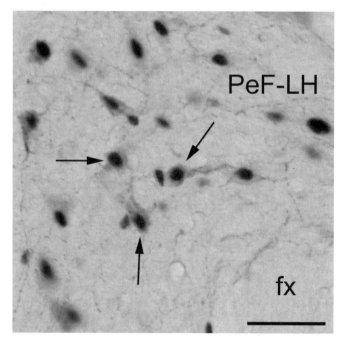

Figure 7.3. Orexin (hypocretin)-containing neurons in the perifornical lateral hypothalamic area (PeF-LH) show increased c-Fos immunoreactivity following systemic injection of a low dose of caffeine (10 mg/kg) in rats. c-Fos-immunoreactivity (appearing in black) is localized to neuronal nuclei, whereas orexin B-immunoreactivity (gray) is localized to the cytoplasm of neurons. Arrows indicate examples of double-labeled neurons. Abbreviation: fx, fornix. Scale bar: 50 μm.

types of stimuli has indicated that orexin neurons are also responsive to a number of environmental and physiological stimuli that increase arousal. These include environmental stressors (España et al., 2003); noxious stimuli (Zhu et al., 2002); treatments that produced hyperphasia, such as administration of orexigenic peptides (e.g., neuropeptide Y; Niimi et al., 2001), energy balance-related manipulations (e.g., insulin-induced hypoglycemia; Moriguchi et al., 1999), and GABA$_A$ receptor activation in the shell region of the accumbens nucleus (Baldo et al., 2004); morphine withdrawal (Georgescu et al., 2003); conditioned place preference for food and drugs (Harris et al., 2005); and stimulants or wake-promoting agents, such as modafinil (Scammell et al., 2000), methamphetamine (Estabrooke et al., 2001), amphetamine (Fadel et al., 2002), dopamine agonists (Bubser et al., 2005), and caffeine (Fig. 7.3; Murphy et al., 2003; Deurveilher et al., in press). Whether single orexin neurons respond to all these stimuli is unclear but unlikely. For example, c-Fos immunoreactivity after the stimulants was greater in orexin neurons in the medial, than in the lateral, perifornical lateral hypothalamus, whereas the reverse was true after treatments that induced consumatory or drug-seeking behaviors (see references above). These findings suggest that medial orexin neurons may be more closely involved in high arousal wake states, whereas lateral orexin neurons may have a role in motivational processes.

In contrast to their high activity during wakefulness, orexin neurons are electrophysiologically virtually silent during slow-wave sleep and REM sleep, with occasional bursts of discharges in association with muscle twitches during REM sleep, in rats (Lee et al., 2005a; Mileykovskiy et al., 2005). In parallel, c-Fos expression in orexin neurons was very low during REM sleep rebound following REM sleep deprivation in rats (Verret et al., 2003). In cats, however, 34% of orexin neurons expressed c-Fos during carbachol-induced REM sleep (Torterolo et al., 2001a). This discrepancy may be due to differences in the species and experimental procedures used. Interestingly, recent c-Fos data suggest that orexin neurons are actively inhibited by GABAergic input during sleep. The blockade of $GABA_A$ receptors in the perifornical region with bicuculline increased wakefulness and induced c-Fos immunoreactivity in orexin neurons (Alam et al., 2004; Goutagny et al., 2005).

The c-Fos technique has also been used to unravel the functional connectivity between orexin neurons and other arousal state-regulatory regions in the forebrain. For example, microinjection of a $GABA_A$ receptor agonist into the basal forebrain (Satoh et al., 2003) or preoptic area (Satoh et al., 2004) produced wakefulness and increased c-Fos expression in orexin neurons ipsilateral to the injection site. These results suggest that orexin neurons receive inhibitory projections, directly and/or indirectly, from the basal forebrain and preoptic area.

Triple-labeling techniques have been used to demonstrate that many orexin neurons that expressed c-Fos during sleep deprivation bore α_{1A} or α_{2A} adrenergic receptors (Modirrousta et al., 2005). Electrophysiological data on the effects of noradrenaline on orexin neurons *in vitro* are inconsistent, with reports of an excitation of intracellularly labeled rat orexin neurons (Bayer et al., 2005), and an inhibition of green fluorescent protein (GFP)-labeled mouse orexin neurons (Li et al., 2002; Yamanaka et al., 2003). Intriguing data indicate that the excitatory action of noradrenaline *in vitro* changed to an inhibitory action when rats were previously sleep-deprived for 2 hours (Grivel et al., 2005). These results suggest that orexin neurons respond to a noradrenergic input with either an excitation or an inhibition depending on the level of arousal during wakefulness or sleep need.

2.7. Histamine Neurons in the Tuberomammillary Hypothalamic Nucleus

Similar to orexin neurons, c-Fos immunohistochemistry has revealed the wake-related activity of histamine neurons. The number of histaminergic neurons that expressed c-Fos was proportional to the amount of time spent awake during the preceding 1 hour (Ko et al., 2003; see also Novak et al., 2000). Electrophysiological and microdialysis evidence has also shown that histamine neurons increased their activity during wakefulness, particularly during wake state with motor activity (Yamatodani et al., 1996; Strecker et al., 2002; Vanni-Mercier et al., 2003; Chu et al., 2004). Interestingly, the pattern of c-Fos activation in histamine neurons following wake periods showed regional heterogeneity, with higher proportions of histamine neurons expressing c-Fos located in the caudal ($\sim$85%) and ventral divisions ($\sim$75%) than in the dorsal division ($\sim$35%; Ko et al., 2003). These regional differences may be related to previously reported differences in the afferent connections of different histaminergic subgroups (Sherin et al., 1998). In addition, c-Fos immunohistochemistry revealed a subset of histamine neurons responding to high arousal states, such as those associated with increased locomotor activity in anticipation of mealtime (Meynard et al., 2005; Valdés et al., 2005), and with various stress challenges (Miklós and Kovács, 2003). Increased c-Fos expression in histaminergic neurons has also been observed during wakefulness following injections of an H_3 histaminergic autoreceptor antagonist, which disinhibits histaminergic neurons directly, in cats (Vanni-Mercier et al., 2003), or following a low dose of caffeine (10 mg/kg, i.p.) in rats (Deurveilher et al., in press). However, amphetamine, methylphenidate, and modafinil in cats (Lin et al., 1996), or higher doses of caffeine in rats (30 and 75 mg/kg; Deurveilher et al., in press), all failed to increase c-Fos immunoreactivity in tuberomammillary neurons. These differences may be due to different mechanisms of action of these stimulants. Administration of the anesthetic dexmedetomidine, on the other hand, reduced c-Fos expression in the tuberomammillary nucleus, and this reduction depended on the integrity of the VLPO (Nelson et al., 2003). This latter finding provides further support for the existence of inhibitory projections from the VLPO to the tuberomammillary nucleus (see 2.1), and illustrates an example of the usefulness of combining lesions with c-Fos immunohistochemistry to examine the functional connectivity between brain regions.

2.8. Issues with the Serotonergic, Noradrenergic, Dopaminergic and Cholinergic Cell Groups

In contrast to orexin and histaminergic neurons, the functional mapping of serotonergic neurons of the dorsal raphe nucleus or noradrenergic neurons of the locus coeruleus using c-Fos immunohistochemistry has met with less success. The increased activity of these neurons during wakefulness, as demonstrated by electrophysiological (McGinty and Harper, 1976; Aston-Jones and Bloom, 1981) and microdialysis studies (Portas et al., 1998; Shouse et al., 2000; Lena et al., 2005), is in stark contrast to the little c-Fos expression observed during wake states (Tononi et al., 1994; Yamuy et al., 1995; Yamuy et al., 1998; Janušonis and Fite, 2001; Lu et al., 2006). The noradrenergic neurons of the locus coeruleus are, however, capable of expressing c-Fos; a majority of these neurons showed increased c-Fos immunoreactivity in response to stress and noxious stimuli (Sved et al., 2002; Takase et al., 2005), or a high dose of caffeine (Deurveilher et al., in press). Serotonergic neurons in the dorsal raphe nucleus may be less reactive in terms of c-Fos expression; only a small proportion of these cells expressed c-Fos under conditions that increase arousal, such as exposure to noxious stimuli and certain stressors (Grahn et al., 1999; Chen et al., 2003; Takase et al., 2004, 2005), or caffeine injection (Abrams et al., 2005; Deurveilher et al., in press).

Contrary to the locus coeruleus noradrenergic and dorsal raphe serotonergic neurons, dopaminergic neurons in the ventral tegmental area and substantia nigra pars compacta are electrophysiologically active during wakefulness and do not change their activity greatly during sleep (Miller et al., 1983; Trulson and Preussler, 1984). High dopamine levels in the forebrain were associated with wakefulness and REM sleep (Lena et al., 2005). However, little or no c-Fos was found in the dopaminergic neurons in the ventral tegmental area and substantia nigra pars compacta during spontaneous or forced wakefulness (Lu et al., 2006), or following motor stimulant doses of caffeine (Deurveilher et al., in press) or an adenosine A_{2A} receptor antagonist (Deurveilher et al., 2005). In contrast, dopaminergic neurons in the ventral periaqueductal gray matter increased c-Fos expression during spontaneous and forced wakefulness (Lu et al., 2006) (but not in response to caffeine administration; Deurveilher et al., in press). These neurons projected to major components of the sleep-wake system, including the basal forebrain, VLPO, laterodorsal tegmental nucleus and locus coeruleus, and specific lesions of these neurons increased total daily sleep (Lu et al., 2006). These dopamine neurons in the periaqueductal gray may be responsible for pharmacological and clinical evidence suggesting an important role of dopamine in the regulation of wakefulness (Lu et al., 2006).

Another example of an arousal-active neuronal group whose firing is not consistently associated with c-Fos expression is the group of cholinergic neurons in the basal forebrain. Immunohistochemically identified cholinergic neurons discharged in bursts during active wake and REM sleep (Lee et al., 2005b). Microdialysis data also indicated that acetylcholine levels in the cerebral cortex were elevated during wakefulness and REM sleep compared to slow-wave sleep (Jasper and Tessier, 1971). However, previous c-Fos studies support this conclusion equivocally. Several studies reported little or no c-Fos in any cholinergic cell groups of the basal forebrain after 3 hours of sleep deprivation (Basheer et al., 1999) or after periods of increased locomotor activity following administration of either caffeine (Deurveilher et al., in press) or an adenosine A_{2A} receptor antagonist (Deurveilher et al., 2005). Other studies found a number of c-Fos-immunoreactive cholinergic neurons after spontaneous waking or 3 or 6 hours of sleep deprivation, but these studies were inconsistent as to the location of these neurons within the basal forebrain (Greco et al., 2000; Modirrousta et al., 2004; Lu et al., 2006).

The c-Fos strategy has also been unsuccessful in revealing REM sleep- or wake-selective activation of the mesopontine cholinergic neurons. Microdialysis studies indicated an increase in acetylcholine levels in the thalamus and pontine reticular formation during REM sleep, and in the thalamus during wakefulness as well, as compared to slow-wave sleep; both regions are innervated by meso-pontine cholinergic neurons (reviewed in Semba, 1999). However, only a small proportion of the mesopontine cholinergic neurons were c-Fos-positive during carbachol microinjection-induced REM sleep in cats (Shiromani et al., 1996; Yamuy et al., 1998; Pose et al., 2000; Torterolo et al., 2001b), or REM sleep rebound after REM sleep deprivation in rats (Maloney et al., 1999; Verret et al., 2005). Similar results were obtained after spontaneous or forced wakefulness (Yamuy et al., 1998; Pose et al., 2000; Torterolo et al., 2001b; Lu et al., 2006), or periods of increased locomotor activity following administration of either caffeine (Deurveilher et al., in press) or an adenosine A_{2A} receptor antagonist (Deurveilher et al., 2005). Presumed mesopontine cholinergic neurons showed a high tonic firing pattern during both wakefulness and REM sleep, and a subset of them displayed a phasic increase in firing during REM sleep (reviewed in Semba, 1999). Although these patterns of discharge would probably be indicative of c-Fos expression, this is apparently not the case.

The reason for the discrepancy between the c-Fos data for these monoaminergic and cholinergic neurons, on one hand, and their electrophysiological and microdialysis data, on the other, is unclear. Evidence indicates that induction of *c-fos* requires a sustained increase in glutamatergic synaptic activity; a strong activation of postsynaptic glutamate receptors, in particular N-methyl-D-aspartate (NMDA) receptors; and a significant increase in intracellular concentrations of calcium and cyclic AMP (Finkbeiner and Greenberg, 1998; Herdegen and Leah, 1998). An increase in intracellular calcium induces *c-fos* expression by activating specific signal transduction pathways, which results in activation of specific transcription factors. These transcription factors include the calcium-cyclic AMP response element binding protein (CREB) and the serum-response factor, which bind to the calcium-cyclic AMP response element (CRE) and serum-response element, respectively, within the *c-fos* promoter (Finkbeiner and Greenberg, 1998; Herdegen and Leah, 1998). In light of this sequence of events leading to *c-fos* expression, several explanations may be offered for the lack of c-Fos expression in the presence of active firing. First, despite their increased firing during wake states and, in case of the cholinergic neurons, REM sleep as well, these neurons may not receive a sufficiently strong and sustained glutamatergic drive during these states to activate fully the calcium-dependent signaling pathway required for c-Fos synthesis. In support of this possibility, the non-specific ionotropic glutamate receptor antagonist kynurenate had no effect on the spontaneous activity of dorsal raphe neurons, but reduced the sensory evoked increase in activity (Levine and Jacobs, 1992). However, the monoaminergic and cholinergic neurons certainly have glutamate input and NMDA receptors (Cherubini et al., 1988; Inglis and Semba, 1996; Sanchez and Leonard, 1996), and it remains difficult to reconcile this with the lack of c-Fos. It is also possible that the calcium-dependent pathways within the stimulus-transcription coupling mechanisms in these neurons are lacking, have a higher threshold for activation, or are desensitized or inactivated during wakefulness. Another possibility is that mechanisms responsible for inhibition of the *c-fos* promoter and arrest of *c-fos* transcription (Sassone-Corsi et al., 1988; Herdegen and Leah, 1998) may be particularly effective in these neurons. In any case, an increase in c-Fos synthesis is a good indicator of increased activity of a neuron, but its absence does not necessarily mean that the cell in question is inactive.

3. Functional Roles of c-Fos in Sleep-Wake Control

Although c-Fos has been used extensively and successfully as a marker of neuronal activation, it is important to remember that c-Fos is a transcription factor. Accordingly, another line of sleep research has been concerned with the function and molecular consequences of *c-fos* expression in neurons that express it during wakefulness or sleep. One successful approach to this question has been to examine how genetic alterations in *c-fos* expression alter sleep and wake states. Mice with a deletion of the *c-fos* gene showed approximately 30% less slow-wave sleep during the day, while REM sleep was not affected; compared to their littermate controls they also took longer to fall asleep after 6-hour sleep deprivation (Shiromani et al., 2000). These data, however, should be interpreted with caution due to the non-specificity of the *c-fos* deletion and several developmental behavioral abnormalities that may affect sleep. An alternative approach is the use of antisense probes. Administration of a *c-fos* antisense oligonucleotide in the preoptic area, which markedly reduced the c-Fos immunostaining normally observed in that area after wakefulness, also decreased both slow-wave sleep and REM sleep on the day after injection (Cirelli et al., 1995b). These studies suggest that c-Fos protein in the preoptic area is involved in promoting sleep, and that loss of its expression decreases sleep.

As a transcription factor, the c-Fos protein can regulate the transcription rates of target genes, commonly referred to as late response genes. To do so, c-Fos must dimerize with a member of the Jun protein family (c-Jun, JunB or JunD) to form the transcription factor activating protein-1 (AP-1), whereas Jun can form a homodimer with itself (Hughes and Dragunow, 1995). This AP-1 factor recognizes either the AP-1 consensus sequence, also known as 12-O-tetradecanoylphorbol-13-acetate (TPA) response elements (TRE), or the CRE (Chinenov and Kerppola, 2001), both in the promoter region of numerous genes. The composition of the AP-1 dimers determines its transcriptional potency; Fos/Jun dimers have stronger transcriptional activity than Jun/Jun dimers in cultured cells (Hughes and Dragunow, 1995).

Of particular interest is the possibility that the composition of the AP-1 transcription factor varies depending on behavioral state. The following observations support this possibility. Like *c-fos* mRNA, *junB* mRNA was dynamically expressed in the brain during wakefulness or in response to sleep deprivation, although with more regional selectivity than *c-fos* (Fig. 7.4; O'Hara et al., 1993; Grassi-Zucconi et al., 1994; Menegazzi et al., 1994; Terao et al., 2003). Similarly, JunB protein immunoreactivity increased in several brain regions in response to 3 or 6 hours of sleep deprivation (Semba et al., 2001). In contrast, both *c-jun* and *junD* mRNAs were expressed constantly at high levels without changes across the light-dark cycle or even after sleep deprivation (see above for references). These findings raise the intriguing possibility that c-Fos/JunB dimers may be prevalent during wakefulness, whereas Jun/Jun dimers may prevail during sleep. It remains to be examined, however, whether IEGs whose expression varies in parallel after wake periods or sleep deprivation co-localize in single neurons.

Behavioral state-dependent changes in the composition of the AP-1 dimers may affect their DNA-binding activity. The AP-1 binding, as measured by electrophoretic mobility shift assay, increased in the basal forebrain and hypothalamus after 3, 6 or 12 hours of sleep deprivation (Basheer et al., 1999; Basheer and

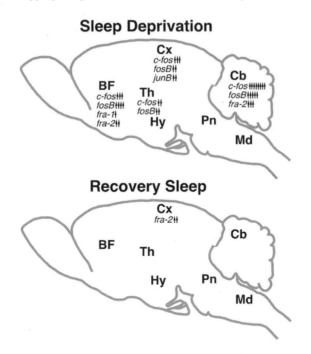

Figure 7.4. mRNA expression of several members of *fos* and *jun* gene families in the mouse brain is upregulated following a 6-hour sleep deprivation (Terao et al., 2003). Sleep deprivation was performed by gentle handling starting at lights on; recovery sleep was allowed for the following 4 hours. Gene expression was quantified by real-time RT-PCR analyses. The number of arrows indicates x-fold increases in magnitude compared to levels in undisturbed control groups. One arrow: 1.5-fold or less increase; Two arrows: 1.5- to 2.5-fold increase; Three arrows: 2.5- to 3.5-fold increase, and so on. Note that there is a widespread activation of *fos/jun* expression during sleep deprivation, whereas very limited changes occur during the subsequent sleep rebound. Abbreviations: BF, basal forebrain; Cb, cerebellum; Cx, cerebral cortex; Hy, hypothalamus; Md, medulla; Pn, pons; Th, thalamus. Reprinted from Neuroscience, Vol. 120, Terao A, Greco MA, Davis RW, Heller HC, Kilduff TS, Region-specific changes in immediate early gene expression in response to sleep deprivation and recovery sleep in the mouse brain, pp. 1115-1124, Copyright (2003), by courtesy of the original authors and with permission from Elsevier.

Shiromani, 2001). Supershift assay and Western blot analyses are necessary to examine the compositional changes in the members of the AP-1 complexes. c-Fos is, however, likely one of the components of the AP-1 dimers observed after sleep deprivation, because *c-fos* knock out mice did not show increased AP-1 binding in the cingulate cortex after 6 hours of sleep deprivation (Shiromani et al., 2000). Changes in DNA-binding by AP-1 may also be linked to interactions with other transcription factors and cofactors, and its phosphorylation status (Herdegen and Leah, 1998; Chinenov and Kerppola, 2001). The functional consequences of changes in the DNA-binding activity of AP-1 after sleep deprivation are yet to be elucidated.

Candidate gene targets for the c-Fos/AP-1 complex are genes bearing AP-1/ CRE elements. Among these diverse genes are several transmitter-related genes present in neurons of the sleep-wake regulatory system, including prodynorphin, glutamic acid decarboxylase, tyrosine hydroxylase and choline acetyltransferase (reviewed in Hughes and Dragunow, 1995; Herdegen and Leah, 1998). It has been shown that the levels of tyrosine hydroxylase mRNA in the locus coeruleus increased after 3 or 5 days of REM sleep deprivation (Porkka-Heiskanen et al., 1995; Basheer et al., 1998). Extensive screening of the mice and rat genome has been undertaken recently to identify genes that are upregulated specifically during sleep and waking (Cirelli and Tononi, 2000a; Cirelli et al., 2004; Terao et al., 2006). However, the relationship between c-Fos and any of its potential target genes remains to be investigated. It is possible that c-Fos regulates different target genes depending on the neuronal type and cellular environment.

The role of c-Fos as a transcription factor in sleep-wake regulation remains unclear. However, it has been proposed that c-Fos induction during waking might represent a neuronal response to increased metabolic activity and cellular stress, and that c-Fos protein might activate signaling pathways to restore equilibrium within a neuron (Sheng and Greenberg, 1990; Cirelli and Tononi, 2000b). Another possible function of c-Fos is a role in neuronal plasticity occurring during wakefulness, in conjunction with sensorimotor experience, learning, and memory (Cirelli and Tononi, 2000b; Kaczmarek, 2002).

4. Conclusions and Future Directions

There is little doubt that c-Fos has been extremely useful in sleep research as a marker of neuronal activation. The major advantages of the c-Fos techniques are the possibility for regional mapping at the cellular level, and its use in combination with other neuroanatomical and neurochemical markers. c-Fos mapping has led to the discovery of prominent clusters of sleep-active neurons in the preoptic area (VLPO and MnPO), the identification of REM-sleep active neurons in several forebrain and brainstem nuclei, and the confirmation of arousal-promoting neuronal groups of the hypothalamus. Using c-Fos and other immunostaining or axonal tracing, it was also possible to identify the neurotransmitters, receptors and projections of a number of these neurons, as well as to characterize their responsiveness to various stimuli. There have been, however, relatively few studies taking advantage of this multi-labeling potential, and the availability of new antibodies for transmitter receptors, for example, would provide attractive opportunities for future investigation.

Despite these major advantages of using c-Fos, several caveats are warranted. One disadvantage is the poor time resolution of c-Fos expression which is due to its time course in the order of tens of minutes. A major limitation of c-Fos mapping, however, is that not all neurons express c-Fos even when they are known to be electrophysiologically active. Examples are cholinergic neurons in the basal forebrain and mesopontine tegmentum and certain monoaminergic neurons. The elucidation of this discrepancy might give insights into the factors that are crucial for c-Fos expression.

Recognizing the limitations of c-Fos as a neuronal marker of activation, several alternative approaches may be proposed. One approach is to use other IEGs as markers of activation, and this is attractive for two reasons. First, subsets of wake- and sleep-active neurons that do not readily express *c-fos* may express other IEGs, for example, nerve growth factor-induced A (NGFI-A; also called Egr-1, ZIF-268, ZENK, krox24, and tis-8). The expression of NGFI-A showed a different time course and regional expression pattern than c-Fos, both during spontaneous behavioral states and in response to sleep deprivation (O'Hara et al., 1993; Pompeiano et al., 1994, 1997; Terao et al., 2003). Second, IEGs with a slower time course of expression and longer persistence than c-Fos, such as chronic Fos related antigen (Fra)-1 and Fra-2 (which are isoforms of ΔFosB, a splice variant of another member of the *fos* gene family, *fosB*; Nestler et al., 1999), may be useful in mapping neuronal activation during long-term sleep deprivation or chronic physiological, behavioral or pharmacological stimulation. Another alternative to c-Fos mapping is to examine activation of signal transduction pathways, using antibodies against phosphorylated activated kinases or CREB, or CRE-LacZ transgenic mice in which the reporter gene *lacZ* is under the control of CREB.

The functional consequence of *c-fos* expression remains unclear requiring further investigation, but several interesting approaches have been developed in recent years. One approach is to characterize the cellular and synaptic properties of neurons *in vitro* in which *c-fos* expression has previously been activated during sleep or wakefulness, using mice that are transgenic for a FosGFP fusion protein that is driven by the *c-fos* promoter (Barth et al., 2004). Another approach is to screen mRNA levels in single c-Fos-expressing neurons to examine functional relationships between the expression of *c-fos* and other specific genes, using laser microdissection and real-time reverse transcription-polymerase chain reaction (RT-PCR; Harthoorn et al., 2005). As alternatives to the conventional gene knock out and antisense technologies which have their own limitations, it will be useful to develop conditional transgenics using Cre/loxP strategy to suppress *c-fos* expression, and dominant negative transgenics to block c-Fos/AP-1 activity, both in a time- and cell-specific manner. Another useful approach will be to use short-interfering RNA to silence the *c-fos* gene in specific brain regions through local injections. It is hoped that these techniques will lead to a better understanding of the role of c-Fos and its functional relationship with other genes.

In view of the major advantages of c-Fos as a marker of neuronal activation, c-Fos will most likely continue to prove useful in expanding the knowledge of neuronal networks involved in the regulation of sleep-wake, and characterizing the impact of various stimuli on sleep and wake regulation at cellular and mole-cular levels. The presence of c-Fos in neurons in association with sleep or wakefulness does not, however, prove by itself that these neurons are responsible for the onset and/or maintenance of that behavioral state, or physiological processes taking place during that behavioral state. Rather, localizing c-Fos-expressing neurons would represent only the first step and it would be necessary to apply neurophysiological, neurochemical, and other techniques complementarily to better understand the role of these neurons. The function of c-Fos after it is expressed is also currently unclear. As we learn more about the target genes of c-Fos and other IEG proteins, it is hoped that a better understanding of the molecular and cellular mechanisms of behavioral state control will ensue, and that new targets for therapeutic intervention for the treatment of sleep and wake disorders will emerge.

Acknowledgements

We thank Drs. Doug Rasmusson, Jennifer Stamp and Raphael Pinaud for helpful comments on an earlier version of this manuscript. Supported by CIHR (MOP14451 and MOP67085) and NSERC (217301-03) grants.

References

Abrams JK, Johnson PL, Hay-Schmidt A, Mikkelsen JD, Shekhar A, Lowry CA (2005). Serotonergic systems associated with arousal and vigilance behaviors following administration of anxiogenic drugs. Neuroscience 133:983-997.

Alam MN, Kumar S, Bashir T, Suntsova N, Methippara MM, Szymusiak R, McGinty D (2004). Evidence for GABA mediated control of hypocretin- but not MCH-immunoreactive neurons during sleep in rats. J Physiol 563:569-582.

Aston-Jones G, Bloom FE (1981). Activity of norepinephrine-containing locus coeruleus neurons in behaving rats anticipates fluctuations in the sleep-waking cycle. J Neurosci 1:876-886.

Baker FC, Shah S, Stewart D, Angara C, Gong H, Szymusiak R, Opp MR, McGinty D (2004). Interleukin 1β enhances non-rapid eye movement sleep and increases c-Fos protein expression in the median preoptic nucleus of the hypothalamus. Am J Physiol Regul Integr Comp Physiol 288:R998-R1005.

Baldo BA, Gual-Bonilla L, Sijapati K, Daniel RA, Landry CF, Kelley AE (2004). Activation of a subpopulation of orexin/hypocretin-containing hypothalamic neurons by $GABA_A$ receptor-mediated inhibition of the nucleus accumbens shell, but not by exposure to a novel environment. Eur J Neurosci 19:376-386.

Barth AL, Gerkin RC, Dean KL (2004). Alteration of neuronal firing properties after *in vivo* experience in a FosGFP transgenic mouse. J Neurosci 24:6466-6475.

Basheer R, Shiromani PJ (2001). Effects of prolonged wakefulness on c-fos and AP1 activity in young and old rats. Brain Res Mol Brain Res 89:153-157.

Basheer R, Magner M, McCarley RW, Shiromani PJ (1998). REM sleep deprivation increases the levels of tyrosine hydroxylase and norepinephrine transporter mRNA in the locus coeruleus. Brain Res Mol Brain Res 57:235-240.

Basheer R, Porkka-Heiskanen T, Stenberg D, McCarley RW (1999). Adenosine and behavioral state control: adenosine increases c-Fos protein and AP1 binding in basal forebrain of rats. Brain Res Mol Brain Res 73:1-10.

Bayer L, Eggermann E, Serafin M, Grivel J, Machard D, Mühlethaler M, Jones BE (2005). Opposite effects of noradrenaline and acetylcholine upon hypocretin/orexin versus melanin concentrating hormone neurons in rat hypothalamic slices. Neuroscience 130:807-811.

Bentivoglio M, Grassi-Zucconi G (1999). Immediate early gene expression in sleep and wakefulness. In: Handbook of Behavioral State Control. Cellular and Molecular Mechanisms (Lydic R, Baghdoyan HA, eds), pp. 235-253. Boca Raton: CRC Press.

Boissard R, Gervasoni D, Schmidt MH, Barbagli B, Fort P, Luppi P-H (2002). The rat ponto-medullary network responsible for paradoxical sleep onset and maintenance: a combined microinjection and functional neuroanatomical study. Eur J Neurosci 16:1959-1973.

Bubser M, Fadel JR, Jackson LL, Meador-Woodruff JH, Jing D, Deutch AY (2005). Dopaminergic regulation of orexin neurons. Eur J Neurosci 21:2993-3001.

Chen T, Dong YX, Li YQ (2003). Fos expression in serotonergic neurons in the rat brainstem following noxious stimuli: an immunohistochemical double-labelling study. J Anat 203:579-588.

Cherubini E, North RA, Williams JT (1988). Synaptic potentials in rat locus coeruleus neurones. J Physiol 406:431-442.

Chinenov Y, Kerppola TK (2001). Close encounters of many kinds: Fos-Jun interactions that mediate transcription regulatory specificity. Oncogene 20:2438-2452.

Chu M, Huang Z-L, Qu W-M, Eguchi N, Yao M-H, Urade Y (2004). Extracellular histamine level in the frontal cortex is positively correlated with the amount of wakefulness in rats. Neurosci Res 49:417-420.

Cirelli C, Tononi G (2000a). Gene expression in the brain across the sleep-waking cycle. Brain Res 885:303-321.

Cirelli C, Tononi G (2000b). On the functional significance of c-fos induction during the sleep-waking cycle. Sleep 23:453-469.

Cirelli C, Pompeiano M, Tononi G (1995a). Sleep deprivation and c-fos expression in the rat brain. J Sleep Res 4:92-106.

Cirelli C, Gutierrez CM, Tononi G (2004). Extensive and divergent effects of sleep and wakefulness on brain gene expression. Neuron 41:35-43.

Cirelli C, Pompeiano M, Arrighi P, Tononi G (1995b). Sleep-waking changes after c-fos antisense injections in the medial preoptic area. Neuroreport 6:801-805.

Collaco AM, Geusz ME (2003). Monitoring immediate-early gene expression through firefly luciferase imaging of HRS/J hairless mice. BMC Physiol 3:8.

Deurveilher S, Huang Z-L, Semba K, Hayaishi O (2005). Behavioral and neuronal activation following systemic injection of the adenosine A_{2A} receptor antagonist MSX-3 in rats. Sleep Suppl 28:41.

Deurveilher S, Lo H, Murphy JA, Burns J, Semba K (in press). Differential c-Fos immunoreactivity in arousal-promoting cell groups following systemic administration of caffeine in rats.

Dragunow M, Faull R (1989). The use of c-fos as a metabolic marker in neuronal pathway tracing. J Neurosci Methods 29:261-265.

Eguchi K, Satoh T (1980a). Characterization of the neurons in the region of solitary tract nucleus during sleep. Physiol Behav 24:99-102.

Eguchi K, Satoh T (1980b). Convergence of sleep-wakefulness subsystems onto single neurons in the region of cat's solitary tract nucleus. Arch Ital Biol 118:331-345.

España RA, Valentino RJ, Berridge CW (2003). Fos immunoreactivity in hypocretin-synthesizing and hypocretin-1 receptor-expressing neurons: effects of diurnal and nocturnal spontaneous waking, stress and hypocretin-1 administration. Neuroscience 121:201-217.

Estabrooke IV, McCarthy MT, Ko E, Chou TC, Chemelli RM, Yanagisawa M, Saper CB, Scammell TE (2001). Fos expression in orexin neurons varies with behavioral state. J Neurosci 21:1656-1662.

Fadel J, Bubser M, Deutch AY (2002). Differential activation of orexin neurons by antipsychotic drugs associated with weight gain. J Neurosci 22:6742-6746.

Farivar R, Zangenehpour S, Chaudhuri A (2004). Cellular-resolution activity mapping of the brain using immediate-early gene expression. Front Biosci 9:104-109.

Finkbeiner S, Greenberg ME (1998). Ca^{2+} channel-regulated neuronal gene expression. J Neurobiol 37:171-189.

Gallopin T, Fort P, Eggermann E, Cauli B, Luppi P-H, Rossier J, Audinat E, Mühlethaler M, Serafin M (2000). Identification of sleep-promoting neurons in vitro. Nature 404:992-995.

Gaus SE, Strecker RE, Tate BA, Parker RA, Saper CB (2002). Ventrolateral preoptic nucleus contains sleep-active, galaninergic neurons in multiple mammalian species. Neuroscience 115:285-294.

Georgescu D, Zachariou V, Barrot M, Mieda M, Willie JT, Eisch AJ, Yanagisawa M, Nestler EJ, DiLeone RJ (2003). Involvement of the lateral hypothalamic peptide orexin in morphine dependence and withdrawal. J Neurosci 23:3106-3111.

Gong H, Szymusiak R, King J, Steininger T, McGinty D (2000). Sleep-related c-Fos protein expression in the preoptic hypothalamus: effects of ambient warming. Am J Physiol Regul Integr Comp Physiol 279:R2079-R2088.

Gong H, McGinty D, Guzman-Marin R, Chew KT, Stewart D, Szymusiak R (2004). Activation of c-fos in GABAergic neurones in the preoptic area during sleep and in response to sleep deprivation. J Physiol 556:935-946.

Goutagny R, Luppi P-H, Salvert D, Gervasoni D, Fort P (2005). GABAergic control of hypothalamic melanin-concentrating hormone-containing neurons across the sleep-waking cycle. Neuroreport 16:1069-1073.

Grahn RE, Will MJ, Hammack SE, Maswood S, McQueen MB, Watkins LR, Maier SF (1999). Activation of serotonin-immunoreactive cells in the dorsal raphe nucleus in rats exposed to an uncontrollable stressor. Brain Res 826:35-43.

Grassi-Zucconi G, Menegazzi M, Carcereri De Prati A, Vescia S, Ranucci G, Bentivoglio M (1994). Different programs of gene expression are associated with different phases of the 24h and sleep-wake cycles. Chronobiologia 21:93-97.

Greco MA, Lu J, Wagner D, Shiromani PJ (2000). c-Fos expression in the cholinergic basal forebrain after enforced wakefulness and recovery sleep. Neuroreport 11:437-440.

Grivel J, Cvetkovic V, Bayer L, Machard D, Tobler I, Mühlethaler M, Serafin M (2005). The wake-promoting hypocretin/orexin neurons change their response to noradrenaline after sleep deprivation. J Neurosci 25:4127-4130.

Gvilia I, Angara C, McGinty D, Szymusiak R (2005). Different neuronal populations of the rat median preoptic nucleus express c-fos during sleep and in response to hypertonic saline or angiotensin-II. J Physiol 569:587-599.

Harris GC, Wimmer M, Aston-Jones G (2005). A role for lateral hypothalamic orexin neurons in reward seeking. Nature 437:556-559.

Harthoorn LF, Sañe A, Nethe M, Van Heerikhuize JJ (2005). Multi-transcriptional profiling of melanin-concentrating hormone and orexin-containing neurons. Cell Mol Neurobiol 25:1209-1223.

Herdegen T, Leah JD (1998). Inducible and constitutive transcription factors in the mammalian nervous system: control of gene expression by Jun, Fos and Krox, and CREB/ATF proteins. Brain Res Brain Res Rev 28:370-490.

Hoffman GE, Lyo D (2002). Anatomical markers of activity in neuroendocrine systems: are we all 'Fos-ed out'? J Neuroendocrinol 14:259-268.

Hughes P, Dragunow M (1995). Induction of immediate-early genes and the control of neurotransmitter-regulated gene expression within the nervous system. Pharmacol Rev 47:133-178.

Inglis WL, Semba K (1996). Colocalization of ionotropic glutamate receptor subunits with NADPH-diaphorase-containing neurons in the rat mesopontine tegmentum. J Comp Neurol 368:17-32.

Janušonis S, Fite KV (2001). Diurnal variation of c-Fos expression in subdivisions of the dorsal raphe nucleus of the Mongolian gerbil (Meriones unguiculatus). J Comp Neurol 440:31-42.

Jasper HH, Tessier J (1971). Acetylcholine liberation from cerebral cortex during paradoxical (REM) sleep. Science 172:601-602.

Jones BE (2005). From waking to sleeping: neuronal and chemical substrates. Trends Pharmacol Sci 26:578-586.

Jouvet M (1962). Recherches sur les structures nerveuses et les mécanismes responsables des différentes phases du sommeil physiologique. Arch Ital Biol 100:125-206.

Kaczmarek L (2002). c-Fos in learning: beyond the mapping of neuronal activity. In: Handbook of Chemical Neuroanatomy, vol. 19: Immediate Early Genes and Inducible Transcription Factors in Mapping of the Central Nervous System Function and Dysfunction (Kaczmarek L, Robertson HJ, eds), pp. 189-215. Amsterdam: Elsevier Science.

Ko EM, Estabrooke IV, McCarthy M, Scammell TE (2003). Wake-related activity of tuberomammillary neurons in rats. Brain Res 992:220-226.

Koyama Y, Hayaishi O (1994). Firing of neurons in the preoptic/anterior hypothalamic areas in rat: its possible involvement in slow wave sleep and paradoxical sleep. Neurosci Res 19:31-38.

Lee MG, Hassani OK, Jones BE (2005a). Discharge of identified orexin/hypocretin neurons across the sleep-waking cycle. J Neurosci 25:6716-6720.

Lee MG, Manns ID, Alonso A, Jones BE (2004). Sleep-wake related discharge properties of basal forebrain neurons recorded with micropipettes in head-fixed rats. J Neurophysiol 92:1182-1198.

Lee MG, Hassani OK, Alonso A, Jones BE (2005b). Cholinergic basal forebrain neurons burst with theta during waking and paradoxical sleep. J Neurosci 25:4365-4369.

Lee RS, Steffensen SC, Henriksen SJ (2001). Discharge profiles of ventral tegmental area GABA neurons during movement, anesthesia, and the sleep-wake cycle. J Neurosci 21:1757-1766.

Lena I, Parrot S, Deschaux O, Muffat-Joly S, Sauvinet V, Renaud B, Suaud-Chagny MF, Gottesmann C (2005). Variations in extracellular levels of dopamine, noradrenaline, glutamate, and aspartate across the sleep-wake cycle in the medial prefrontal cortex and nucleus accumbens of freely moving rats. J Neurosci Res 81:891-899.

Levine ES, Jacobs BL (1992). Neurochemical afferents controlling the activity of serotonergic neurons in the dorsal raphe nucleus: microiontophoretic studies in the awake cat. J Neurosci 12:4037-4044.

Li Y, Gao XB, Sakurai T, van den Pol AN (2002). Hypocretin/Orexin excites hypocretin neurons via a local glutamate neuron-A potential mechanism for orchestrating the hypothalamic arousal system. Neuron 36:1169-1181.

Lin J-S, Hou Y, Jouvet M (1996). Potential brain neuronal targets for amphetamine-, methylphenidate-, and modafinil-induced wakefulness, evidenced by *c-fos* immunocytochemistry in the cat. Proc Natl Acad Sci USA 93:14128-14133.

Lu J, Chou TC, Saper CB (2006). Identification of wake-active dopaminergic neurons in the ventral periaqueductal gray matter. J Neurosci 26:193-202.

Lu J, Greco MA, Shiromani P, Saper CB (2000). Effect of lesions of the ventrolateral preoptic nucleus on NREM and REM sleep. J Neurosci 20:3830-3842.

Lu J, Bjorkum AA, Xu M, Gaus SE, Shiromani PJ, Saper CB (2002). Selective activation of the extended ventrolateral preoptic nucleus during rapid eye movement sleep. J Neurosci 22:4568-4576.

Lydic R, Baghdoyan HA (1999). Handbook of Behavioural State Control. Cellular and Molecular Mechanisms. Boca Raton: CRC Press.

Maloney KJ, Mainville L, Jones BE (1999). Differential c-Fos expression in cholinergic, monoaminergic, and GABAergic cell groups of the pontomesencephalic tegmentum after paradoxical sleep deprivation and recovery. J Neurosci 19:3057-3072.

Maloney KJ, Mainville L, Jones BE (2000). c-Fos expression in GABAergic, serotonergic, and other neurons of the pontomedullary reticular formation and raphe after paradoxical sleep deprivation and recovery. J Neurosci 20:4669-4679.

Maloney KJ, Mainville L, Jones BE (2002). c-Fos expression in dopaminergic and GABAergic neurons of the ventral mesencephalic tegmentum after paradoxical sleep deprivation and recovery. Eur J Neurosci 15:774-778.

Martínez GS, Smale L, Nunez AA (2002). Diurnal and nocturnal rodents show rhythms in orexinergic neurons. Brain Res 955:1-7.

McGinty DJ, Harper RM (1976). Dorsal raphe neurons: depression of firing during sleep in cats. Brain Res 101:569-575.

Menegazzi M, Carcereri De Prati AC, Zucconi GG (1994). Differential expression pattern of *jun B* and *c-jun* in the rat brain during the 24-h cycle. Neurosci Lett 182:295-298.

Merchant-Nancy H, Vazquez J, Garcia F, Drucker-Colin R (1995). Brain distribution of *c-fos* expression as a result of prolonged rapid eye movement (REM) sleep period duration. Brain Res 681:15-22.

Meynard MM, Valdés JL, Recabarren M, Serón-Ferré M, Torrealba F (2005). Specific activation of histaminergic neurons during daily feeding anticipatory behavior in rats. Behav Brain Res 158:311-319.

Miklós IH, Kovács KJ (2003). Functional heterogeneity of the responses of histaminergic neuron subpopulations to various stress challenges. Eur J Neurosci 18:3069-3079.

Mileykovskiy BY, Kiyashchenko LI, Siegel JM (2005). Behavioral correlates of activity in identified hypocretin/orexin neurons. Neuron 46:787-798.

Miller JD, Farber J, Gatz P, Roffwarg H, German DC (1983). Activity of mesencephalic dopamine and non-dopamine neurons across stages of sleep and walking in the rat. Brain Res 273:133-141.

Modirrousta M, Mainville L, Jones BE (2004). GABAergic neurons with α_2-adrenergic receptors in basal forebrain and preoptic area express c-Fos during sleep. Neuroscience 129:803-810.

Modirrousta M, Mainville L, Jones BE (2005). Orexin and MCH neurons express c-Fos differently after sleep deprivation vs. recovery and bear different adrenergic receptors. Eur J Neurosci 21:2807-2816.

Morales FR, Sampogna S, Yamuy J, Chase MH (1999). *c-fos* expression in brainstem premotor interneurons during cholinergically induced active sleep in the cat. J Neurosci 19:9508-9518.

Morgan JI, Curran T (1991). Stimulus-transcription coupling in the nervous system: involvement of the inducible proto-oncogenes *fos* and *jun*. Annu Rev Neurosci 14:421-451.

Moriguchi T, Sakurai T, Nambu T, Yanagisawa M, Goto K (1999). Neurons containing orexin in the lateral hypothalamic area of the adult rat brain are activated by insulin-induced acute hypoglycemia. Neurosci Lett 264:101-104.

Murphy JA, Deurveilher S, Semba K (2003). Stimulant doses of caffeine induce c-Fos activation in orexin/hypocretin-containing neurons in rat. Neuroscience 121:269-275.

Nelson LE, Guo TZ, Lu J, Saper CB, Franks NP, Maze M (2002). The sedative component of anesthesia is mediated by $GABA_A$ receptors in an endogenous sleep pathway. Nat Neurosci 5:979-984.

Nelson LE, Lu J, Guo T, Saper CB, Franks NP, Maze M (2003). The α_2-adrenoceptor agonist dexmedetomidine converges on an endogenous sleep-promoting pathway to exert its sedative effects. Anesthesiology 98:428-436.

Nestler EJ, Kelz MB, Chen J (1999). ΔFosB: a molecular mediator of long-term neural and behavioral plasticity. Brain Res 835:10-17.

Niimi M, Sato M, Taminato T (2001). Neuropeptide Y in central control of feeding and interactions with orexin and leptin. Endocrine 14:269-273.

Nitz D, Siegel J (1997a). GABA release in the dorsal raphe nucleus: role in the control of REM sleep. Am J Physiol 273:R451-455.

Nitz D, Siegel JM (1997b). GABA release in the locus coeruleus as a function of sleep/wake state. Neuroscience 78:795-801.

Novak CM, Smale L, Nunez AA (2000). Rhythms in Fos expression in brain areas related to the sleep-wake cycle in the diurnal *Arvicanthis niloticus*. Am J Physiol Regul Integr Comp Physiol 278:R1267-1274.

O'Hara BF, Young KA, Watson FL, Heller HC, Kilduff TS (1993). Immediate early gene expression in brain during sleep deprivation: preliminary observations. Sleep 16:1-7.

Pace-Schott EF, Hobson JA (2002). The neurobiology of sleep: genetics, cellular physiology and subcortical networks. Nat Rev Neurosci 3:591-605.

Peterfi Z, Churchill L, Hajdu I, Obal Jr F, Krueger JM, Parducz A (2004). Fos-immunoreactivity in the hypothalamus: dependency on the diurnal rhythm, sleep, gender, and estrogen. Neuroscience 124:695-707.

Pompeiano M, Cirelli C, Tononi G (1994). Immediate-early genes in spontaneous wakefulness and sleep: expression of *c-fos* and NGFI-A mRNA and protein. J Sleep Res 3:80-96.

Pompeiano M, Cirelli C, Ronca-Testoni S, Tononi G (1997). NGFI-A expression in the rat brain after sleep deprivation. Brain Res Mol Brain Res 46:143-153.

Porkka-Heiskanen T, Smith SE, Taira T, Urban JH, Levine JE, Turek FW, Stenberg D (1995). Noradrenergic activity in rat brain during rapid eye movement sleep deprivation and rebound sleep. Am J Physiol 268:R1456-1463.

Portas CM, Bjorvatn B, Fagerland S, Grønli J, Mundal V, Sørensen E, Ursin R (1998). On-line detection of extracellular levels of serotonin in dorsal raphe nucleus and frontal cortex over the sleep/wake cycle in the freely moving rat. Neuroscience 83:807-814.

Pose I, Sampogna S, Chase MH, Morales FR (2000). Cuneiform neurons activated during cholinergically induced active sleep in the cat. J Neurosci 20:3319-3327.

Sanchez R, Leonard CS (1996). NMDA-receptor-mediated synaptic currents in guinea pig laterodorsal tegmental neurons in vitro. J Neurophysiol 76:1101-1111.

Sassone-Corsi P, Sisson JC, Verma IM (1988). Transcriptional autoregulation of the proto-oncogene *fos*. Nature 334:314-319.

Satoh S, Matsumura H, Nakajima T, Nakahama K, Kanbayashi T, Nishino S, Yoneda H, Shigeyoshi Y (2003). Inhibition of rostral basal forebrain neurons promotes wakefulness and induces FOS in orexin neurons. Eur J Neurosci 17:1635-1645.

Satoh S, Matsumura H, Fujioka A, Nakajima T, Kanbayashi T, Nishino S, Shigeyoshi Y, Yoneda H (2004). FOS expression in orexin neurons following muscimol perfusion of preoptic area. Neuroreport 15:1127-1131.

Scammell T, Gerashchenko D, Urade Y, Onoe H, Saper CB, Hayaishi O (1998). Activation of ventrolateral preoptic neurons by the somnogen prostaglandin D_2. Proc Natl Acad Sci USA 95:7754-7759.

Scammell TE, Estabrooke IV, McCarthy MT, Chemelli RM, Yanagisawa M, Miller MS, Saper CB (2000). Hypothalamic arousal regions are activated during modafinil-induced wakefulness. J Neurosci 20:8620-8628.

Scammell TE, Gerashchenko DY, Mochizuki T, McCarthy MT, Estabrooke IV, Sears CA, Saper CB, Urade Y, Hayaishi O (2001). An adenosine A2a agonist increases sleep and induces Fos in ventrolateral preoptic neurons. Neuroscience 107:653-663.

Semba K (1999). The mesopontine cholinergic system: a dual role in REM sleep and wakefulness. In: Handbook of Behavioral State Control. Cellular and Molecular Mechanisms (Lydic R, Baghdoyan H, eds), pp. 161-180. Boca Raton: CRC Press.

Semba K, Pastorius J, Wilkinson M, Rusak B (2001). Sleep deprivation-induced *c-fos* and *junB* expression in the rat brain: effects of duration and timing. Behav Brain Res 120:75-86.

Sheng M, Greenberg ME (1990). The regulation and function of *c-fos* and other immediate early genes in the nervous system. Neuron 4:477-485.

Sherin J, Shiromani P, McCarley R, Saper C (1996). Activation of ventrolateral preoptic neurons during sleep. Science 271:216-219.

Sherin JE, Elmquist JK, Torrealba F, Saper CB (1998). Innervation of histaminergic tuberomammillary neurons by GABAergic and galaninergic neurons in the ventrolateral preoptic nucleus of the rat. J Neurosci 18:4705-4721.

Shiromani PJ, Winston S, McCarley RW (1996). Pontine cholinergic neurons show Fos-like immunoreactivity associated with cholinergically induced REM sleep. Brain Res Mol Brain Res 38:77-84.

Shiromani PJ, Basheer R, Thakkar J, Wagner D, Greco MA, Charness ME (2000). Sleep and wakefulness in *c-fos* and *fos B* gene knockout mice. Brain Res Mol Brain Res 80:75-87.

Shouse MN, Staba RJ, Saquib SF, Farber PR (2000). Monoamines and sleep: microdialysis findings in pons and amygdala. Brain Res 860:181-189.

Strecker RE, Nalwalk J, Dauphin LJ, Thakkar MM, Chen Y, Ramesh V, Hough LB, McCarley RW (2002). Extracellular histamine levels in the feline preoptic/anterior hypothalamic area during natural sleep-wakefulness and prolonged wakefulness: an *in vivo* microdialysis study. Neuroscience 113:663-670.

Suntsova N, Szymusiak R, Alam MN, Guzman-Marin R, McGinty D (2002). Sleep-waking discharge patterns of median preoptic nucleus neurons in rats. J Physiol 543:665-677.

Sved AF, Cano G, Passerin AM, Rabin BS (2002). The locus coeruleus, Barrington's nucleus, and neural circuits of stress. Physiol Behav 77:737-742.

Szymusiak R, McGinty D (1986). Sleep-related neuronal discharge in the basal forebrain of cats. Brain Res 370:82-92.

Szymusiak R, Alam N, Steininger T, McGinty D (1998). Sleep-waking discharge patterns of ventrolateral preoptic/anterior hypothalamic neurons in rats. Brain Res 803:178-188.

Takase LF, Nogueira MI, Baratta M, Bland ST, Watkins LR, Maier SF, Fornal CA, Jacobs BL (2004). Inescapable shock activates serotonergic neurons in all raphe nuclei of rat. Behav Brain Res 153:233-239.

Takase LF, Nogueira MI, Bland ST, Baratta M, Watkins LR, Maier SF, Fornal CA, Jacobs BL (2005). Effect of number of tailshocks on learned helplessness and activation of serotonergic and noradrenergic neurons in the rat. Behav Brain Res 162:299-306.

Terao A, Greco MA, Davis RW, Heller HC, Kilduff TS (2003). Region-specific changes in immediate early gene expression in response to sleep deprivation and recovery sleep in the mouse brain. Neuroscience 120:1115-1124.

Terao A, Wisor JP, Peyron C, Apte-Deshpande A, Wurts SW, Edgar DM, Kilduff TS (2006). Gene expression in the rat brain during sleep deprivation and recovery sleep: an Affymetrix GeneChip® study. Neuroscience 137:593-605.

Tononi G, Pompeiano M, Cirelli C (1994). The locus coeruleus and immediate-early genes in spontaneous and forced wakefulness. Brain Res Bull 35:589-596.

Torterolo P, Sampogna S, Morales FR, Chase MH (2002). Gudden's dorsal tegmental nucleus is activated in carbachol-induced active (REM) sleep and active wakefulness. Brain Res 944:184-189.

Torterolo P, Yamuy J, Sampogna S, Morales FR, Chase MH (2000). GABAergic neurons of the cat dorsal raphe nucleus express *c-fos* during carbachol-induced active sleep. Brain Res 884:68-76.

Torterolo P, Yamuy J, Sampogna S, Morales FR, Chase MH (2001a). Hypothalamic neurons that contain hypocretin (orexin) express *c-fos* during active wakefulness and carbachol-induced active sleep. Sleep Res Online 4:25-32.

Torterolo P, Yamuy J, Sampogna S, Morales FR, Chase MH (2001b). GABAergic neurons of the laterodorsal and pedunculopontine tegmental nuclei of the cat express *c-fos* during carbachol-induced active sleep. Brain Res 892:309-319.

Torterolo P, Yamuy J, Sampogna S, Morales FR, Chase MH (2003). Hypocretinergic neurons are primarily involved in activation of the somatomotor system. Sleep 26:25-28.

Trulson ME, Preussler DW (1984). Dopamine-containing ventral tegmental area neurons in freely moving cats: activity during the sleep-waking cycle and effects of stress. Exp Neurol 83:367-377.

Valdés JL, Farías P, Ocampo-Garcés A, Cortés N, Serón-Ferré M, Torrealba F (2005). Arousal and differential Fos expression in histaminergic neurons of the ascending arousal system during a feeding-related motivated behaviour. Eur J Neurosci 21:1931-1942.

van den Pol AN, Acuna-Goycolea C, Clark KR, Ghosh PK (2004). Physiological properties of hypothalamic MCH neurons identified with selective expression of reporter gene after recombinant virus infection. Neuron 42:635-652.

Vanni-Mercier G, Gigout S, Debilly G, Lin J-S (2003). Waking selective neurons in the posterior hypothalamus and their response to histamine H_3-receptor ligands: an electrophysiological study in freely moving cats. Behav Brain Res 144:227-241.

Verret L, Léger L, Fort P, Luppi P-H (2005). Cholinergic and noncholinergic brainstem neurons expressing Fos after paradoxical (REM) sleep deprivation and recovery. Eur J Neurosci 21:2488-2504.

Verret L, Goutagny R, Fort P, Cagnon L, Salvert D, Léger L, Boissard R, Salin P, Peyron C, Luppi P-H (2003). A role of melanin-concentrating hormone producing neurons in the central regulation of paradoxical sleep. BMC Neurosci 4:19.

Wagner D, Salin-Pascual R, Greco M, Shiromani P (2000). Distribution of hypocretin-containing neurons in the lateral hypothalamus and c-fos-immunoreactive neurons in the VLPO. Sleep Research Online 3:35-42.

Yamanaka A, Muraki Y, Tsujino N, Goto K, Sakurai T (2003). Regulation of orexin neurons by the monoaminergic and cholinergic systems. Biochem Biophys Res Commun 303:120-129.

Yamatodani A, Mochizuki T, Mammoto T (1996). New vistas on histamine arousal hypothesis: microdialysis study. Meth Find Exp Clin Pharmacol 18(Suppl A):113-117.

Yamuy J, Mancillas JR, Morales FR, Chase MH (1993). c-fos expression in the pons and medulla of the cat during carbachol-induced active sleep. J Neurosci 13:2703-2718.

Yamuy J, Sampogna S, Morales FR, Chase MH (1998). c-fos expression in mesopontine noradrenergic and cholinergic neurons of the cat during carbachol-induced active sleep: a double-labeling study. Sleep Res Online 1:28-40.

Yamuy J, Ramos O, Torterolo P, Sampogna S, Chase MH (2005). The role of tropomyosin-related kinase receptors in neurotrophin-induced rapid eye movement sleep in the cat. Neuroscience 135:357-369.

Yamuy J, Sampogna S, López-Rodríguez F, Luppi P-H, Morales FR, Chase MH (1995). Fos and serotonin immunoreactivity in the raphe nuclei of the cat during carbachol-induced active sleep: a double-labeling study. Neuroscience 67:211-223.

Zhu L, Onaka T, Sakurai T, Yada T (2002). Activation of orexin neurones after noxious but not conditioned fear stimuli in rats. Neuroreport 13:1351-1353.

8

c-Fos and Zif268 in Learning and Memory—Studies on Expression and Function

ROBERT K. FILIPKOWSKI, EWELINA KNAPSKA and LESZEK KACZMAREK

Nencki Institute, Warsaw, Poland

1. Introduction

c-Fos and Zif268[1] are both transcription factors (TF) implicated in brain plasticity and activated in the vertebrate brain by a variety of stimuli, including learning and memory. While the expression of *zif268* is elevated by on-going synaptic activity in the adult brain, the expression of *c-fos* is usually very low, only to be stimulated in selected areas, involved in a particular process.

Since c-Fos and Zif268 are both TF, they bind with specific DNA sequences to regulate the activity of target genes and thus may be the key players in guiding the cell's response to various stimuli. c-Fos interacts with Jun proteins to form a dimer, AP-1 (activatory protein-1; Angel and Karin, 1991). There are three Juns identified: c-Jun, JunB and JunD. All interact with c-Fos bringing about differential effects on DNA binding and gene expression (Kaminska et al., 2000; Szabowski et al., 2000). c-Fos can be phosphorylated and glycosylated (Barber et al., 1987; Barber and Verma, 1987; Muller et al., 1987; Reason et al., 1992). Moreover, there are at least four additional members of Fos family: FosB, ΔFosB, Fra-1, and Fra-2 which can participate in AP-1 complex, binding a set of closely related gene regulatory sequences known also as AP-1 or TRE (TPA-responsive element). The *zif268* gene (also called Egr-1, NGFI-A, Krox-24, TIS8 or ZENK; Lau and Nathans, 1987; Milbrandt, 1987) codes for a zinc finger TF. Zif268 presents some variability in the possible sequences of binding sites (Swirnoff and Milbrandt, 1995) and was also found to be phosphorylated and glycosylated (Cao et al., 1990; Lemaire et al., 1990). Both c-Fos and Zif268 interact with an array of other TF (Herdegen and Leah, 1998; Knapska and Kaczmarek, 2004). The

---

[1] Throughout this review, we use the standard convention of denoting genes and their mRNA products in italics (*c-fos*, *zif268*) and the protein encoded by them with a single capital letter (c-Fos, Zif268).

R. Pinaud, L.A. Tremere (Eds.), Immediate Early Genes in Sensory Processing, Cognitive Performance and Neurological Disorders, 137–158, ©2006 Springer Science + Business Media, LLC

composition of the AP-1, modifications of both proteins and their interactions with other TF further expand DNA-binding properties of c-Fos and Zif268 and contribute to their regulatory potential.

Owing to the great diversity of possible target genes and numerous interactions of AP-1 and Zif268 with other TF, different potential functions of both TF can be expected. They may maintain the homeostasis of neural cells, replenish key cellular components, guide and sustain plastic changes and/or contribute to information integration (see section 4). In this chapter, we discuss the data on patterns of *c-fos* and *zif268* expression in the mammalian brain[2] along with functional studies involving genetically modified animals, hoping that such a synthesis may shed some light on a possible function of both TF.

2. c-Fos

2.1. Correlative Studies (c-fos Expression in Learning)

The expression of *c-fos* is usually very low, only to be stimulated in selected areas, involved in a particular process (see Fig. 8.1). Some of the first studies on *c-fos* expression showed its activation following behavioral training. Also, this expression correlated with learning (Kaczmarek, 1990; Kaczmarek and Nikolajew, 1990; Maleeva et al., 1990; Maleeva et al., 1989; Tischmeyer et al., 1990). These experiments—on instrumental conditioning with aversive reinforcement—laid the foundation for the field and were followed by those on instrumental and classical conditioning, motor, spatial and recognintion learning in (mostly) rats and mice (for reviews see Anokhin, 1997; Dragunow, 1996; Herdegen and Leah, 1998; Hughes and Dragunow, 1995; Kaczmarek, 1993, 2002; Kaczmarek and Chaudhuri, 1997; Kovacs, 1998; Tischmeyer and Grimm, 1999).

As mentioned above, the first studies on *c-fos* expression in learning relied on instrumental conditioning with aversive reinforcement and employed two-way (active) avoidance paradigm, where an animal is placed in one of the two compartments and is supposed to learn to avoid the US (footshock) by reacting to CS (tone, light) and moving to the other compartment. Using this paradigm, our group reported a marked increase in *c-fos* mRNA and protein levels in rat hippocampus, sensory and limbic cortex (Kaczmarek and Nikolajew, 1990; Lukasiuk et al., 1999; Nikolaev et al., 1992a, b). Later, in detailed analysis of c-Fos expression in the amygdala, we applied factor analysis to separate and group a number of behavioral components (Savonenko et al., 1999). Surprisingly, no correlation was found between c-Fos levels in any of the 13 amygdalar subdivisions and either avoidance reaction or sum of the shock received. On the other hand, c-Fos expression in cortical amygdala correlated with grooming behavior (reflecting the lack of fear), whereas c-Fos levels in basolateral and medial amygdala correlated with anticipatory anxiety. Levels of *c-fos*/c-Fos expression were found to be elevated in relevant brain structures in several further paradigms of avoidance learning as well as following other behavioral tasks (listed in Table 8.1).

---

[2] Zif268 and c-Fos expression in birds was investigated repeatedly and it is reviewed elsewhere; see: Chapter 3 of this volume, Rose (2000); Mello (2002); Mello et al. (2004).

Table 8.1. Behavioral tests with induced *c-fos*/c-Fos expression in the brain. Brain regions are shown where *c-fos* expression was augmented (underlined when observed repeatedly)

Tests used	Structures with *c-fos* induction	References
Two-way avoidance	Sensory cortex, limbic cortex, hippocampus, amygdala	(Maleeva et al., 1989, 1990; Kaczmarek, 1990; Kaczmarek and Nikolajew, 1990; Tischmeyer et al., 1990; Nikolaev et al., 1992a, b; Lukasiuk et al., 1999; Savonenko et al., 1999; Radwanska et al., 2002)
Other paradigms of avoidance learning	Cerebral cortex, hippocampus, medial prefrontal, restrosplenial, cingulate, and ventrolateral orbital cortices, taenia tecta, nucleus accumbens, bed nucleus of stria terminalis, paraventricular nucleus, amygdala, lateral septum, locus coeruleus, inferior colliculus, central grey, cuneiform nucleus, cerebellum	(Maleeva et al., 1989, 1990; Tischmeyer et al., 1990; Castro-Alamancos et al., 1992; Duncan et al., 1996; Grimm and Tischmeyer, 1997; Anokhin et al., 2000)
Passive avoidance	Nucleus basalis of Meynert, hippocampus	(Cammarota et al., 2000; Zhang et al., 2000)
Instrumental conditioning (appetitve reinforcement)	Hippocampus, cingulate cortex, occipital cortex, parietal cortex, subiculum, amygdala, visual cortex, superior colliculus, olfactory bulb	(Heurteaux et al., 1993; Bertaina and Destrade, 1995; Hess et al., 1995a, b, 1997; Gall et al., 1998; Bertaina-Anglade et al., 2000)
Classical conditioning (appetitive reinforcement paired with odors)	Olfactory bulb	(Johnson et al., 1995; Woo et al., 1996; Allingham et al., 1999)
Classical conditioning (a nonphotic stimulus paired with light)	Hypothalamic suprachiasmatic nucleus	(Amir and Stewart, 1996, 1998a, b, c, 1999)
Classical fear conditioning (footshock alone)	Amygdala, hippocampus, paraventricular nucleus, parietal, frontal, occipital, temporal and cingular cortex, locus coeruleus, nucleus of the solitary tract, ventral lateral medulla, dorsal and ventral subdivisions of the periaqueductal gray, dorsal raphe nuclei	(Campeau et al., 1991; Smith et al., 1992; Pezzone et al., 1992, 1993; Li et al., 1996; Carrive et al., 1997b; Milanovic et al., 1998; Radulovic et al., 1998; Rosen et al., 1998; Bubser and Deutch, 1999; Morrow et al., 1999)
Classical fear conditioning (novel environment or tone only)	Amygdala, cingulate, parietal and piriform cortex	(Smith et al., 1992; Rosen et al., 1998; Filipkowski et al., 2000)

Table 8.1. Continued.

Tests used	Structures with *c-fos* induction	References
Classical fear conditioning (re-exposure)	Cingulate, piriform, infralimbic, and retrosplenial cortex, anterior olfactory nucleus, claustrum, endopiriform nucleus, nucleus accumbens shell, lateral septal nucleus, amygdala, hippocampus, paraventricular thalamic nucleus, ventral lateral geniculate nucleus, hypothalamus, ventral tegmental area, supramammillary area	(Campeau et al., 1991; Beck and Fibiger, 1995; Carrive et al., 1997b; Milanovic et al., 1998; Radulovic et al., 1998; Rosen et al., 1998; Strekalova et al., 2003; Scicli et al., 2004)
Classical fear conditioning (delayed and trace)	Hippocampus, anterior cingulate cortex	(Weitemier and Ryabinin, 2004)
Classical fear conditioning (odor stimulus)	Olfactory bulb, accessory olfactory nucleus; infralimbic, orbital, and perirhinal-entorhinal cortex, amygdala	(Funk and Amir, 2000)
Classical fear conditioning (extinction)	Medial prefrontal cortex, amygdala	(Herry and Mons, 2004; Santini et al., 2004)
Classical conditioning of rabbit nictitating membrane reflex	Trigeminal and auditory nuclei, raphe nuclei, ventrolateral medulla, locus coeruleus	(Irwin et al., 1992; Carrive et al., 1997a)
Conditioned taste aversion	Nucleus of the solitary tract, paraventricular nucleus, supraoptic nuclei of the hypothalamus, amygdala, area postrema, parabrachial nucleus, hypoglossal nucleus, cingulate and parietal cortex	(Gu et al., 1993; Swank and Bernstein, 1994; Houpt et al., 1994, 1996a, b, 1997; Yamamoto et al., 1994, 1997; Lamprecht and Dudai, 1995; Schafe et al., 1995; Swank et al., 1995; Schafe and Bernstein, 1996, 1998; Thiele et al., 1996; Sakai and Yamamoto, 1997; Yanagawa et al., 1997; Cubero et al., 1999; Montag-Sallaz et al., 1999; Spray et al., 2000; Swank, 2000; Koh et al., 2003)
Conditioned taste aversion (extinction)	Nucleus of the solitary tract, amygdala, gustatory cortex	(Mickley et al., 2004)
Motor learning	Motor cortex	(Kleim et al., 1996)
Spatial task (eight-arm radial maze)	Hippocampus, entorhinal and postrhinal cortex	(Vann et al., 2000)
Morris water maze	Hippocampus, subiculum, entorhinal cortex, medial mammillary nucleus, supramammillary nucleus	(Jenkins et al., 2003; Santin et al., 2003)

Table 8.1. Continued.

Tests used	Structures with *c-fos* induction	References
Visual recognition	Occipital visual association cortex, area TE of temporal cortex, perirhinal cortex, entorhinal cortex, anterior cingulate cortex, diagonal band of Broca, hippocampus, mediodorsal nucleus and geniculate nucleus of the thalamus	(Zhu et al., 1995, 1996, 1997)
Two-odor discrimination	Orbital and infralimbic cortex	(Roullet et al., 2004)
Sexual recognition	Olfactory bulb	(Brennan et al., 1992)
Maternal recognition	Medial preoptic area, medial and cortical amygdala, nucleus accumbens	(Fleming et al., 1994; Fleming and Walsh, 1994)
Acquisition of copulatory reactions	Medial preoptic area, bed nucleus of stria terminalis, medial amygdala, central tegmental field, parieto-occipital cortex	(Bialy et al., 1992, 2000; Bialy and Kaczmarek, 1996)

2.2. Functional Studies (c-fos in Learning as Indicated by the Interventive Approaches)

Paylor et al. (1994) carried out a behavioral characterization of mice lacking the *c-fos* gene (knock-out mice). The majority of these animals were impaired in spatial learning, as revealed by exposure to the Morris water maze. However, this poor performance in spatial learning was highly correlated with their performance in a non-spatial version of the task. These findings suggested that these animals displayed a behavioral impairment that interrupted their ability to perform adequately on both versions of the task, as was the case with wild-type and heterozygous littermates. On the other hand, *c-fos* mutants were not impaired in a simple left/right discrimination in a T-maze. Furthermore, the mutants were also characterized by sensory (e.g., auditory) deficiencies, which made interpretations of the observed behavioral deficits very difficult.

Recently, Fleischmann et al. (2003) have generated mice with nervous system-specific *c-fos* knock-out. The mutant mice displayed a reduction of a long-term potentiation (LTP) in hippocampal CA3-CA1 synapses, however, the magnitude of LTP was restored by a repeated tetanization procedure. Moreover, the knock-outs showed impairment in spatial learning in the Morris water maze as well as in contextual fear conditioning. However, surprisingly, the observed learning deficits were rather subtle and limited only to the hippocampus-dependent tasks. These findings suggest a possibility for a developmental compensation in these mutant animals.

Recent developments in brain area-specific and inducible transgenes or gene knock-outs may offer a technological breakthrough allowing overcoming the development-based deficiencies and their compensations in knock-out mice. However, the number of experiments employing these emerging technologies is still very limited. Moreover, they have not yet been applied successfully to investigate the role of *c-fos* expression in learning and memory formation.

On the other hand, since the first antisense application in the brain (Chiasson et al., 1992) there were a number of studies employing *c-fos* antisense oligodeoxynucleotides (ODNs). However, one has to be aware of all the limitations in interpreting the results of these experiments. The works on cell cultures clearly show that there is a multitude of potential caveats and, therefore, a necessity to use plentitude of controls (Neumann, 1997; Stein and Krieg, 1994; Stein, 1996). Suffice to say that a reliable experiment should employ several different ODNs, directed against various target mRNA regions, each time matched with missense, and nonsense counterparts. This greatly increases the number of animals required to test the function of a gene in learning and makes such studies very difficult to be carried out. Indeed, it appears that no such complete investigation has apparently been described for *c-fos* (for review see Szklarczyk and Kaczmarek, 1999). Nevertheless, the experiments described below, despite all the aforementioned reservations, support the hypothesis that *c-fos* expression plays a pivotal role in memory consolidation.

Lamprecht and Dudai (1996) applied amygdalar microinjection of *c-fos* antisense ODNs several hours before conditioned taste aversion (CTA) training and found this treatment to impair CTA memory tested 3–5 days after conditioning. In contrast, injection of the antisense ODNs several days before training or testing, or into the basal ganglia, or injection of *c-fos* sense ODNs, had no effect on CTA memory. Notably, inhibition of translation by amygdalar microinjection of anisomycin shortly before, as well as during, CTA training, but not several days before training or shortly before testing, also impaired CTA memory. Using similar training paradigms in mice, Swank et al. (1996) found that blockade of US-induced c-Fos translation in another brain structure, brain stem, by antisense ODNs (injected into the fourth ventricle) specifically blocked both acquisition and extinction of CTA, but did not impair sensory processing of either CS or US, suggesting that c-Fos antisense blocked associative events supporting CTA.

Grimm et al. (1997) used antisense ODNs to suppress the expression of c-Fos in rat brain evoked by footshock-motivated brightness discrimination in a Y-maze. Intrahippocampal application of the ODNs, 10 hrs and 2 hrs before starting the training, drastically reduced c-Fos induction in limbic and cortical areas. Acquisition of the discrimination reaction was not affected by this treatment. However, in a re-learning session, retention of the discrimination reaction was specifically impaired in ODNs-treated rats when compared with animals pretreated with control ODNs or saline.

Morrow et al. (1999) examined inducible c-Fos immunoreactivity in subregions of the prefrontal cortex during conditioned fear paradigm. During acquisition, the rats were conditioned by pairing tone with footshock. Animals were then tested for fearful behavior by re-exposure to the tone without additional footshock. Antisense ODNs against *c-fos* suppressed c-Fos production without altering behavior when injected 12 but not 72 hrs before training (into the infralimbic/prelimbic cortex). Three days after the acquisition session, rats were tested for fearful behavior; blockade of c-Fos production triggered by antisense ODN application during acquisition was associated with less fearful response during the test session.

2.3. Summary on c-fos Results

Results of the experiments described above clearly show that behavioral training activates *c-fos* expression in the brain. Especially striking is the very high number of brain regions displaying elevated c-Fos expression in a context of behavioral experience. Notably, in many cases, induction of c-Fos was observed in brain regions known to subserve the learning tasks under study. Furthermore, a number of studies with unilateral sensory occlusion or impairment of lateralized brain structures resulted in lateralized c-Fos response in downstream brain regions. These results suggest that c-Fos activation depends on specific neuronal connections and not on widespread, non-specific responses. In addition, the changes in *c-fos* expression were often seen following the first training session (Maleeva et al., 1989, 1990; Kaczmarek and Nikolajew, 1990; Nikolaev et al., 1992a, b; Lukasiuk et al., 1999; Anokhin et al., 2000; Bertaina-Anglade et al., 2000).

Interestingly, in the reviewed literature an apparent controversy has emerged, on whether the elevated *c-fos* expression correlates with either *learning* or *performing* an already acquired task. In particular, in a number of studies authors claimed that *c-fos* expression could be detected following the testing session of the already learned behavior. However, a very careful scrutiny of these reports reveals that in each case the testing session differed from the training ones (e.g., either in its duration or additional elements being included; Castro-Alamancos et al., 1992; Duncan et al., 1996). Thus, it is clear that the "testing" involved a learning component of novel aspects of the procedure. Notably, answers to this issue has already been provided in an early report, showing increased *c-fos* mRNA accumulation after a single, but not multiple, training sessions of two-active avoidance reaction in rats (Nikolaev et al., 1992b). Also, in the majority of cases, repeated exposure to the same stimulus ceases to evoke initial *c-fos* activation (e.g., Maleeva et al., 1989, 1990; Hess et al., 1995a, b; 1997; Gall et al., 1998; Anokhin et al., 2000; Bertaina-Anglade et al., 2000). Moreover, animals which acquired particular behaviors more rapidly display lower *c-fos* mRNA levels than slow learners (which still acquire the task) (Maleeva et al., 1989, 1990; Anokhin et al., 2000). Finally, the correlation between the higher performance scores and *c-fos* induction were reported repeatedly (Sakai and Yamamoto, 1997; Radulovic et al., 1998; Bertaina-Anglade et al., 2000).

Special considerations should also be given to the effects of post-training (in an absence of the US) and exposure of animals to the conditioned stimuli (e.g. sweet taste in the CTA paradigm or training environment in which footshocks were delivered during conditioned fear training). The *c-fos*/c-Fos expression was repeatedly found to be very pronounced after such experience. Most interestingly, in the CTA procedure, c-Fos labeling of the nucleus of the solitary tract could be revealed only after treatment with the conditioned taste, but not just after exposure to the same taste, not previously paired with the US. This clearly shows that learning modifies spatial pattern of the gene expression in the brain. However, as already pointed out by Grimm and Tischmeyer (1997), the gene activation in the re-exposure condition may be related to a number of phenomena such as memory recall, expression of the learned behavior, re-learning and extinction. The last possibility is especially important, as it is well established that the extinction processes comprise a condition of a new learning experience, not involving the very same circuits that subserved original acquisition of the trace.

Further misconception surrounds the use of pseudoconditioning (random presentation of a number of CSs and USs, not allowing for learning any association between them). This is a very good control procedure for acquisition of the CS-US association. However, it does not preclude *any learning* experience and indeed it has been shown that it results in learning of the *lack of association*. Existence of such learning can be proven by difficulties to train an appropriate CS-US contingency in the previously pseudoconditioned animals (Rescorla, 1967). Thus, an elevated *c-fos* expression in the pseudotrained or yoked animals (e.g., Tischmeyer et al., 1990; Grimm and Tischmeyer, 1997; Zhang et al., 2000) does not disprove its role in learning. On the other hand, the observation of a selective *c-fos* activation in a given brain region in animals exposed to appropriate CS-US contingency, but not in the pseudotrained ones, may be indicative of the specific involvement of this brain area in forming memory for the trained response.

Another critical issue is whether acquisition-related *c-fos* expression could be attributed to the learning process or to the stress accompanying this phenomenon. This is a very difficult issue to tackle. There is probably no learning without a component of stress, and by the same token every stressful procedure (at least such as context conditioning and habituation) results in learning. Indeed a number of studies have shown that *c-fos* expression accompanies a stressful experience (Campeau et al., 1991; Melia et al., 1994; Chen and Herbert, 1995; Cullinan et al., 1995; Kovacs, 1998). Interestingly, Melia et al. (1994) investigated effects of acute and repeated restraint stress on *c-fos* expression in rat brain. The authors found that acute stress results in massive accumulation of *c-fos* mRNA in various brain regions. However, this response was completely habituated by the ninth restraint stress session. Furthermore, this habituation was stressor-specific as significant *c-fos* expression was observed in brains of the animals that were exposed to acute swim stress after they were treated to the repeated immobilization. Notably, adrenalectomy that completely abolished stress-related increases in plasma corticosterone levels did not affect *c-fos* activation following acute stress (Melia et al., 1994).

The correlative results described in chapter 2.1 concerning *c-fos* expression in learning raise the problem of biological significance of this phenomenon. One way to approach it is to continue the expression pattern studies, especially through extensive analysis of the complex animal behavior and thorough correlative examination of the c-Fos expression in various brain regions (Savonenko et al., 1999). Another important approach is through functional alterations in c-Fos expression and/or activity. However, the brain has proven to be a notoriously difficult organ to apply strategies that alter gene expression *in vivo* and the use of *c-fos* knock-out mice illustrates problems with this approach (see discussion above).

In conclusion, the c-Fos mapping studies provide paramount support for the existence of a brain-wide network that is involved in information processing (or even underlying widespread plastic changes) during learning. However, at the same time, careful behavioral analyses into correlations between various animal reactions and c-Fos expression patterns reveal that specific aspects of information processing (plastic changes) are encoded by well determined brain structures (such as emotions in the amygdala).

3. Zif268

3.1. Correlative Studies (zif268 Expression in Learning)

Contrary to the low basal expression of *c-fos* in the brain, Zif268 has a relatively high expression maintained by normal ongoing neuronal activity (Worley et al., 1991; Beckmann and Wilce, 1997; Herdegen and Leah, 1998). High basal levels of *zif268* and its protein have repeatedly been observed in the visual cortex (Kaczmarek and Chaudhuri, 1997), somatosensory cortex (Mack and Mack, 1992; Steiner and Gerfen, 1994; Melzer and Steiner, 1997; Filipkowski, 2000; Filipkowski et al., 2001; Bisler et al., 2002), hippocampus (Hughes et al., 1992; Cullinan et al., 1995; Okuno et al., 1995; Desjardins et al., 1997) as well as other brain regions (Herdegen et al., 1990; Cullinan et al., 1995; Herdegen et al., 1995; Okuno et al., 1995; see Fig. 8.1).

Nikolaev et al. (1992a) provided the first demonstration of the increased levels of *zif268* mRNA in the hippocampus and visual cortex after one session of the two-way avoidance training. However, most of the studies on *zif268* expression and behavioral training involved fear conditioning. According to some studies (Rosen et al., 1998; Malkani and Rosen, 2000a, b, 2001), the increase in *zif268* expression was limited exclusively to the lateral nucleus of the amygdala and to the group of rats that exhibited freezing reaction. On the other hand, Hall et al. (2000) reported non-specific induction of *zif268* following contextual fear conditioning, i.e., the expression of *zif268* was increased in the hippocampus and amygdala in the experimental group and also in all control groups exposed to the training chamber. Moreover, Weitemier and Ryabinin (2004) have recently shown a lack of difference in Zif268 expression between fear conditioned or naive C57BL/6J mice after the training phase in the septum, amygdala, hippocampus and anterior cingulate cortex. The expression of *zif268* and its protein has been repeatedly observed in other behavioral tasks. The results of these experiments are, however, mostly inconsistent (see Table 8.2 and discussion in Knapska and Kaczmarek, 2004).

3.2. Functional Studies (*zif268* in Learning as Indicated by the Interventive Approaches)

Jones et al. (2001) investigated behavior of mice with a targeted disruption of *zif268* in the variety of different tests. Following CTA, the mice were supposed to choose between water and sucrose solution. The normal mice avoided drinking the sucrose solution, but neither heterozygous nor homozygous mutants exhibited significant taste aversion.

The social transmission of food preference test is a one-trial, olfactory discrimination task that measures the preference of rodents for novel food which has recently been smelled on the breath of another animal. 24 hrs following presentation, only the normal mice showed preference for the demonstrated food.

The *zif268*-deficient mice were also tested in the Morris water maze, which requires long-term memory. The performance of the mutant mice and the heterozygous mice was markedly impaired in comparison to the performance of the normal ones following massive training. However, when the extended and distributed training was applied, all mice showed normal acquisition and long-term recall. Short-term memory remained intact in mutants (it was also examined by testing spontaneous alternation in a T-maze). These mutant mice were also tested in a hippocampus-dependent novel object recognition test. They were

Table 8.2. Behavioral tests in which the expression of *zif268* and/or its protein was investigated; observed increase (+) or no difference in expression (−) is indicated

Tests used	Brain structure	Increase (+) or no increase (−) of expression	References
Contextual fear memory retrieval	Hippocampus	+	(Hall et al., 2001)
	Hippocampus	−	(Malkani and Rosen, 2000a)
	Neocortex	−	
	Amygdala	−	
	Hippocampus	+ (after 1 d)	(Weitemier and Ryabinin, 2004)
	Hippocampus (CA1)	+ (after 1 d)	(Frankland et al., 2004)
	Hippocampus (CA1)	− (after 36 d)	
	Nucleus accumbens core	+	(Thomas et al., 2002)
	Nucleus accumbens shell	+	
	Anterior cingulate cortex	+	
Cued fear memory retrieval	Nucleus accumbens core	+	(Thomas et al., 2002)
	Nucleus accumbens shell	−	
	Anterior cingulate cortex	−	
Visual pair-association task	Inferior temporal gyrus (perirhinal cortex)	+	(Okuno and Miyashita, 1996; Tokuyama et al., 2002)
Morris water maze; spatial and cued version	Dorsal hippocampus	+	(Guzowski et al., 2001)
Conditioned taste aversion or lithium chloride injection itself	Nucleus of the solitary tract	+	(Lamprecht and Dudai, 1995)
	Parabrachial nucleus	+	
	Hypothalamic paraventricular nucleus	+	
	Central nucleus of amygdala	+	
Memory of pheromones	Accessory olfactory bulb	+	(Brennan et al., 1992, 1999)
	Accessory olfactory bulb	−	(Polston and Erskine, 1995)
	Medial amygdala	+	
	Preoptic area	+	
	Bed nucleus of stria terminalis	+	
	Ventromedial and paraventricular nuclei of the hypothalamus	+	

allowed to explore two objects for 20 minutes and then, after a delay, one of the familiar objects was replaced with a novel object and the time spent exploring the novel and the familiar objects was measured. After the 10 min delay, when only the short-term memory was required, all genotypes explored the novel object significantly longer than the familiar one. After the 24-hr delay, when the long-term memory was engaged, the normal and heterozygous mice still explored the novel object for longer time than the familiar one, in contrast to the homozygous mutants.

Bozon et al. (2002) extended the studies on a task-dependent *zif268* dosage effect in the above mentioned heterozygous mice. These mice showed about 50% reduction in *zif268* expression in comparison to the wild-type mice. Bozon and associates modified the object recognition task in order to distinguish between the spatial and non-spatial component in the same type of learning. They exposed mice to three objects, with a cue card placed on one of the sidewalls of the open field. The long-term memory was tested 24 hrs later with one of the familiar objects moved to a new location in the arena (the place-change task). On the third day, the mice were given a training session with three new objects and the remaining conditions identical to those used on the first day. On the following day, the mice were tested with one of the familiar objects replaced by a novel object (the object-change task). The mutant mice did not express long-term memory of the objects, in contrast to both wild type and heterozygous mice. However, after changing the spatial configuration of the objects the heterozygous mice, as well as the mutant mice, exhibited a severe deficit in the long-term memory in contrast to the wild-type mice.

Because of the impaired acquisition and the severe deficit in long-term spatial memory of the mutant mice (Jones et al., 2001) during massive training, one could believe that Zif268 is required either for memory consolidation or retrieval. However, there are two interpretations of the fact that the mutant mice managed to overcome the deficits in the extended and distributed training. (1) This kind of training recruited other unidentified signaling pathways that allow information to enter long-term memory storage or (2) Zif268 is not necessary for neuronal plasticity, but for the maintenance or replenishment of some aspects of cell homeostasis, which may be required for the creation of plastic changes. In this case, other signaling pathways which re-establish homeostasis would be activated. Since the impairment caused by deficiency in *zif268* expression is restricted to the spatial version of the object recognition task (Bozon et al., 2002), it should be noted that the spatial and non-spatial versions of the object recognition task engage different brain structures. For example, Mumby et al. (2002) observed that hippocampal damage impaired memory for contextual and spatial aspects of experience, whereas memory for objects was left intact. The efficacy of compensating processes may vary between different brain structures, thereby accounting for the observed differences.

The infusion of *zif268* antisense ODNs into the amygdala prior to contextual fear conditioning impaired memory formation (Malkani et al., 2004). This result might support the hypothesis that Zif268 plays an important role in fear memory formation. However, taking into account the limitations in interpreting the results of the experiments with ODNs (see section 2.2), one should be cautious in drawing final conclusions.

3.3. Summary on zif268 Results

Most of our present knowledge regarding *zif268* induction is based on correlating expression patterns with various types of stimulation. Extensive review of the literature on this topic suggests that there is a close link between *zif268* mRNA and protein expression with neuronal activity. However, this correlation has been observed in studies on brain structures that are usually implicated in neuronal plasticity as well. It is conspicuous that very little is known about Zif268 activation in, for example, the thalamus, which is also known to be engaged in intense neuronal activity during processing of sensory information. Hence, it seems that plasticity-linked neuronal activity is particularly effective in driving expression of Zif268.

At present, the correlative approach allows for drawing the most credible suggestions about Zif268 functions. However, in light of the possible influences of various inputs (e.g., sensory information processing, arousal, motivation, emotion, stress responses), careful scrutiny of gene expression patterns in the context of learning has to be applied. Such a scrutiny is a necessary step before any learning-related roles could be ascribed to Zif268.

The correlative method does not allow for distinguishing between learning and the processes engaged in cellular homeostatic maintenance or, especially, in the replenishment of metabolic components required for normal cell physiology. Therefore, it is impossible to conclude about a direct or indirect engagement of Zif268 in neuronal plasticity with the use of this approach. The fact that the Zif268 recognition element was found in genes coding for such neuronal proteins, as in the case of the synapsins and glutamate dehydrogenase, underscores the role of this TF in critical neuronal functions. Identifying additional Zif268-regulated genes should be an important avenue for the following investigations. Recent applications of knock-out mice have also contributed to our knowledge about Zif268 in the brain. However, this approach can not exclude the possibility that observed learning deficits in *zif268* KO animals results from the disturbance of basic metabolic functions that are revealed upon strong neuronal stimulation occurring during behavioral training. Thus, ultimately, the full description of Zif268 function in the brain cells will require a combination of all the aforementioned approaches as well as the use of new transgenic methods allowing for inducible gene down- and up-regulation.

4. Summary

Though several genes whose expression in the brain can be controlled by Zif268 and c-Fos have been identified (see recent reviews by Knapska and Kaczmarek, 2004; Rylski and Kaczmarek, 2004), most of them almost certainly still remain unknown. Moreover, full understanding of the roles of AP-1 and

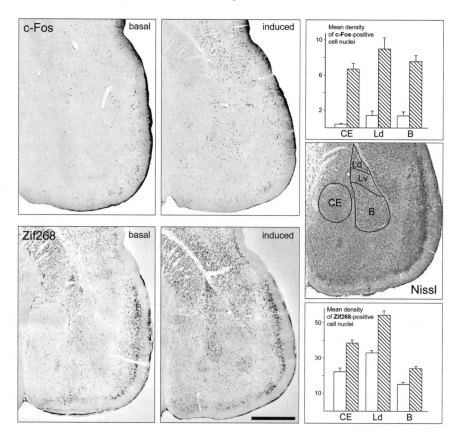

Figure 8.1. Behavioral training induces c-Fos and, to a lesser extent, Zif268 in the brain, as shown for the rat amygdala. The *basal* (control) expression is shown in the animals taken directly from their home-cages and the *induced* rats had already acquired bar-pressing response to partial food reinforcement and were further trained to learn that an acoustic stimulus signaled continuous food reinforcement. The rats were sacrificed after the first session of such training (see Knapska et al., 2006 for details). Coronal sections of the brain were immunostained with either c-Fos or Zif268 specific antibody and visualized by the peroxidase-DAB reaction. Mean density was calculated for central (CE), lateral (dorsal part) (Ld) and basal (B) nuclei of amygdala (see the Nissl-stained section for the demarcation of the structures) in both basal (empty bars) and induced conditions (hatched bars). Scale bar = 1 mm. Please note that the number of Zif268-positive neuronal nuclei is much higher in both basal and trained brains. Furthermore, whereas the increase in number of c-Fos positive cells is several-fold, the induction of Zif268 is markedly lower in magnitude. Also of note is, the general pattern of protein expression changes that is different between c-Fos and Zif268 in all three nuclei investigated.

Zif268 will not be uncovered without the knowledge of the mechanisms by which they interact with other TF.

General potential neuronal functions of *c-fos* and *zif268* can be suggested based on the complex mechanisms controlling their expression, as well as their interactions with other TF and target genes (for details see Kaczmarek, 2000, 2002). We propose that the first and simplest hypothesis explaining the role of

TF in the neuronal cell is *homeostatic maintenance of neural functioning*, for example, making up for the proteins that are lost during physiological metabolic turnover. The second possible function is given by *the replenishment hypothesis*. The concept of replenishment implies that the neuronal activity-evoked TF expression is involved in the metabolic recovery triggered by depletion of key cellular components (e.g., synaptic release machinery, metabolic enzymes) during neuronal activation and thus serves to reinstate the same situation as before the training. The third potential role of TF in neurons is *maintenance of plastic changes*, for example, when they serve to produce the proteins whose function is to maintain the plastically reorganized neuronal connections. These proteins should be targeted to specific synapses to support their newly gained functions. The fourth possibility is *information integration*. This hypothesis proposes that the regulatory regions of the genes encoding the proteins directly subserving synaptic reorganization may act as "coincidence detectors", thereby allowing convergence of information provided by various TF, activated by different signaling pathways of behavioral relevance, such as sensory information, arousal and motivation (Kaczmarek, 1993, 1995).

In light of the results we summarized above, it is now impossible to determine which of the functions are provided by c-Fos and Zif268. However, *c-fos* expression seems to be connected with neuronal plasticity, while Zif268 appears to be primarily involved in maintaining neuronal activity.

References

Allingham K, Brennan PA, Distel H, Hudson R (1999). Expression of c-fos in the main olfactory bulb of neonatal rabbits in response to garlic as a novel and conditioned odour. Behav Brain Res 104:157-167.

Amir S, Stewart J (1996). Resetting of the circadian clock by a conditioned stimulus. Nature 379:542-545.

Amir S, Stewart J (1998a). Conditioned fear suppresses light-induced resetting of the circadian clock. Neuroscience 86:345-351.

Amir S, Stewart J (1998b). Conditioning in the circadian system. Chronobiol Int 15:447-456.

Amir S, Stewart J (1998c). Induction of Fos expression in the circadian system by unsignaled light is attenuated as a result of previous experience with signaled light: a role for Pavlovian conditioning. Neuroscience 83:657-661.

Amir S, Stewart J (1999). Conditioned and unconditioned aversive stimuli enhance light-induced fos expression in the primary visual cortex. Neuroscience 89:323-327.

Angel P, Karin M (1991). The role of Jun, Fos and the AP-1 complex in cell-proliferation and transformation. Biochim Biophys Acta 1072:129-157.

Anokhin KV (1997). Towards synthesis of systems and molecular genetics approaches to memory consolidation. Journal of Higher Nervous Activity 47:157-169.

Anokhin KV, Riabinin AE, Sudakov KV (2000). [The expression of the c-fos gene in the brain of mice in the dynamic acquisition of defensive behavioral habits]. Zh Vyssh Nerv Deiat Im I P Pavlova 50:88-94.

Barber JR, Sassone-Corsi P, Verma IM (1987). Proto-oncogene fos: factors affecting expression and covalent modification of the gene product. Ann NY Acad Sci 511:117-130.

Barber JR, Verma IM (1987). Modification of fos proteins: phosphorylation of c-fos, but not v-fos, is stimulated by 12-tetradecanoyl-phorbol-13-acetate and serum. Mol Cell Biol 7:2201-2211.

Beck CH, Fibiger HC (1995). Conditioned fear-induced changes in behavior and in the expression of the immediate early gene c-fos: with and without diazepam pretreatment. J Neurosci 15:709-720.

Beckmann AM, Wilce PA (1997). Egr transcription factors in the nervous system. Neurochem Int 31:477-510; discussion 517-476.

Bertaina-Anglade V, Tramu G, Destrade C (2000). Differential learning-stage dependent patterns of c-Fos protein expression in brain regions during the acquisition and memory consolidation of an operant task in mice. Eur J Neurosci 12:3803-3812.

Bertaina V, Destrade C (1995). Differential time courses of c-fos mRNA expression in hippocampal subfields following acquisition and recall testing in mice. Brain Res Cogn Brain Res 2:269-275.

Bialy M, Kaczmarek L (1996). c-Fos expression as a tool to search for the neurobiological base of the sexual behaviour of males. Acta Neurobiol Exp (Wars) 56:567-577.

Bialy M, Nikolaev E, Beck J, Kaczmarek L (1992). Delayed c-fos expression in sensory cortex following sexual learning in male rats. Brain Res Mol Brain Res 14:352-356.

Bialy M, Rydz M, Kaczmarek L (2000). Precontact 50-kHz vocalizations in male rats during acquisition of sexual experience. Behav Neurosci 114:983-990.

Bisler S, Schleicher A, Gass P, Stehle JH, Zilles K, Staiger JF (2002). Expression of c-Fos, ICER, Krox-24 and JunB in the whisker-to-barrel pathway of rats: time course of induction upon whisker stimulation by tactile exploration of an enriched environment. J Chem Neuroanat 23:187-198.

Bozon B, Davis S, Laroche S (2002). Regulated transcription of the immediate-early gene Zif268: mechanisms and gene dosage-dependent function in synaptic plasticity and memory formation. Hippocampus 12:570-577.

Brennan PA, Hancock D, Keverne EB (1992). The expression of the immediate-early genes c-fos, egr-1 and c-jun in the accessory olfactory bulb during the formation of an olfactory memory in mice. Neuroscience 49:277-284.

Brennan PA, Schellinck HM, Keverne EB (1999). Patterns of expression of the immediate-early gene egr-1 in the accessory olfactory bulb of female mice exposed to pheromonal constituents of male urine. Neuroscience 90:1463-1470.

Bubser M, Deutch AY (1999). Stress induces Fos expression in neurons of the thalamic paraventricular nucleus that innervate limbic forebrain sites. Synapse 32:13-22.

Cammarota M, Bevilaqua LR, Ardenghi P, Paratcha G, Levi de Stein M, Izquierdo I, Medina JH (2000). Learning-associated activation of nuclear MAPK, CREB and Elk-1, along with Fos production, in the rat hippocampus after a one-trial avoidance learning: abolition by NMDA receptor blockade. Brain Res Mol Brain Res 76:36-46.

Campeau S, Hayward MD, Hope BT, Rosen JB, Nestler EJ, Davis M (1991). Induction of the c-fos proto-oncogene in rat amygdala during unconditioned and conditioned fear. Brain Res 565:349-352.

Cao XM, Koski RA, Gashler A, McKiernan M, Morris CF, Gaffney R, Hay RV, Sukhatme VP (1990). Identification and characterization of the Egr-1 gene product, a DNA-binding zinc finger protein induced by differentiation and growth signals. Mol Cell Biol 10:1931-1939.

Carrive P, Kehoe EJ, Macrae M, Paxinos G (1997a). Fos-like immunoreactivity in locus coeruleus after classical conditioning of the rabbit's nictitating membrane response. Neurosci Lett 223:33-36.

Carrive P, Leung P, Harris J, Paxinos G (1997b). Conditioned fear to context is associated with increased Fos expression in the caudal ventrolateral region of the midbrain periaqueductal gray. Neuroscience 78:165-177.

Castro-Alamancos MA, Borrell J, Garcia-Segura LM (1992). Performance in an escape task induces fos-like immunoreactivity in a specific area of the motor cortex of the rat. Neuroscience 49:157-162.

Chen X, Herbert J (1995). Regional changes in c-fos expression in the basal forebrain and brainstem during adaptation to repeated stress: correlations with cardiovascular, hypothermic and endocrine responses. Neuroscience 64:675-685.

Chiasson BJ, Hooper ML, Murphy PR, Robertson HA (1992). Antisense oligonucleotide eliminates in vivo expression of c-fos in mammalian brain. Eur J Pharmacol 227:451-453.

Cubero I, Thiele TE, Bernstein IL (1999). Insular cortex lesions and taste aversion learning: effects of conditioning method and timing of lesion. Brain Res 839:323-330.

Cullinan WE, Herman JP, Battaglia DF, Akil H, Watson SJ (1995). Pattern and time course of immediate early gene expression in rat brain following acute stress. Neuroscience 64:477-505.

Desjardins S, Mayo W, Vallee M, Hancock D, Le Moal M, Simon H, Abrous DN (1997). Effect of aging on the basal expression of c-Fos, c-Jun, and Egr-1 proteins in the hippocampus. Neurobiol Aging 18:37-44.

Dragunow M (1996). A role for immediate-early transcription factors in learning and memory. Behav Genet 26:293-299.

Duncan GE, Knapp DJ, Breese GR (1996). Neuroanatomical characterization of Fos induction in rat behavioral models of anxiety. Brain Res 713:79-91.

Filipkowski RK (2000). Inducing gene expression in barrel cortex–focus on immediate early genes. Acta Neurobiol Exp 60:411-418.

Filipkowski RK, Rydz M, Berdel B, Morys J, Kaczmarek L (2000). Tactile experience induces c-fos expression in rat barrel cortex. Learn Mem 7:116-122.

Filipkowski RK, Rydz M, Kaczmarek L (2001). Expression of c-Fos, Fos B, Jun B, and Zif268 transcription factor proteins in rat barrel cortex following apomorphine-evoked whisking behavior. Neuroscience 106:679-688.

Fleischmann A, Hvalby O, Jensen V, Strekalova T, Zacher C, Layer LE, Kvello A, Reschke M, Spanagel R, Sprengel R, Wagner EF, Gass P (2003). Impaired long-term memory and NR2A-type NMDA receptor-dependent synaptic plasticity in mice lacking c-Fos in the CNS. J Neurosci 23:9116-9122.

Fleming AS, Suh EJ, Korsmit M, Rusak B (1994). Activation of Fos-like immunoreactivity in the medial preoptic area and limbic structures by maternal and social interactions in rats. Behav Neurosci 108:724-734.

Fleming AS, Walsh C (1994). Neuropsychology of maternal behavior in the rat: c-fos expression during mother-litter interactions. Psychoneuroendocrinology 19:429-443.

Frankland PW, Bontempi B, Talton LE, Kaczmarek L, Silva AJ (2004). The involvement of the anterior cingulate cortex in remote contextual fear memory. Science 304:881-883.

Funk D, Amir S (2000). Enhanced fos expression within the primary olfactory and limbic pathways induced by an aversive conditioned odor stimulus. Neuroscience 98:403-406.

Gall CM, Hess US, Lynch G (1998). Mapping brain networks engaged by, and changed by, learning. Neurobiol Learn Mem 70:14-36.

Grimm R, Schicknick H, Riede I, Gundelfinger ED, Herdegen T, Zuschratter W, Tischmeyer W (1997). Suppression of c-fos induction in rat brain impairs retention of a brightness discrimination reaction. Learn Mem 3:402-413.

Grimm R, Tischmeyer W (1997). Complex patterns of immediate early gene induction in rat brain following brightness discrimination training and pseudotraining. Behav Brain Res 84:109-116.

Gu Y, Gonzalez MF, Chin DY, Deutsch JA (1993). Expression of c-fos in brain subcortical structures in response to nauseant lithium chloride and osmotic pressure in rats. Neurosci Lett 157:49-52.

Guzowski JF, Setlow B, Wagner EK, McGaugh JL (2001). Experience-dependent gene expression in the rat hippocampus after spatial learning: a comparison of the immediate-early genes Arc, c-fos, and zif268. J Neurosci 21:5089-5098.

Hall J, Thomas KL, Everitt BJ (2000). Rapid and selective induction of BDNF expression in the hippocampus during contextual learning. Nat Neurosci 3:533-535.

Hall J, Thomas KL, Everitt BJ (2001). Cellular imaging of zif268 expression in the hippocampus and amygdala during contextual and cued fear memory retrieval: selective activation of hippocampal CA1 neurons during the recall of contextual memories. J Neurosci 21:2186-2193.

Herdegen T, Kovary K, Buhl A, Bravo R, Zimmermann M, Gass P (1995). Basal expression of the inducible transcription factors c-Jun, JunB, JunD, c-Fos, FosB, and Krox-24 in the adult rat brain. J Comp Neurol 354:39-56.

Herdegen T, Leah JD (1998). Inducible and constitutive transcription factors in the mammalian nervous system: control of gene expression by Jun, Fos and Krox, and CREB/ATF proteins. Brain Res Brain Res Rev 28:370-490.

Herdegen T, Walker T, Leah JD, Bravo R, Zimmermann M (1990). The KROX-24 protein, a new transcription regulating factor: expression in the rat central nervous system following afferent somatosensory stimulation. Neurosci Lett 120:21-24.

Herry C, Mons N (2004). Resistance to extinction is associated with impaired immediate early gene induction in medial prefrontal cortex and amygdala. Eur J Neurosci 20:781-790.

Hess US, Gall CM, Granger R, Lynch G (1997). Differential patterns of c-fos mRNA expression in amygdala during successive stages of odor discrimination learning. Learn Mem 4:262-283.

Hess US, Lynch G, Gall CM (1995a). Changes in c-fos mRNA expression in rat brain during odor discrimination learning: differential involvement of hippocampal subfields CA1 and CA3. J Neurosci 15:4786-4795.

Hess US, Lynch G, Gall CM (1995b). Regional patterns of c-fos mRNA expression in rat hippocampus following exploration of a novel environment versus performance of a well-learned discrimination. J Neurosci 15:7796-7809.

Heurteaux C, Messier C, Destrade C, Lazdunski M (1993). Memory processing and apamin induce immediate early gene expression in mouse brain. Brain Res Mol Brain Res 18:17-22.

Houpt TA, Berlin R, Smith GP (1997). Subdiaphragmatic vagotomy does not attenuate c-Fos induction in the nucleus of the solitary tract after conditioned taste aversion expression. Brain Res 747:85-91.

Houpt TA, Philopena JM, Joh TH, Smith GP (1996a). c-Fos induction in the rat nucleus of the solitary tract by intraoral quinine infusion depends on prior contingent pairing of quinine and lithium chloride. Physiol Behav 60:1535-1541.

Houpt TA, Philopena JM, Joh TH, Smith GP (1996b). c-Fos induction in the rat nucleus of the solitary tract correlates with the retention and forgetting of a conditioned taste aversion. Learn Mem 3:25-30.

Houpt TA, Philopena JM, Wessel TC, Joh TH, Smith GP (1994). Increased c-fos expression in nucleus of the solitary tract correlated with conditioned taste aversion to sucrose in rats. Neurosci Lett 172:1-5.

Hughes P, Dragunow M (1995). Induction of immediate-early genes and the control of neurotransmitter-regulated gene expression within the nervous system. Pharmacol Rev 47:133-178.

Hughes P, Lawlor P, Dragunow M (1992). Basal expression of Fos, Fos-related, Jun, and Krox 24 proteins in rat hippocampus. Brain Res Mol Brain Res 13:355-357.

Irwin KB, Craig AD, Bracha V, Bloedel JR (1992). Distribution of c-fos expression in brainstem neurons associated with conditioning and pseudo-conditioning of the rabbit nictitating membrane reflex. Neurosci Lett 148:71-75.

Jenkins TA, Amin E, Harold GT, Pearce JM, Aggleton JP (2003). Distinct patterns of hippocampal formation activity associated with different spatial tasks: a Fos imaging study in rats. Exp Brain Res 151:514-523.

Johnson BA, Woo CC, Duong H, Nguyen V, Leon M (1995). A learned odor evokes an enhanced Fos-like glomerular response in the olfactory bulb of young rats. Brain Res 699:192-200.

Jones MW, Errington ML, French PJ, Fine A, Bliss TV, Garel S, Charnay P, Bozon B, Laroche S, Davis S (2001). A requirement for the immediate early gene Zif268 in the expression of late LTP and long-term memories. Nat Neurosci 4:289-296.

Kaczmarek L (1990). Molecular biology of long lasting memory formation. ESF Scientific Networks: Zeist 1-3 December, 1989.

Kaczmarek L (1993). Glutamate receptor-driven gene expression in learning. Acta Neurobiol Exp 53:187-196.

Kaczmarek L (1995). Towards understanding of the role of transcription factors in learning processes. Acta Biochim Pol 42:221-226.

Kaczmarek L (2000). Gene expression in learning processes. Acta Neurobiol Exp 60:419-424.

Kaczmarek L (2002). In: Handbook of Chemical Neuroanatomy (Kaczmarek L, Robertson HA, eds), p. 370. Amsterdam: Elsevier.

Kaczmarek L, Chaudhuri A (1997). Sensory regulation of immediate-early gene expression in mammalian visual cortex: implications for functional mapping and neural plasticity. Brain Res Brain Res Rev 23:237-256.

Kaczmarek L, Nikolajew E (1990). c-fos protooncogene expression and neuronal plasticity. Acta Neurobiol Exp (Wars) 50:173-179.

Kaminska B, Pyrzynska B, Ciechomska I, Wisniewska M (2000). Modulation of the composition of AP-1 complex and its impact on transcriptional activity. Acta Neurobiol Exp (Wars) 60:395-402.

Kleim JA, Lussnig E, Schwarz ER, Comery TA, Greenough WT (1996). Synaptogenesis and Fos expression in the motor cortex of the adult rat after motor skill learning. J Neurosci 16:4529-4535.

Knapska E, Kaczmarek L (2004). A gene for neuronal plasticity in the mammalian brain: Zif268/Egr-1/NGFI-A/Krox-24/TIS8/ZENK? Prog Neurobiol 74:183-211.

Knapska A, Walasek G, Nikolaev E, Neuhäusser-Wespy F, Lipp H-P, Kaczmarek L, Tomasz Werka T (2006). Differential involvement of the central amygdala in appetitive versus aversive learning. Learn Mem 13:192-200.

Koh MT, Wilkins EE, Bernstein IL (2003). Novel tastes elevate c-fos expression in the central amygdala and insular cortex: implication for taste aversion learning. Behav Neurosci 117:1416-1422.

Kovacs KJ (1998). c-Fos as a transcription factor: a stressful (re)view from a functional map. Neurochem Int 33:287-297.

Lamprecht R, Dudai Y (1995). Differential modulation of brain immediate early genes by intraperitoneal LiCl. Neuroreport 7:289-293.

Lamprecht R, Dudai Y (1996). Transient expression of c-Fos in rat amygdala during training is required for encoding conditioned taste aversion memory. Learn Mem 3:31-41.

Lau LF, Nathans D (1987). Expression of a set of growth-related immediate early genes in BALB/c 3T3 cells: coordinate regulation with c-fos or c-myc. Proc Natl Acad Sci USA 84:1182-1186.

Lemaire P, Vesque C, Schmitt J, Stunnenberg H, Frank R, Charnay P (1990). The serum-inducible mouse gene Krox-24 encodes a sequence-specific transcriptional activator. Mol Cell Biol 10:3456-3467.

Li HY, Ericsson A, Sawchenko PE (1996). Distinct mechanisms underlie activation of hypothalamic neurosecretory neurons and their medullary catecholaminergic afferents in categorically different stress paradigms. Proc Natl Acad Sci USA 93:2359-2364.

Lukasiuk K, Savonenko A, Nikolaev E, Rydz M, Kaczmarek L (1999). Defensive conditioning-related increase in AP-1 transcription factor in the rat cortex. Brain Res Mol Brain Res 67:64-73.

Mack KJ, Mack PA (1992). Induction of transcription factors in somatosensory cortex after tactile stimulation. Brain Res Mol Brain Res 12:141-147.

Maleeva NE, Bikbulatova LS, Ivolgina GL, Anokhin KV, Limborskaia SA, Kruglikov RI (1990). [Activation of the c-fos proto-oncogene in different structures of the rat brain during training and pseudoconditioning]. Dokl Akad Nauk SSSR 314:762-764.

Maleeva NE, Ivolgina GL, Anokhin KV, Limborskaia SA (1989). [Analysis of the expression of the c-fos proto-oncogene in the rat cerebral cortex during learning]. Genetika 25:1119-1121.

Malkani S, Rosen JB (2000a). Differential expression of EGR-1 mRNA in the amygdala following diazepam in contextual fear conditioning. Brain Res 860:53-63.

Malkani S, Rosen JB (2000b). Specific induction of early growth response gene 1 in the lateral nucleus of the amygdala following contextual fear conditioning in rats. Neuroscience 97:693-702.

Malkani S, Rosen JB (2001). N-Methyl-D-aspartate receptor antagonism blocks contextual fear conditioning and differentially regulates early growth response-1 messenger RNA expression in the amygdala: implications for a functional amygdaloid circuit of fear. Neuroscience 102:853-861.

Malkani S, Wallace KJ, Donley MP, Rosen JB (2004). An egr-1 (zif268) antisense oligodeoxynucleotide infused into the amygdala disrupts fear conditioning. Learn Mem 11:617-624.

Melia KR, Ryabinin AE, Schroeder R, Bloom FE, Wilson MC (1994). Induction and habituation of immediate early gene expression in rat brain by acute and repeated restraint stress. J Neurosci 14:5929-5938.

Mello CV (2002). Immediate-early gene (IEG) expression mapping of vocal communication areas in the avian brain. In: Immediate Early Genes and Inducible Transcription Factors in Mapping of the Central Nervous System Function and Dysfunction (Kaczmarek L, Robertson HA, eds), pp. 59-101. Amsterdam: Elsevier Science.

Mello CV, Velho TA, Pinaud R (2004). Song-induced gene expression: a window on song auditory processing and perception. Ann NY Acad Sci 1016:263-281.

Melzer P, Steiner H (1997). Stimulus-dependent expression of immediate-early genes in rat somatosensory cortex. J Comp Neurol 380:145-153.

Mickley GA, Kenmuir CL, McMullen CA, Yocom AM, Valentine EL, Dengler-Crish CM, Weber B, Wellman JA, Remmers-Roeber DR (2004). Dynamic processing of taste aversion extinction in the brain. Brain Res 1016:79-89.

Milanovic S, Radulovic J, Laban O, Stiedl O, Henn F, Spiess J (1998). Production of the Fos protein after contextual fear conditioning of C57BL/6N mice. Brain Res 784:37-47.

Milbrandt J (1987). A nerve growth factor-induced gene encodes a possible transcriptional regulatory factor. Science 238:797-799.

Montag-Sallaz M, Welzl H, Kuhl D, Montag D, Schachner M (1999). Novelty-induced increased expression of immediate-early genes c-fos and arg 3.1 in the mouse brain. J Neurobiol 38:234-246.

Morrow BA, Elsworth JD, Inglis FM, Roth RH (1999). An antisense oligonucleotide reverses the footshock-induced expression of fos in the rat medial prefrontal cortex and the subsequent expression of conditioned fear-induced immobility. J Neurosci 19:5666-5673.

Muller R, Bravo R, Muller D, Kurz C, Renz M (1987). Different types of modification in c-fos and its associated protein p39: modulation of DNA binding by phosphorylation. Oncogene Res 2:19-32.

Mumby DG, Gaskin S, Glenn MJ, Schramek TE, Lehmann H (2002). Hippocampal damage and exploratory preferences in rats: memory for objects, places, and contexts. Learn Mem 9:49-57.

Neumann I (1997). Antisense oligonucleotides in neuroendocrinology: enthusiasm and frustration. Neurochem Int 31:363-378.

Nikolaev E, Kaminska B, Tischmeyer W, Matthies H, Kaczmarek L (1992a). Induction of expression of genes encoding transcription factors in the rat brain elicited by behavioral training. Brain Res Bull 28:479-484.

Nikolaev E, Werka T, Kaczmarek L (1992b). C-fos protooncogene expression in rat brain after long-term training of two-way active avoidance reaction. Behav Brain Res 48:91-94.

Okuno H, Miyashita Y (1996). Expression of the transcription factor Zif268 in the temporal cortex of monkeys during visual paired associate learning. Eur J Neurosci 8:2118-2128.

Okuno H, Saffen DW, Miyashita Y (1995). Subdivision-specific expression of ZIF268 in the hippocampal formation of the macaque monkey. Neuroscience 66:829-845.

Paylor R, Johnson RS, Papaioannou V, Spiegelman BM, Wehner JM (1994). Behavioral assessment of c-fos mutant mice. Brain Res 651:275-282.

Pezzone MA, Lee WS, Hoffman GE, Pezzone KM, Rabin BS (1993). Activation of brainstem catecholaminergic neurons by conditioned and unconditioned aversive stimuli as revealed by c-Fos immunoreactivity. Brain Res 608:310-318.

Pezzone MA, Lee WS, Hoffman GE, Rabin BS (1992). Induction of c-Fos immunoreactivity in the rat forebrain by conditioned and unconditioned aversive stimuli. Brain Res 597:41-50.

Polston EK, Erskine MS (1995). Patterns of induction of the immediate-early genes c-fos and egr-1 in the female rat brain following differential amounts of mating stimulation. Neuroendocrinology 62:370-384.

Radulovic J, Kammermeier J, Spiess J (1998). Relationship between fos production and classical fear conditioning: effects of novelty, latent inhibition, and unconditioned stimulus preexposure. J Neurosci 18:7452-7461.

Radwanska K, Nikolaev E, Knapska E, Kaczmarek L (2002). Differential response of two subdi-visions of lateral amygdala to aversive conditioning as revealed by c-Fos and P-ERK mapping. Neuroreport 13:2241-2246.

Reason AJ, Morris HR, Panico M, Marais R, Treisman RH, Haltiwanger RS, Hart GW, Kelly WG, Dell A (1992). Localization of O-GlcNAc modification on the serum response transcription factor. J Biol Chem 267:16911-16921.

Rescorla RA (1967). Pavlovian conditioning and its proper control procedures. Psychol Rev 74:71-80.

Rose SP (2000). God's organism? The chick as a model system for memory studies. Learn Mem 7:1-17.

Rosen JB, Fanselow MS, Young SL, Sitcoske M, Maren S (1998). Immediate-early gene expression in the amygdala following footshock stress and contextual fear conditioning. Brain Res 796:132-142.

Roullet F, Datiche F, Lienard F, Cattarelli M (2004). Cue valence representation studied by Fos immunocytochemistry after acquisition of a discrimination learning task. Brain Res Bull 64:31-38.

Rylski M, Kaczmarek L (2004). Ap-1 targets in the brain. Front Biosci 9:8-23.

Sakai N, Yamamoto T (1997). Conditioned taste aversion and c-fos expression in the rat brainstem after administration of various USs. Neuroreport 8:2215-2220.

Santin LJ, Aguirre JA, Rubio S, Begega A, Miranda R, Arias JL (2003). c-Fos expression in supramammillary and medial mammillary nuclei following spatial reference and working memory tasks. Physiol Behav 78:733-739.

Santini E, Ge H, Ren K, Pena de Ortiz S, Quirk GJ (2004). Consolidation of fear extinction requires protein synthesis in the medial prefrontal cortex. J Neurosci 24:5704-5710.

Savonenko A, Filipkowski RK, Werka T, Zielinski K, Kaczmarek L (1999). Defensive conditioning-related functional heterogeneity among nuclei of the rat amygdala revealed by c-Fos mapping. Neuroscience 94:723-733.

Schafe GE, Bernstein IL (1996). Forebrain contribution to the induction of a brainstem correlate of conditioned taste aversion: I. The amygdala. Brain Res 741:109-116.

Schafe GE, Bernstein IL (1998). Forebrain contribution to the induction of a brainstem correlate of conditioned taste aversion. II. Insular (gustatory) cortex. Brain Res 800:40-47.

Schafe GE, Seeley RJ, Bernstein IL (1995). Forebrain contribution to the induction of a cellular correlate of conditioned taste aversion in the nucleus of the solitary tract. J Neurosci 15:6789-6796.

Scicli AP, Petrovich GD, Swanson LW, Thompson RF (2004). Contextual fear conditioning is associated with lateralized expression of the immediate early gene c-fos in the central and basolateral amygdalar nuclei. Behav Neurosci 118:5-14.

Smith MA, Banerjee S, Gold PW, Glowa J (1992). Induction of c-fos mRNA in rat brain by conditioned and unconditioned stressors. Brain Res 578:135-141.

Spray KJ, Halsell CB, Bernstein IL (2000). c-Fos induction in response to saccharin after taste aversion learning depends on conditioning method. Brain Res 852:225-227.

Stein CA (1996). Phosphorothioate antisense oligodeoxynucleotides: questions of specificity. Trends Biotechnol 14:147-149.

Stein CA, Krieg AM (1994). Problems in interpretation of data derived from in vitro and in vivo use of antisense oligodeoxynucleotides. Antisense Res Dev 4:67-69.

Steiner H, Gerfen CR (1994). Tactile sensory input regulates basal and apomorphine-induced immediate- early gene expression in rat barrel cortex. J Comp Neurol 344:297-304.

Strekalova T, Zorner B, Zacher C, Sadovska G, Herdegen T, Gass P (2003). Memory retrieval after contextual fear conditioning induces c-Fos and JunB expression in CA1 hippocampus. Genes Brain Behav 2:3-10.

Swank MW (2000). Conditioned c-Fos in mouse NTS during expression of a learned taste aversion depends on contextual cues. Brain Res 862:138-144.

Swank MW, Bernstein IL (1994). c-Fos induction in response to a conditioned stimulus after single trial taste aversion learning. Brain Res 636:202-208.

Swank MW, Ellis AE, Cochran BN (1996). c-Fos antisense blocks acquisition and extinction of conditioned taste aversion in mice. Neuroreport 7:1866-1870.

Swank MW, Schafe GE, Bernstein IL (1995). c-Fos induction in response to taste stimuli previously paired with amphetamine or LiCl during taste aversion learning. Brain Res 673:251-261.

Swirnoff AH, Milbrandt J (1995). DNA-binding specificity of NGFI-A and related zinc finger transcription factors. Mol Cell Biol 15:2275-2287.

Szabowski A, Maas-Szabowski N, Andrecht S, Kolbus A, Schorpp-Kistner M, Fusenig NE, Angel P (2000). c-Jun and JunB antagonistically control cytokine-regulated mesenchymal-epidermal interaction in skin. Cell 103:745-755.

Szklarczyk AW, Kaczmarek L (1999). Brain as a unique antisense environment. Antisense Nucleic Acid Drug Dev 9:105-116.

Thiele TE, Roitman MF, Bernstein IL (1996). c-Fos induction in rat brainstem in response to ethanol- and lithium chloride-induced conditioned taste aversions. Alcohol Clin Exp Res 20:1023-1028.

Thomas KL, Hall J, Everitt BJ (2002). Cellular imaging with zif268 expression in the rat nucleus accumbens and frontal cortex further dissociates the neural pathways activated following the retrieval of contextual and cued fear memory. Eur J Neurosci 16:1789-1796.

Tischmeyer W, Grimm R (1999). Activation of immediate early genes and memory formation. Cell Mol Life Sci 55:564-574.

Tischmeyer W, Kaczmarek L, Strauss M, Jork R, Matthies H (1990). Accumulation of c-fos mRNA in rat hippocampus during acquisition of a brightness discrimination. Behav Neural Biol 54:165-171.

Tokuyama W, Okuno H, Hashimoto T, Li YX, Miyashita Y (2002). Selective zif268 mRNA induction in the perirhinal cortex of macaque monkeys during formation of visual pair-association memory. J Neurochem 81:60-70.

Vann SD, Brown MW, Erichsen JT, Aggleton JP (2000). Fos imaging reveals differential patterns of hippocampal and parahippocampal subfield activation in rats in response to different spatial memory tests. J Neurosci 20:2711-2718.

Weitemier AZ, Ryabinin AE (2004). Subregion-specific differences in hippocampal activity between Delay and Trace fear conditioning: an immunohistochemical analysis. Brain Res 995:55-65.

Woo CC, Oshita MH, Leon M (1996). A learned odor decreases the number of Fos-immunopositive granule cells in the olfactory bulb of young rats. Brain Res 716:149-156.

Worley PF, Christy BA, Nakabeppu Y, Bhat RV, Cole AJ, Baraban JM (1991). Constitutive expression of zif268 in neocortex is regulated by synaptic activity. Proc Natl Acad Sci USA 88:5106-5110.

Yamamoto T, Sako N, Sakai N, Iwafune A (1997). Gustatory and visceral inputs to the amygdala of the rat: conditioned taste aversion and induction of c-fos-like immunoreactivity. Neurosci Lett 226:127-130.

Yamamoto T, Shimura T, Sakai N, Ozaki N (1994). Representation of hedonics and quality of taste stimuli in the parabrachial nucleus of the rat. Physiol Behav 56:1197-1202.

Yanagawa Y, Kobayashi T, Kamei T, Ishii K, Nishijima M, Takaku A, Tamura S (1997). Structure and alternative promoters of the mouse glutamic acid decarboxylase 67 gene. Biochem J 326:573-578.

Zhang YQ, Ji YP, Mei J (2000). Behavioral training-induced c-Fos expression in the rat nucleus basalis of Meynert during aging. Brain Res 879:156-162.

Zhu XO, Brown MW, McCabe BJ, Aggleton JP (1995). Effects of the novelty or familiarity of visual stimuli on the expression of the immediate early gene c-fos in rat brain. Neuroscience 69:821-829.

Zhu XO, McCabe BJ, Aggleton JP, Brown MW (1996). Mapping visual recognition memory through expression of the immediate early gene c-fos. Neuroreport 7:1871-1875.

Zhu XO, McCabe BJ, Aggleton JP, Brown MW (1997). Differential activation of the rat hippocampus and perirhinal cortex by novel visual stimuli and a novel environment. Neurosci Lett 229:141-143.

9

Immediate Early Genes and the Mapping of Environmental Representations in Hippocampal Neural Networks

JOHN F. GUZOWSKI

Department of Neurobiology & Behavior and Center for the Neurobiology of Learning and Memory, University of California, Irvine, CA, USA

1. Introduction

1.1. Hippocampus in the Encoding, Consolidation, and Retrieval of Long-Term Memories: Plasticity of Neurons, Plasticity of Circuits

The hippocampus has received much attention by researchers interested in understanding the nature of "declarative" or "explicit" memory—the memory for discrete events, which requires conscious recollection of disparate facts. These memories, such as remembering one's wedding day or last vacation, serve to define us as individuals, and consequently are of critical importance in human life. The ability to rapidly form explicit memories, such as remembering the location of a food source or an encounter with a predator, can also be necessary for the survival of an animal. In studies from rodents, primates, and humans, the hippocampus has been shown repeatedly to be important for the formation and consolidation of declarative memories (Zola-Morgan and Squire, 1993; Suzuki and Eichenbaum, 2000). For this reason, the hippocampus is arguably the most studied region of the brain. The disparate approaches used to understanding hippocampal function include the study of neural plasticity in hippocampal slice and dissociated neural cultures, lesion/pharmacological experiments in behaving rodents, electrophysiological recordings in rodents and primates, and human functional neuroimaging. Despite years of research across these domains, we still have much to learn as to how neurons of the hippocampus facilitate the formation and consolidation of long-term memories.

It is well established that memories are not acquired instantaneously, but rather that the consolidation of recent experiences into long-term memories (LTM)

R. Pinaud, L.A. Tremere (Eds.), Immediate Early Genes in Sensory Processing, Cognitive Performance and Neurological Disorders, 159–176, ©2006 Springer Science + Business Media, LLC

requires time (McGaugh, 2000). The time-dependent processes involved in consolidating LTM happen at vastly different scales, including biochemical changes in single cells that occur within seconds to hours and interactions of neuronal ensembles throughout the brain that occur over milliseconds to years (Dudai, 1997). A convenient distinction for separating the processes associated with long-term memory formation is to divide them into "synaptic plasticity" and "systems plasticity". Synaptic plasticity, which can involve the growth of new synaptic connections or the modification of pre-existing synaptic connections, occurs on the time scale of seconds to hours after the neural activity associated with learning. The electrophysiological phenomena of long-term potentiation (LTP) and long-term depression (LTD), both well studied in hippocampus, provide examples of the types of molecular/cellular changes associated with synaptic plasticity (reviewed in Bear and Malenka, 1994; Malenka, 1994). Systems level plasticity, on the other hand, occurs in the days to potentially years (in humans) after learning and is characterized by a change in the relative contribution of different brain regions in supporting the memory "trace" (reviewed in Ribeiro and Nicolelis, 2004; Frankland and Bontempi, 2005). It is believed that hippocampal neurons use LTP/LTD-like synaptic plasticity mechanisms to rapidly encode explicit memories, and then during offline periods (e.g., sleep), these modified neural networks play an important role in driving changes in neocortical networks to support permanent memory storage (i.e. systems level plasticity) (Sutherland and McNaughton, 2000; Ribeiro and Nicolelis, 2004; Frankland and Bontempi, 2005). I will revisit this idea in greater detail in the Discussion.

1.2. Physiology of Rodent Hippocampal Neurons: Place Cells and Population Codes

One line of investigation has shown that rodents with lesions (cellular damage or genetic perturbations) of the hippocampus are unable to form spatial memories, such as in radial arm mazes or in open field spatial tasks (e.g., the Morris water maze) (O'Keefe and Nadel, 1978; Nakazawa et al., 2004). Interestingly, hippocampal neurons have been shown to reliably encode spatial information as rats randomly forage in open environments or navigate tracks (O'Keefe and Dostrovsky, 1971; Wilson and McNaughton, 1993). The presence of hippocampal "place cells" has been used to explain the deficits of hippocampal lesioned animals in spatial memory tasks. In recent years, however, it has become increasingly clear that hippocampal neurons encode more than space, such as conjunctive representations of salient stimuli (e.g., place and reward; Young et al., 1994; Markus et al., 1995; Gothard et al., 1996; Wood et al., 1999). Such findings are congruent with other lines of research that animals (and humans) with hippocampal dysfunction have impairments in memories that are episodic (unique events) or conjunctive in nature (reviewed in Eichenbaum, 2001; O'Reilly and Rudy, 2001).

The neural computations involved in storing and retrieving memories, however, are not the domain of single neurons, but are rather the property of large number of neurons—neural ensembles or assemblies. Thus, it is the "population code" of hippocampal neural ensembles that must be understood to better define the role of the hippocampus in learning and memory. This fact has led to advances in parallel cell recording methods for use in both rodent and primate subjects. Although the temporal resolution of these methods is without peer, there are limits to the numbers of neurons, and the number of brain regions, that can be analyzed within an animal. We have developed two related immediate-early gene (IEG) based imaging approaches that provides both cellular and temporal resolution for many regions of the brain and can be used to map neural circuits associated with specific behaviors (Guzowski et al., 1999; Guzowski and Worley, 2001; Vazdarjanova and Guzowski, 2004).

1.3. IEGs in Mapping Behaviorally-Relevant Brain Circuits: A few Considerations and Caveats

Following synaptic activity associated with learning, the expression of several immediate-early genes (IEGs) is dramatically and rapidly increased in neurons of many regions of the brain (Morgan et al., 1987; Cole et al., 1989). Moreover, this experience-dependent gene expression can be critical for transforming recent experience into long-term memory (Tischmeyer and Grimm, 1999; Guzowski, 2002). Operationally, IEGs can been divided into two classes: (1) transcription factor IEGs (e.g., zif268, c-fos, c-jun, ICER, etc.) and (2) effector IEGs (*Arc*, Homer 1a, Narp, cox-2, BDNF, etc.) (Lanahan and Worley, 1998; Guzowski, 2002). As both transcription factor IEGs and effector IEGs have been shown to be important for synaptic plasticity and memory consolidation, it seems likely that IEGs, working in an orchestrated fashion with each other and existing synaptic proteins, are essential for protein synthesis-dependent synaptic plasticity. The studies of the current chapter examine expression of two effector IEGs, *Arc* and *Homer 1a* (*H1a*). Both *Arc* and *H1a* are components of the postsynaptic density and can play prominent roles in enabling lasting synaptic modification. *Arc* (Lyford et al., 1995), also identified as *Arg3.1* (Link et al., 1995), is a component of NMDA receptor complexes (Husi et al., 2000) and can be targeted to active dendrites (Steward et al., 1998). Inhibiting Arc protein expression in hippocampus disrupts long-term potentiation and long-term spatial memory (Guzowski et al., 2000). *H1a* is a synaptic activity regulated isoform of the *Homer 1* gene, which can act as a dominant negative regulator of constitutively expressed *Homer 1* isoforms (*Homer 1b/c*) in coupling membrane events to intracellular calcium release (Xiao et al., 2000; Yuan et al., 2003). Moreover, double labeling studies have shown that the expression of *Arc* and *H1a* is strongly induced in the same population of neurons of the rat hippocampus and neocortex after exploration of a novel environment (Vazdarjanova et al., 2002), suggesting that they may function in concert in the structural and functional modifications of synapses responsible for storing memories.

Since the discovery that IEGs are dynamically regulated in the brain by neural activity, there has been a substantial research focus using IEGs as neural activity markers in studies of behavior and cognition. In a typical experiment, separate groups of animals perform a behavioral task or are exposed to control conditions and are sacrificed at an appropriate time after handling. Brain sections are processed using either immunhistochemical methods or *in situ* hybridization to detect IEG protein or RNA, respectively. Levels of gene product or numbers of positive cells are then compared across groups. IEG approaches have been useful in mapping brain areas engaged by single experiences, but the specificity of the response has been repeatedly questioned. To address this concern, researchers have attempted to use separate "non-learning" control groups, which are designed to control for stress and arousal/attentional components, motor activity, and sensory features experienced by animals performing the task of interest. In essence, these controls are designed to factor out everything but the specific cognitive component of the behavioral task. Invariably, however, learning occurs in such non-learning groups (i.e., latent learning) (Tolman, 1932), undermining such efforts. Furthermore, an underlying assumption of this approach is that the behavioral task must activate a greater number of cells as compared to the control condition for it to be concluded that the region of interest is specifically involved in task performance. More often, however, neural coding involves variation in which elements of a population are activated rather than how many, and a crucial question in the study of neural coding involves how much overlap there is in the populations activated by two different experiences. Such determinations are not possible using conventional IEG methods. As a result, brain structures in which the same numbers of neurons, but different populations, are activated in the control and experimental conditions may be overlooked as important contributors to the cognitive process under study.

An important, but often overlooked, consideration in the use of IEGs as neuronal activity markers in behavioral studies is whether RNA or protein product is being detected. Although there are many instances where the coupling between RNA expression (including transcriptional and posttranscriptional regulation of steady-state mRNA levels) and protein expression are high, there are also many examples in biology where the two are not tightly coupled. Indeed, there is a rather active research focus in understanding posttranscriptional regulation of gene expression in neurons, including the regulation of polyadenylation of neural transcripts and mechanisms for translational regulation (Klann et al., 2004). For IEGs, which are essentially undetectable in quiescent neurons, regulation of functional protein expression requires activation of transcription. As such, IEG mRNA expression is a more direct "readout" of neural activity than IEG protein expression. Furthermore, for a given IEG, the half-life of the mRNA is typically much shorter than that of the protein, giving mRNA detection methods a higher signal to noise ratio. In our imaging methods (described below), we take this even one step closer to the inducing event by measuring cells actively transcribing specific IEG RNAs at the time of sacrifice. This transcription based approach removes differences in mRNA stability from consideration, which could influence steady-state mRNA levels as much as new synthesis (i.e., transcription).

2. catFISH: A Novel IEG Method to Map Hippocampal Network Functions

While developing a sensitive fluorescent *in situ* hybridization (FISH) assay to study co-localization of different IEG mRNAs, my colleagues and I made a rather serendipitous finding. This discovery would ultimately enable a novel approach for mapping hippocampal neuronal responses to different experiences with cellular *and* temporal resolution. The principle finding was that both primary (unprocessed) RNA transcript in the nucleus (at the genomic alleles) and mature mRNA in the cytoplasm could be detected with the use of sensitive FISH and confocal microscopy (Fig. 9.1a). Time course analyses using a temporally precise strong stimulus, maximal electroconvulsive shock (MECS), showed that *Arc* RNA transcript was detected within 1–2 minutes at intense intranuclear foci (INF) and that these INF disappeared within ~15 minutes (Guzowski et al., 1999). Several control experiments (RNase pretreatment of samples, use of sense riboprobes, combined use of intron and exon probes) have lead us to conclude that these INF represent the sites of transcription at the genomic alleles for specific IEGs. Subsequently, from ~15–45 min a prominent labeling was observed in perinuclear/cytoplasmic/dendritic regions. After that, *Arc* mRNA could be observed in dendritic regions. This time course of appearance/disappearance of *Arc* signal holds for the pyramidal cell layers of the hippocampus and for neurons in many cortical and subcortical regions. One region that displays different kinetics is the granule cell layer of the dentate gyrus. In these neurons, the appearance of *Arc* transcription foci is as rapid, but transcription continues to persist for greater than 4 and less than 8 hours after MECS. Thus, in granule cells *Arc* mRNA levels reach much higher levels as compared to CA1 and CA3 pyramidal neurons, due to this prolonged period of active transcription.

The kinetics of *Arc* mRNA appearance/disappearance were then examined using natural stimuli—a brief (5 min) exposure to a novel environmental context. In initial (Guzowski et al., 1999) and then follow-up studies (Vazdarjanova et al., 2002), a maximum of *Arc* transcriptionally active cells was detected in rats sacrificed immediately after the 5 min exploration session. Within 16 min after removal from the environment, the proportion of *Arc* INF-positive cells was back to baseline levels. Thus, a brief "burst" of *Arc* transcription accompanies learning about a novel environment. *Arc* mRNA was then detected in the cytoplasm and dendrites ~20–60 min after removal from environment (Guzowski et al., 1999; Vazdarjanova et al., 2002). Because the time course of the nuclear signal is distinct from the cytoplasmic signal, the activity history of individual neurons at two different times can be inferred through the use of *Arc* FISH and confocal microscopy (Fig. 9.2b).

The ability to define the time of activation of single neurons based on the subcellular distribution of *Arc* RNA suggested that it might be possible to expose rats to two different behavioral experiences, separated by a rest interval, and visualize the population of cells activated during each experience. We termed this new imaging approach "cellular analysis of temporal activity by fluorescence *in situ* hybridization", or "catFISH". In concept, catFISH provided an alternative means to address whether activation of IEG transcription was specifically related to information processing or merely associated with increased behavioral arousal, motor activity, stress, etc, typically associated with different learning paradigms. With this approach, neural activity "maps" for two discrete experiences by a single animal can be visualized and compared. The within-subject comparisons provided by catFISH contrast strongly with "standard approach" for IEG imaging, in which changes in the proportion of activated neurons are compared across different groups of animals (between-subjects). With the between-subjects design of standard IEG experiments, a brain area can be inferred to be specifically associated with a behavioral task only if a different percentage of cells are active as compared to the "learning control" group. Although this type of "coding by numbers" can be important in cortical regions, the hippocampus is engaged rather readily when an animal is attending to various stimuli, increasing the likelihood of it being "factored out" in the standard approach, especially if the behavior is very well controlled.

3. Environment-Specific Expression of the IEG Arc in Hippocampal CA1 Neuronal Ensembles

Previous studies using parallel ensemble recording methods indicate that a fraction of CA1 neurons are specifically activated during exploration of a given environment (typically, 30–50% depending on environment size (Kubie and Ranck, 1983; Wilson and McNaughton, 1993; Gothard et al., 1996). Interestingly, a similar proportion of CA1 neurons activate transcription of *Arc* during exposure to a novel context (Guzowski et al., 1999; Vazdarjanova et al., 2002). The correspondence between the proportions of activated neurons detected by *Arc* catFISH and by single unit recording studies in similar experimental conditions led us to speculate that IEG expression may be induced in CA1 neurons following periods of intense, natural activity, such as when a cell expresses a place field. If true, the cells detected by *Arc* catFISH should meet predictions derived from place cell studies. Exploiting the temporal and cellular resolution of catFISH, we examined *Arc* activation patterns in rats exposed sequentially to two different environments, or to the same environment twice (Guzowski et al., 1999). This paradigm, combined with the use of catFISH, allowed us to examine whether *Arc* gene induction in CA1 neurons was a generalized response or whether it was related to information processing. Neurophysiological recording methods indicate that a fraction of CA1 neurons are activated during exploration of a given environment (Kubie and Ranck, 1983; Wilson and McNaughton, 1993). Moreover, rats reintroduced into the same environment typically reactivate the same cells (Thompson and Best, 1990), while different cells are activated if the rat is placed in a new environment. Activation of "place" cells in a given environment appears to be stochastically determined, in the sense that if the fraction of cells active in environment A is p_A and the fraction of cells active in environment B is p_B, one can predict the percentage of cells active in one or both environments as:

$$A \text{ only} = p_A$$
$$B \text{ only} = p_B$$
$$A \text{ and } B = p_A \times p_B$$
$$\text{neither} = (1 - p_A) \times (1 - p_B).$$

Thus, the stability and population specificity of CA1 place cell representations in normal rats provided an ideal system to examine the specificity of *Arc* gene induction using catFISH.

Rats were first placed in environment "A" for 5 minutes and returned to their home cage for 20 minutes. Rats were then returned to the same environment A (A/A group) or placed in a different environment "B" in an adjacent room (A/B group) for 5 minutes and then immediately sacrificed (Fig. 9.2a). The only difference between these groups was the nature of the environment experienced during the second session. In the A/A group, most cells activated in the first exploration session were again activated in the second exploration session (Fig. 9.2c; note the high level of "double" positive cells—those containing both *Arc* INF and cytoplasmic labeling). This single population of cells represented ~40% of neurons in CA1. By contrast, the A/B group showed 3 populations of neurons (cytoplasmic only, nuclear foci only or double-positive) that each represented 22%, 23% and 16% of total CA1 neurons, respectively (Fig. 9.2c). The experimentally observed cell staining frequencies of the A/B group were consistent with the conclusion that environments A and B activated statistically independent populations of neurons. The high degree of reactivation of neurons in the A/A condition and the completely dissimilar activation profile in the A/B condition supports the conclusion that *Arc* transcription in CA1 neurons is linked to information processing, and is not a nonspecific response to stress, novelty, or motor activity. These data, which are quantitatively consistent with electrophysiological studies of hippocampal place cells recorded in similar behavioral paradigms (Kubie and Ranck, 1983; Thompson and Best, 1990), support the hypothesis that *Arc* catFISH provides visualization of cell ensembles activated by two sequential experiences. Additionally, these results suggest that the cell firing associated with the expression of a place field provides a sufficient signal to activate *Arc* transcription in hippocampal neurons.

4. Population Response of CA3 and CA1 Neuronal Ensembles to Changes in an Environmental Context

As mentioned previously, the hippocampus is critical for acquiring, consolidating, and retrieving declarative memories (reviewed in Squire and Knowlton, 1994; Eichenbaum, 2001; Nadel and Moscovitch, 2001). The relative contributions of the hippocampal subregions (including CA3 and CA1) to these memory processes are only partially understood, despite being an area of active research (reviewed in Kesner et al., 2004). Theoretical considerations first developed by Marr (Marr,

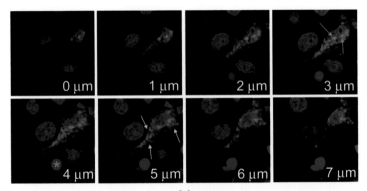

(a)

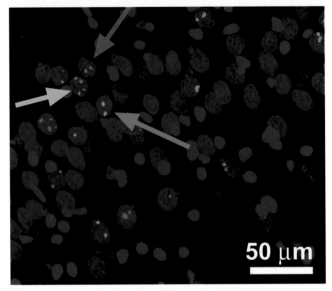

(b)

Figure 9.1. Detection of IEG RNA in hippocampal CA3 neurons using FISH and confocal microscopy. (a) z-series confocal images from the CA3 region of rat hippocampus showing a cell positive for intranuclear foci and cytoplasmic labeling using a digoxigenin-labeled antisense RNA probe to the IEG *Arc*. Nuclei are stained with the DNA stain DAPI (blue color) and *Arc* RNA is detected with cyanine 3 (red color). Optical sections were taken at one-μm intervals. *Arc* INF are indicated in the 3-μm image by green arrows. *Arc* cytoplasmic labeling is indicated in the 5-μm image by yellow arrows. The green asterisk in the 4-μm image indicates a small, intensely staining glial cell nucleus. (From Guzowski, 2002). (b) A confocal projection image (of 20 one-micron optical slices compressed into a single plane) of area CA3 from an A/B rat. The section was processed for double label catFISH for *Arc* RNA (red color) and *H1a* RNA (green color). Nuclei were counterstained and are shown as blue. The green arrow points to a *H1a*-positive neuron which has only *H1a* INF and thus was activated only during epoch 1; the red arrow points to a *Arc*-positive neuron which has only *Arc* INF and thus was activated only during epoch 2; and the yellow arrow points to a *Arc/H1a*-positive neuron that contains both *Arc* and *H1a* INF and thus was activated during both epoch 1 and epoch 2. Neurons that did not contain either *Arc* or *H1a* INF were classified as negative. Please note that confocal microscope settings were optimized for the detection of *Arc* and H1a INF, thus minimizing the appearance of cytoplasmic mRNA staining. Scale bar is 50 μm. (From Vazdarjanova and Guzowski, 2004).

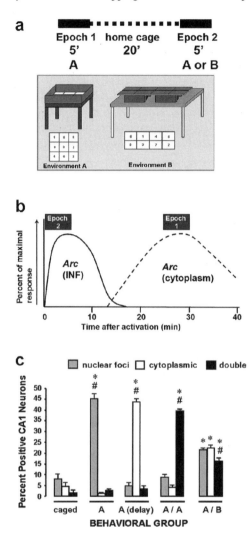

Figure 9.2. Environment-specific expression of the IEG *Arc* in hippocampal CA1 neuronal ensembles. (a) Behavioral timeline (upper figure) and drawing of environments A and B (lower figure). Rats were allowed to explore environment A for 5 minutes (epoch 1). After a 20-minute rest period in the home cage, the rats were exposed to either environment A again or to environment B (epoch 2). At the end of the epoch 2, the rats were sacrificed and brains were processed for *Arc* catFISH. (b) The approximate time course of detection for *Arc* INF and cytoplasmic *Arc* mRNA following neuronal activation. These estimated curves are based on data from (Guzowski et al., 1999; Vazdarjanova et al., 2002). The behavioral epochs for which *Arc* INF or cytoplasmic *Arc* RNA provide "readout" are shown above the curves. (c) Rats explored environments designated A and B as described in the text and panel a. The distinct staining profiles seen in A/A and A/B groups demonstrate that the induction of *Arc* transcription in CA1 is highly specific to the nature of the behavioral experience. The A/immediate, A/delay, and A/B/delay groups define the temporal resolution properties of *Arc* catFISH. $N = 3$ rats per group. The percentage of positive cells for each staining profile was determined for each individual rat; the reported values indicate the group mean. *, $p < 0.002$ relative to caged controls, ANOVA with Scheffe post hoc analysis. #, $p < 0.05$, paired t-test, relative to the other 2 cell populations for that group. (From Guzowski et al., 1999).

1971) have placed emphasis on the sparsity of coding necessary to store the large number of representations that the hippocampus encodes into memory (reviewed in Skaggs and McNaughton, 1992). Sparsely encoded patterns are proposed to originate in CA3 after highly correlated patterns of activity arriving from the entorhinal cortex have been filtered extensively by the dentate gyrus (McNaughton and Nadel, 1990; Treves and Rolls, 1994; Vinogradova, 2001). Additionally, because of its extensive network of recurrent collaterals, CA3 is proposed to be an autoassociator (McNaughton and Nadel, 1990; Treves and Rolls, 1994) and comparator (Vinogradova, 2001) capable of retrieving entire patterns from partial or degraded input and comparing them with processed sensory input arriving from the entorhinal cortex. Thus, CA3 can act as a filter controlling the information transmitted to CA1. The role of CA1 in pattern generation/modification, however, is less well understood.

We used a modified catFISH approach to examine how the ensemble code of both CA3 and CA1 neurons changes in response to defined alterations of an environmental context. In *Arc/H1a* catFISH the sites of transcription for two different immediate-early genes (IEGs), *Arc* and *Homer 1a* (*H1a*), provides a histological readout for the activity history of hippocampal and neocortical neuronal networks (Vazdarjanova et al., 2002). This method capitalizes on the different structure of the primary transcripts for *Arc* and *H1a*, and the precise onset and shutoff of synaptic activity-dependent IEG transcription in neurons. Whereas *Arc* mRNA is derived from a short primary transcript ($\sim$3.5 kb), *H1a* mRNA is generated from a longer and more complex primary transcript ($\sim$55 kb; Bottai et al., 2002). By performing double label FISH with a *H1a* 3'UTR specific riboprobe and a full length *Arc* riboprobe, cells activated $\sim$2–15 min before sacrifice are labeled with *Arc* intranuclear foci of transcription (INF), and cells activated $\sim$25–40 min earlier are labeled with *H1a* INF (Vazdarjanova et al., 2002; Figs 9.1a and 9.3b). Compared to the original catFISH method, the use of *H1a* INF as a readout for activity 30 min earlier provides a more reliable and less subjective marker, as compared to cytoplasmic *Arc* mRNA.

To examine how discrete changes to an environmental context influence the ensemble activation patterns of CA3 and CA1 neuronal networks, 3-month old male rats were exposed sequentially to two similar, but modified, environments, separated by a rest interval of 20 min (Fig. 9.2a). As controls, separate groups of rats ($n = 4$ per group) were exposed sequentially to the same environment twice (A/A group), or to two entirely different environments (A/B group). The environments were defined by both local ("intramaze") and distal ("extramaze") cues. In addition, four rats were sacrificed directly from the home cage, at different times throughout the course of the experiment, without any new behavioral experience, to determine the basal levels of *Arc* and *Homer 1a* (*H1a*) expression. Exploration sessions were videotaped and later analyzed for the number of zone transitions, object approaches, and rearings of each rat. The cell populations active during each exploration session were detected using *Arc/H1a* catFISH (Vazdarjanova et al., 2002).

Analyses of the behavioral data failed to show any significant differences between rats of the various groups. Importantly, all measures of exploratory activity were high for both sessions, in rats of the different groups. Despite the lack of detectable behavioral differences, *Arc/H1a* catFISH analyses revealed highly significant differences in the neural ensemble activity patterns for the rats from the different behavioral groups. Confocal image stacks were acquired and analyzed for the presence or absence of INF for both *Arc* and *H1a* in individual neuronal nuclei. The resulting raw data gives percentages (per rat, per region) of neurons that were negative, positive only for *Arc* INF, positive only for *H1a* INF, or positive for both *Arc* and *H1a* INF. This raw data can then be processed to determine the percentage of CA3 or CA1 neurons active in Epoch 1 (*H1a*-positive neurons plus *Arc/H1a* double-positive neurons/total neurons) or Epoch 2 (*Arc*-positive neurons plus *Arc/H1a* double-positive neurons/total neurons). An additional derived value was also calculated—the similarity score. In *Arc/H1a* catFISH imaging experiments involving two behavioral experiences (Epochs), it is helpful to reduce the complex cell staining data to a value that can be used to compare cell activity patterns across multiple brain regions. The similarity score takes the 4 measured cell-staining values (negative, *Arc*-positive, *H1a*-positive, and *Arc/H1a*-positive) and reduces them to a single value. With this method, a value of 1 represents a single neuronal population faithfully activated in both epochs. A value of 0 indicates that two statistically independent cell populations were activated during the two epochs.

In both CA3 and CA1, exploring the same environment during both behavioral epochs activated largely the same neuronal ensemble, as evidenced by a similarity score close to 1 for the A/A group (Fig. 9.2c). Exploring two very different environments during epoch 1 and epoch 2 (A/B group), on the other hand, activated two statistically independent CA3 neuronal ensembles, as evidenced by a similarity score of essentially 0. Like CA3, the similarity score for CA1 of the A/B group was dramatically lower than that of the A/A group, although the observed value was significantly greater than 0 ($p = 0.02$). Interestingly, although the similarity scores of the A/Aobj, A/Aconf and A/Ab groups were different from both the A/A and A/B groups (ANOVA for CA3: $F(4,15) = 91.95$, $p < 0.0001$; and for CA1: $F(4,15) = 19.75$, $p < 0.0001$, all appropriate post-hoc comparisons had $p < 0.02$), they were not different from each other ($p > 0.82$ for all three comparisons). Thus, changing the type or configuration of intramaze cues, as well as changing the extramaze cues, of environment A activated CA3 and CA1 neuronal ensembles that only partially overlapped with the respective neuronal ensembles activated by the original environment A. Furthermore, any of these perturbations of environment A resulted in a similar degree of overlap with the neuronal ensembles initially activated by environment A.

Despite the similarities in staining profiles and overlap ratios (similarity scores) observed for CA3 and CA1, there were three notable differences. First, the proportion of active cells comprising the CA3 ensemble was significantly less than that of CA1 for all groups for both epochs ($p < 0.005$; Fig. 9.2b). Second, the similarity scores of the A/Aobj, A/Aconf and A/Ab groups combined (collectively termed A/A') were significantly greater for CA3 than those for CA1 ($p = 0.001$; Fig. 9.2c). Thus, the partially modified environments were treated as more similar to the original environment in CA3 than in CA1, where differences between the environments were emphasized. Third, the similarity score of the A/B group was significantly lower in CA3 compared to CA1 ($p = 0.016$), suggesting that the representation of environment B was more orthogonal in CA3 as compared to CA1.

At the time we were conducting the above experiment, two parallel cell recording studies were underway in the labs of James Knierim (Lee et al., 2004) and Edvard Moser (Leutgeb et al., 2004) to explicitly compare CA3 and CA1 neural responses, within the same subjects, to alterations of an environmental context. If our "working hypothesis"—that the expression of at least some IEGs, like *Arc* and *H1a*, is tightly coupled to place cell firing—is correct, then aspects of our findings should have also been seen in these recording experiments. Although discussing differences in the behavioral designs of the 3 studies is beyond the scope of this chapter, it is gratifying that the three studies produced similar, and complementary, findings (Guzowski et al., 2004). In both recording studies, as with our IEG imaging study, the proportion of activated CA1 neurons was significantly higher than the proportion of CA3 neurons. This is to say that the CA3 ensemble representation of a context is considerable more "sparse" than in CA1. In Lee et al. (Lee et al., 2004), CA3 neural ensembles of rats subjected to a "double rotation" paradigm (of intramaze and extramaze cues) on a circular track tended to produce more coherent responses as compared to CA1 ensembles, consistent with the idea of CA3 neural ensembles performing a "pattern completion" function. By contrast, Leutgeb and co-workers (Leutgeb et al., 2004) found that in rats exposed to two similar box environments, but placed in two different rooms (i.e., different "extramaze cues"), the CA3 neuronal ensembles activated were completely independent (orthogonal), whereas the CA1 neural ensembles were more related than predicted by chance. Such orthogonalization of similar, but yet distinct, input patterns is termed "pattern separation".

As described above, the Lee et al. study provided evidence for pattern completion in CA3 neural ensembles and the Leutgeb et al. study provided evidence for pattern separation in CA3 ensembles. At first glance, the Lee and Leutgeb studies would seem at odds. However, and as noted earlier, a defining characteristic of CA3 neurons is their robust recurrent connections. Such recurrent networks have been hypothesized to be biological attractor networks, which can respond nonlinearly to changes in input patterns. As such, the input-output responses of CA3 neurons have been modeled as a sigmoidal function, whereas those of CA1 neurons have been modeled as a linear function (McClelland and Goddard, 1996; Rolls and Treves, 1998). Such behavior for CA3 and CA1 ensembles was observed in our catFISH study. When confronted with two different environments (A/B group), CA3 produced orthogonal representations for each environment (pattern separation). However, when input was sufficiently similar (A/A and A/A' groups), pattern completion in CA3 de-emphasized the mild perturbations of the environment. These two functions produced the discontinuous distribution of similarity scores (Fig. 9.3c). By contrast, the varying environmental modifications were represented continuously by CA1 ensembles (Fig. 9.3c).

The previous findings suggested that CA3 and CA1 neuronal ensembles may perform distinct functions in the processing of spatial and contextual information. We next investigated the possible interaction between CA3 and CA1 by examining whether overall cell activity levels in these regions were correlated within a behavioral epoch. Such a finding would be consistent with a unitary "hippocampal activation" involving both CA3 and CA1. The cells active during epoch 1 were defined as the total percentage of $H1a$-positive neurons, irrespective of Arc staining. Conversely, the cells active during epoch 2 were defined as the total percentage of Arc-positive cells, irrespective of $H1a$ staining. For both epoch 1 and epoch 2, CA3 and CA1 activity levels were not correlated (epoch 1, $r^2 = 0.04$, $p = 0.43$; epoch 2, $r^2 = 0.01$, $p = 0.66$), arguing against a simple co-activation of these regions during behavior. By contrast, activity levels within a rat for both CA3 and CA1 were modestly correlated across behavioral epochs ($r^2 = 0.46$ for CA3, $p < 0.001$; $r^2 = 0.41$ for CA1, $p < 0.002$), indicating that for each rat a similar proportion of CA3 or CA1 cells were active during two closely spaced exploration sessions.

We also examined whether the ratio of CA3/CA1 activity was constant across behavioral epochs within a rat. This value was generated by dividing the proportion of CA3 cells active per epoch by the proportion of CA1 cells active within that same epoch. Thus, a higher number indicates a greater bias towards CA3 activity, and a lower value indicates a bias towards CA1 activity. The CA3/CA1 ratio within a rat was highly correlated across epochs (Fig. 9.2d; $r^2 = 0.60$, $p < 0.0001$, slope $= 0.97$). This correlation was not simply related to exploratory activity because exploratory activity was not correlated within an animal across epochs ($r^2 = 0.18$, $p > 0.05$) and the distribution of CA3/CA1 ratios for the different groups did not cluster in any way for epoch 2 (Fig. 9.3d), despite the fact that the input was widely different and detected in the ensemble responses of both CA3 and CA1. Our finding that the relative balance of CA3/CA1 network activity was rat specific suggests an interaction between these regions, a finding consistent with an earlier study by Hess and co-workers (Hess et al., 1995). In that study, Hess and colleagues used $in\ situ$ hybridization to c-fos mRNA to show that the balance of labeling between CA3 and CA1 changed at different stages of learning an odor discrimination task. Based on these two IEG imaging studies, it would appear that both intrinsic and extrinsic factors can influence the balance of CA3 and CA1 neural activity in the hippocampal network, although the factors that govern this balance are not clear at present.

5. Use of IEGs to Map Hippocampal Networks: Future Directions and Questions

The finding that the transcription of the IEGs Arc and $Homer\ 1a$ is tightly linked to the neural activity associated with the expression of a "place field" in hippocampal neurons has several important implications for understanding hippocampal function in memory and provides a point of convergence for other lines

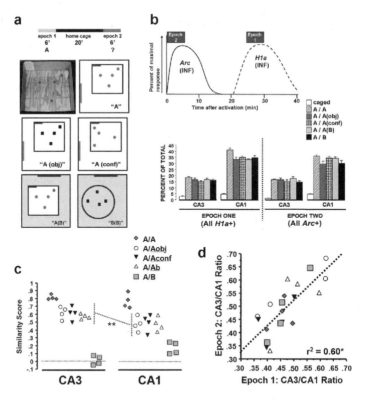

Figure 9.3. Response of hippocampal CA1 and CA3 ensembles to modifications of an environmental context. (a) Schematic representation of the time course (upper figure) and types of environments (lower figures) used in the experiment. Animals were allowed to freely explore the novel environment A (top row of panel b) for 6 minutes (epoch 1). After a 20-minute rest period in the home cage, the rats were exposed to either environment A again, or to one of the other four environments schematically represented in panel b (epoch 2). Three of these environments were modified A environments: in Aobj the local cues were changed from balls to cubes; in Aconf the same local cues (balls) were rearranged in a different cue configuration; and in Ab the entire local environment was moved into a different room with different distal cues. Environment B was in the second room and was as different as possible from environment A, yet contained the same number of local cues to ensure a similar level of complexity. (b) The approximate time course of detection for *Arc* and *H1a* INF and following neuronal activation. These estimated curves are based on data from (Guzowski et al., 1999; Vazdarjanova et al., 2002). The behavioral epochs for which *Arc* INF or *H1a* INF provide "readout" are shown above the curves. (c) Analysis of similarity scores of rats from the different behavioral groups. A similarity score of 1 indicates a complete overlap of the neuronal ensembles activated by epoch 1 and epoch 2 (no remapping), while a similarity score of 0 indicates no overlap beyond that predicted by chance (complete "remapping"). In both CA3 and CA1, the similarity scores from rats of the A/Aobj, A/Aconf, and A/Ab groups were significantly different from the A/A and A/B groups ($p < 0.02$), but were not different from each other ($p > 0.82$). Thus, in both CA3 and CA1 all alterations to environment A (here termed A′ and comprising from groups A/Aobj, A/Aconf, and A/Ab) produced a similar effect on the neural ensemble response in that cell population. Notably, the similarity score of the A/A′ rats was greater in CA3 as compared to CA1 (**, $p = 0.001$). (d) The within-rat ratio of CA3 to CA1 activity was strongly correlated across behavioral epochs. The fact that the CA3 and CA1 activity per epoch was not correlated across rats (not shown), but that the CA3/CA1 within rat ratio was correlated across sessions, suggests that the balance of CA3 to CA1 activity is highly specific to the individual rat. Panels (a), (c), and (d) are from Vazdarjanova and Guzowski (2004).

of research in this area. First, studies by Kentros and colleagues have shown that the stabilization of place field "maps" of CA1 neurons requires NMDA receptor function (Kentros et al., 1998) and protein synthesis (Agnihotri et al., 2004). In these two studies, the NMDA antagonist CPP or the protein synthesis inhibitor anisomycin did not affect the formation of place field maps or their stability at short intervals (i.e., 1 hr) when rats where exposed to a novel environment, but prevented the long-term stabilization of these maps. This NMDA receptor- and protein synthesis-dependent stabilization was evidenced when the rats were re-introduced to the environment 24 hours later and a new map was activated. In vehicle treated controls, hippocampal maps for the same environment were stabile at this 24 hour interval. Interestingly, neither of these drugs influenced the stability of map activation in familiar environments. Thus, stabilization of a place cell network representation of a new experience requires NMDA receptor activation and protein synthesis. These findings are consistent with the notion that expression of IEG proteins such as *Arc* and *H1a*, which require NMDA receptor activation for transcriptional activation (Lyford et al., 1995; Brakeman et al., 1997) and an intact translation apparatus for functional protein expression, is critical for the stabilization of place field maps. This hypothesis is supported by the finding that inhibition of *Arc* protein expression can block the maintenance phase of long-term potentiation (without affecting its induction) and impair long-term memory in a spatial task (without affecting learning or short-term memory) (Guzowski et al., 2000).

The advent of catFISH will facilitate advanced investigations into the nature of how information is processed within distinct circuits of the hippocampal formation. Because catFISH is a histological method, it will be possible to map neural activity for distinct experiences across many regions of the brain including the hippocampus proper as well as its efferent and afferent structures. Furthermore, combining catFISH with techniques such as pharmacological re-versible neural activation, investigators will be able to test directly hypotheses regarding functional neural circuits associated with specific behaviors. At present, a significant limitation of such broad investigations using catFISH has been caused by difficulties in quantifying gene expression in 3 dimensional space from confocal image stacks. This limitation may soon be overcome, however, with recent developments in automated, quantitative 3 dimensional imaging algorithms and software (Chawla et al., 2004; Lin et al., 2005).

References

Agnihotri NT, Hawkins RD, Kandel ER, Kentros C (2004). The long-term stability of new hippocam-pal place fields requires new protein synthesis. Proc Natl Acad Sci USA 101:3656-3661.

Bear MF, Malenka RC (1994). Synaptic plasticity: LTP and LTD. Curr Opin Neurobiol 4:389-399.

Bottai D, Guzowski JF, Schwarz MK, Kang SH, Xiao B, Lanahan A, Worley PF, Seeburg PH (2002). Synaptic activity-induced conversion of intronic to exonic sequence in Homer 1 immediate-early gene expression. J Neurosci 22:167-175.

Brakeman PR, Lanahan AA, O'Brien R, Roche K, Barnes CA, Huganir RL, Worley PF (1997). Homer: A protein that selectively binds metabotropic glutamate receptors. Nature 386:284-288.

Chawla MK, Lin G, Olson K, Vazdarjanova A, Burke SN, McNaughton BL, Worley PF, Guzowski JF, Roysam B, Barnes CA (2004). 3D-catFISH: a system for automated quantitative three-dimensional compartmental analysis of temporal gene transcription activity imaged by fluorescence in situ hybridization. J Neurosci Methods 139:13-24.

Cole AJ, Saffen DW, Baraban JM, Worley PF (1989). Rapid increase of an immediate early gene messenger RNA in hippocampal neurons by synaptic NMDA receptor activation. Nature 340:474-476.

Dudai Y (1997). Time to remember. Neuron 18:179-182.

Eichenbaum H (2001). The hippocampus and declarative memory: cognitive mechanisms and neural codes. Behav Brain Res 127:199-207.

Frankland PW, Bontempi B (2005). The organization of recent and remote memories. Nat Rev Neurosci 6:119-130.

Gothard KM, Skaggs WE, Moore KM, McNaughton BL (1996). Binding of hippocampal CA1 neural activity to multiple reference frames in a landmark-based navigation task. J Neurosci 16:823-835.

Guzowski JF (2002). Insights into immediate-early gene function in hippocampal memory consolidation using antisense oligonucleotide and fluorescent imaging approaches. Hippocampus 12:86-104.

Guzowski JF, Worley PF (2001). Cellular compartment analysis of temporal activity by fluorescence *in situ* hybridization (catFISH). In: Current Protocols in Neuroscience, pp. 1.8.1-1.8.16: John Wiley & Sons, Inc.

Guzowski JF, Knierim JJ, Moser EI (2004). Ensemble dynamics of hippocampal regions CA3 and CA1. Neuron 44:581-584.

Guzowski JF, McNaughton BL, Barnes CA, Worley PF (1999). Environment-specific expression of the immediate-early gene *Arc* in hippocampal neuronal ensembles. Nat Neurosci 2:1120-1124.

Guzowski JF, Lyford GL, Stevenson GD, Houston FP, McGaugh JL, Worley PF, Barnes CA (2000). Inhibition of activity-dependent arc protein expression in the rat hippocampus impairs the maintenance of long-term potentiation and the consolidation of long-term memory. J Neurosci 20:3993-4001.

Hess US, Lynch G, Gall CM (1995). Changes in c-fos mRNA expression in rat brain during odor discrimination learning: differential involvement of hippocampal subfields CA1 and CA3. J Neurosci 15:4786-4795.

Husi H, Ward MA, Choudhary JS, Blackstock WP, Grant SG (2000). Proteomic analysis of NMDA receptor-adhesion protein signaling complexes. Nat Neurosci 3:661-669.

Kentros C, Hargreaves E, Hawkins RD, Kandel ER, Shapiro M, Muller RV (1998). Abolition of long-term stability of new hippocampal place cell maps by NMDA receptor blockade. Science 280:2121-2126.

Kesner RP, Lee I, Gilbert P (2004). A behavioral assessment of hippocampal function based on a subregional analysis. Rev Neurosci 15:333-351.

Klann E, Antion MD, Banko JL, Hou L (2004). Synaptic plasticity and translation initiation. Learn Mem 11:365-372.

Kubie JL, Ranck JBJ (1983). Sensory-behavioral correlates in individual hippocampal neurons in three situations: space and context. In: Neurobiology of the Hippocampus (Seifert W, ed.), pp. 433-447. New York: Academic Press.

Lanahan A, Worley P (1998). Immediate-early genes and synaptic function. Neurobiol Learn Mem 70:37-43.

Lee I, Yoganarasimha D, Rao G, Knierim JJ (2004). Comparison of population coherence of place cells in hippocampal subfields CA1 and CA3. Nature 430:456-459.

Leutgeb S, Leutgeb JK, Treves A, Moser MB, Moser EI (2004). Distinct Ensemble Codes in Hippocampal Areas CA3 and CA1. Science.

Lin G, Chawla MK, Olson K, Guzowski JF, Barnes CA, Roysam B (2005). Hierarchical, model-based merging of multiple fragments for improved three-dimensional segmentation of nuclei. Cytometry A 63:20-33.

Link W, Konietsko U, Kauselmann G, Krug M, Schwanke B, Frey U, Kuhl D (1995). Somatodendritic expression of an immediate-early gene is regulated by synaptic activity. Proc Natl Acad Sci USA 92(12):5734-5738.

Lyford GL, Yamagata K, Kaufmann WE, Barnes CA, Sanders LK, Copeland NG, Gilbert DJ, Jenkins NA, Lanahan AA, Worley PF (1995). Arc, a growth factor and activity-regulated gene, encodes a novel cytoskeleton-associated protein that is enriched in neuronal dendrites. Neuron 14:433-445.

Malenka RC (1994). Synaptic plasticity in the hippocampus: LTP and LTD. Cell 78:535-538.

Markus EJ, Qin YL, Leonard B, Skaggs WE, McNaughton BL, Barnes CA (1995). Interactions between location and task affect the spatial and directional firing of hippocampal neurons. J Neurosci 15:7079-7094.

Marr D (1971). Simple memory: a theory for archicortex. Philos Trans R Soc Lond B Biol Sci 262:23-81.

McClelland JL, Goddard NH (1996). Considerations arising from a complementary learning systems perspective on hippocampus and neocortex. Hippocampus 6:654-665.

McGaugh JL (2000). Memory—a century of consolidation. Science 287:248-251.

McNaughton BL, Nadel L (1990). Hebb-Marr networks and the neurobiological representation of action in space. In: Neuroscience and Connectionist Theory (Gluck MA, Rumelhart DE, eds), pp. 1-64. Hillsdale, NJ: Erlbaum.

Morgan JI, Cohen DR, Hempstead JL, Curran T (1987). Mapping patterns of c-fos expression in the central nervous system after seizure. Science 237:192-197.

Nadel L, Moscovitch M (2001). The hippocampal complex and long-term memory revisited. Trends Cogn Sci 5:228-230.

Nakazawa K, McHugh TJ, Wilson MA, Tonegawa S (2004). NMDA receptors, place cells and hippocampal spatial memory. Nat Rev Neurosci 5:361-372.

O'Keefe J, Dostrovsky J (1971). The hippocampus as a spatial map. Preliminary evidence from unit activity in the freely-moving rat. Brain Res 34:171-175.

O'Keefe J, Nadel L (1978). The Hippocampus as a Cognitive Map. Oxford: Clarendon Press.

O'Reilly RC, Rudy JW (2001). Conjunctive representations in learning and memory: principles of cortical and hippocampal function. Psychol Rev 108:311-345.

Ribeiro S, Nicolelis MA (2004). Reverberation, storage, and postsynaptic propagation of memories during sleep. Learn Mem 11:686-696.

Rolls ET, Treves A (1998). Neural Networks and Brain Function. Oxford: Oxford University Press.

Skaggs WE, McNaughton BL (1992). Computational approaches to hippocampal function. Curr Opin Neurobiol 2:209-211.

Squire LR, Knowlton BJ (1994). Memory, hippocampus, and brain systems. In: The Cognitive Neurosciences (Gazzaniga M, ed.), pp. 825-837. Cambridge: MIT Press.

Steward O, Wallace CS, Lyford GL, Worley PF (1998). Synaptic activation causes the mRNA for the IEG Arc to localize selectively near activated postsynaptic sites on dendrites. Neuron 21:741-751.

Sutherland GR, McNaughton B (2000). Memory trace reactivation in hippocampal and neocortical neuronal ensembles. Curr Opin Neurobiol 10:180-186.

Suzuki WA, Eichenbaum H (2000). The neurophysiology of memory. Ann NY Acad Sci 911:175-191.

Thompson LT, Best PJ (1990). Long-term stability of the place-field activity of single units recorded from the dorsal hippocampus of freely behaving rats. Brain Res 509:299-308.

Tischmeyer W, Grimm R (1999). Activation of immediate early genes and memory formation. Cell Mol Life Sci 55:564-574.

Tolman EC (1932). Purposive Behavior in Animals and Men. New York: Century.

Treves A, Rolls ET (1994). Computational analysis of the role of the hippocampus in memory. Hippocampus 4:374-391.

Vazdarjanova A, Guzowski JF (2004). Differences in hippocampal neuronal population responses to modifications of an environmental context: evidence for distinct, yet complementary, functions of CA3 and CA1 ensembles. J Neurosci 24:6489-6496.

Vazdarjanova A, McNaughton BL, Barnes CA, Worley PF, Guzowski JF (2002). Experience-dependent coincident expression of the effector immediate-early genes *Arc* and *Homer 1a* in hippocampal and neocortical neuronal networks. J Neurosci 22:10067-10071.

Vinogradova OS (2001). Hippocampus as comparator: role of the two input and two output systems of the hippocampus in selection and registration of information. Hippocampus 11:578-598.

Wilson MA, McNaughton BL (1993). Dynamics of the hippocampal ensemble code for space. Science 261(5124):1055-1058.

Wood ER, Dudchenko PA, Eichenbaum H (1999). The global record of memory in hippocampal neuronal activity. Nature 397:613-616.

Xiao B, Tu JC, Worley PF (2000). Homer: a link between neural activity and glutamate receptor function. Curr Opin Neurobiol 10:370-374.

Young BJ, Fox GD, Eichenbaum H (1994). Correlates of hippocampal complex-spike cell activity in rats performing a nonspatial radial maze task. J Neurosci 14:6553-6563.

Yuan JP, Kiselyov K, Shin DM, Chen J, Shcheynikov N, Kang SH, Dehoff MH, Schwarz MK, Seeburg PH, Muallem S, Worley PF (2003). Homer binds TRPC family channels and is required for gating of TRPC1 by IP3 receptors. Cell 114:777-789.

Zola-Morgan S, Squire LR (1993). Neuroanatomy of memory. Annu Rev Neurosci 16:547-563.

10

Neuronal Dysfunction and Cognitive Impairment Resulting from Inactivation of the Egr-Family Transcription Factor zif268

SABRINA DAVIS, MEGAN LIBBEY and SERGE LAROCHE

Laboratoire de Neurobiologie de l'Apprentissage, de la Mémoire et de la Communication, CNRS UMR 8620, Université Paris-Sud, 91405 Orsay, France

1. Introduction

A leading theory explaining how the brain encodes, stores and represents information, postulates that memories for events are represented as specific spatio-temporal patterns of cellular activity within distributed networks, and that changes take place at the cellular level in the activated neurons to store these representations. This is based theoretically on the notion that activity-driven modifications of synaptic strength lead to functional and structural remodelling of neural networks such that the specific distribution of activity for a given memory can form a subsequent 'readout' pattern during recall. This prevailing view evolved out of the discovery, thirty years ago, of an enduring form of activity-dependent synaptic plasticity in the mammalian brain, known as long-term potentiation, or LTP (Bliss and Lomo, 1973). Since then, considerable attention has been directed towards determining the cellular and molecular mechanisms that underlie synaptic plasticity and its contribution to memory storage. The results of the past three decades of research have firmly established that the type of synaptic change that is brought about by LTP is a key player in memory function and, when dysfunctional, results in a dysfunction in memory processes, even if there are still issues about the exact role of LTP in memory storage that are the subject of intensive debate. In parallel, as a wealth of information concerning the underlying cellular and molecular mechanisms of LTP has been made available, the beginnings of a new era has emerged, exploring the cellular

R. Pinaud, L.A. Tremere (Eds.), Immediate Early Genes in Sensory Processing, Cognitive Performance and Neurological Disorders, 177–195, ©2006 Springer Science + Business Media, LLC

and molecular mechanisms of memory under both normal conditions and in disease states. This has been driven not only by the increasing knowledge of the molecular mechanisms of LTP, but also by the development of novel technologies specifically designed to exploit our means of analysing molecular changes within a behavioral context. Among the many important advances that have been made in discovering some of these molecular mechanisms, one of the most intriguing is the realisation that the mechanisms underlying the longer-lasting phases of both LTP and memory storage set in motion the genetic programs of neurons and result in the synthesis of new proteins. Evidence has accumulated over the past years, showing that the persistence of LTP over days as well as the stabilization of long-term, but not short-term, memory requires transcription of genes and *de novo* synthesis of proteins (Davis and Squire, 1984; Krug et al., 1984; Otani et al., 1989; Nguyen et al., 1994; Frey and Morris, 1997; Meiri et al., 1998). There is also much evidence that the induction of LTP and learning themselves induce transcriptional regulation of a variety of genes. However, although many genes have been shown to be up- and down-regulated in a finely tuned and co-ordinated manner in correlation with specific aspects of LTP and memory processes, the full genomic response of neurons associated with these processes is still largely unknown and is currently the subject of intensive investigation using newly available large-scale screening methods. An important issue in this context is to understand how synaptic events signal gene regulation and the synthesis of proteins. Capitalizing on studies investigating gene regulation mechanisms in response to a variety of stimuli in cell models, the search for molecules underlying synapse-to-nuclear signalling in LTP has provided important inroads into discovering molecular mechanisms of memory. In LTP, as in many other aspects of cell function involving regulated gene transcription, a critical step in triggering the genomic response of neurons is the expression of a class of immediate early genes (IEGs) encoding inducible transcription factors which interact with promoter regulatory elements of a host of downstream effector genes. This response generally occurs within minutes after membrane receptor activation and is mediated via the activation of multiple kinase cascades converging onto constitutively expressed transcriptional regulators. These, when phosphorylated in the nucleus, bind to specific response elements in the promoter region of IEGs and induce their expression. This constitutes an early genomic response required for triggering the mechanisms underlying persistent cell modification.

To date, several lines of evidence suggest that rapid regulation of the expression of IEGs can be a key mechanism underlying the enduring modification of neural networks required for the stabilisation of memories. At its most fundamental level, it suggests a molecular process implemented in the computations that neurons are capable of making, in order to change their functional properties and the properties of the circuits in which they are embedded, based on activation of transcriptional switches initiating the regulation of effector genes. To illustrate this molecular process and its potential role in long-term memory formation and synaptic plasticity, this article will focus on one such activity-dependent immediate early gene encoding a transcriptional regulator, *zif268*. *Zif268*, also known as *Egr-1*, *Krox-24*, *NGFI-A*, *Tis-8* or *Zenk*, is of particular interest in the context of neural plasticity and memory as its activation has been closely associated with NMDA receptor-dependent synaptic plasticity and with different forms of learning in a variety of animal species. Readers are referred to recent reviews for thorough survey of these studies (Abraham et al., 1991; O'Donovan et al., 1999; Tischmeyer and Grimm, 1999; Bozon et al., 2002; Guzowski, 2002; Davis et al., 2003; Knapska and Kaczmarek, 2004; Pinaud, 2004). In this chapter, we will limit ourselves to the consequences of *zif268* inactivation to neural and cognitive functions. We begin with a brief overview of the characteristics and properties of *zif268*. We then summarize results of several experiments in which we examined synaptic plasticity and learning and memory in mutant mice with a targeted inactivation of the *zif268* gene, and discuss these results together with those of more recent experiments using an antisense strategy in rats.

2. The Immediate Early Gene *zif268* and the *Egr* Family of Transcription Factors

The *zif268* gene was originally identified in the late 1980s as a nerve growth factor response gene product in PC12 cells and an immediate-early serum response gene product in fibroblasts (Milbrandt, 1987; Christy et al., 1988; Lemaire et al., 1988; Sukhatme et al., 1988). It belongs to a multi-gene family encoding the closely related transcription factors, *Egr-2* (*Krox-20*), *Egr-3* (*Pilot*) and *Egr-4* (*NGFI-C*), which have been implicated in diverse processes in a variety of cell types, including cell growth, differentiation and apoptosis in response to extracellular stimuli (Gashler and Sukhatme, 1995). The *zif268* gene encodes a 80–82 kDa protein containing a DNA binding domain with three zinc-finger motifs shared by all four members of the *Egr* family. The protein binds to cognate GC-rich sequences in the DNA containing the consensus 9-base-pair binding site GCG(G/T)GGGCG, the EGR response element, to regulate downstream expression of late-response genes (Christy and Nathans, 1989; Swirnoff and Milbrandt, 1995; O'Donovan et al., 1999). The gene is evolutionary conserved (Burmeister and Fernald, 2005); it is present in the mammalian genome as a single copy gene on chromosome 18 in the mouse. In brain, the mRNA and protein are expressed in several areas of the neocortex, hippocampus, entorhinal cortex, amygdala, striatum and cerebellar cortex (Christy et al., 1988; Mack et al., 1990; Worley et al., 1991). In the hippocampus, its expression gradually increases in the second week after birth and remains elevated in the CA regions, but it is only transient in the dentate gyrus where its constitutive expression fades after three weeks (Watson and Milbrandt, 1990; Herms et al., 1994). As an inducible transcription factor, *zif268* can be rapidly and transiently turned on by a variety of pharmacological and physiological stimuli including neurotransmitters, growth factors, peptides, depolarisation, seizures, ischemia and brain injury or cellular stress (for reviews see Gashler and Sukhatme, 1995; Beckmann and Wilce, 1997).

The origin of *zif268* research in neuronal plasticity can be traced back to the discovery in the early 1990s that the induction of LTP in the dentate gyrus of the hippocampus is associated with a rapid and robust transcription of *zif268* in the activated granule cells (Cole et al., 1989; Wisden et al., 1990). The induction of *zif268* mRNA occurs between ten minutes and two hours following LTP-inducing stimulation and is dependent on activation of the NMDA receptor. Among the many IEGs of the *Fos*, *Jun* or *Egr* family that are regulated after LTP, *zif268* has attracted particular attention because its transcription is reliably associated with the expression of the protein synthesis-dependent late phase of LTP, and appears to correlate more with the persistence than the induction of LTP (Abraham et al., 1991, 1993; Richardson et al., 1992; Worley et al., 1993; Dragunow, 1996). In addition to the expression of *zif268*, others members of the *Egr* family such as *Egr-2* and *Egr-3* are also up-regulated after LTP in the dentate gyrus (Yamagata et al., 1994; Williams et al., 1995). Basal expression of *zif268* mRNA/protein is rapidly and dramatically reduced in the brain by systemic administration of NMDA receptor antagonists as well as in the primary visual cortex after monocular deprivation or dark-adaptation, suggesting that naturally occurring afferent synaptic activity regulates the steady-state basal expression of *zif268* (Worley et al., 1991; Chandhuri et al., 1995). Characterization of the cellular signals implicated in LTP-dependent transcription of *zif268* in dentate granule cells has shown the requirement for MAPK/ERK activation and the current model suggests transcriptional regulation via MAPK/ERK-dependent phosphorylation of CREB and Elk-1, binding to CRE and SRE response elements present on the *zif268* promoter (Davis et al., 2000).

In parallel, as exemplified in several chapters of this volume and in many recent reviews (Tischmeyer and Grimm, 1999; Bozon et al., 2002; Guzowski, 2002; Davis et al., 2003; Knapska and Kaczmarek, 2004) a number of studies have provided evidence that the expression of *zif268* is sensitive to information gained following exposure to novel environments or in a learning context. For example, increases in the expression of *zif268* mRNA or protein has been found in specific brain structures after active avoidance learning (Nikolaev et al., 1992), brightness discrimination (Grimm and Tischmeyer, 1997), visual paired associate learning in monkeys (Okuno and Miyashita, 1996; Tokuyama et al., 2002), song learning in birds (Mello et al., 1992; Jarvis et al., 1995; Bolhuis et al., 2000), learning and/or retrieval of contextual and cued fear learning (Hall et al., 2000, 2001; Malkani and Rosen, 2000; Frankland et al., 2004; Weitemier and Ryabinin, 2004) and spatial learning (Guzowski et al., 2001). In general, as in synaptic plasticity paradigms, the induction of *zif268* occurs rapidly after learning or exposure to learning-associated cues, and is transient, suggesting a role in the transition from short- to long-term memory.

Although these correlational studies point to a possible role of *zif268* in processes of learning and memory formation, they do not determine whether *zif268* is necessary for these neural and cognitive functions and if it is, what functional role *zif268* plays in specific forms of plasticity or memory processes. Such issues were only approachable after methods for inactivating *zif268* expression *in vivo* were made available.

3. Neuronal Dysfunction Resulting from Inactivation of the *zif268* Gene

In the mid-1990s, two independent groups generated mutant mice with a null *zif268* allele (Lee et al., 1996; Topilko et al., 1998). The initial characterisation of these mice by the two groups revealed different degrees of endocrine defect associated with reduced reproductive capability in homozygous mice due to deficiency in the synthesis of luteinizing hormone β in the pituitary-gonadal axis and a slightly reduced body size, although heterozygous mice are phenotypically normal in terms of size and fertility. At that time, no other phenotype had been reported (for review see O'Donovan et al., 1999).

The evidence that *zif268* is required for both the expression of late LTP and for long-term memory emerged from collaborative experiments with Tim Bliss and his colleagues in which we examined LTP in the dentate gyrus and memory abilities of *zif268* mutant mice (Jones et al., 2001). In our experiments, we used the mutant mice generated by Topilko and colleagues (1998). The mutation involves the insertion of a *lacZ-neo* cassette between the promoter and the coding sequence to prevent the transcription of the gene, with the addition of a frame-shift mutation into the coding sequence at the level of a *Nde*I restriction site corresponding to the beginning of the DNA-binding motif (Topilko et al., 1998). The mice were generated in a hybrid 129SV/C57Bl/6J background, and were then backcrossed onto a C57Bl/6J background.

Histochemical examination of the homozygous mutant mice using cellular, neuronal and presynaptic markers confirmed similar cell densities and hippocampal architecture in wild-type and mutant mice, showing that disruption of *zif268* had no gross effects on hippocampal circuitry. Electrophysiological analyses in the dentate gyrus were first conducted in anesthetised mice. We found that basal synaptic transmission and forms of short-term plasticity such as paired-pulse facilitation and paired-pulse depression were normal, suggesting *zif268* inactivation does not affect synaptic transmission, cell excitability, recurrent inhibition and disinhibition, or forms of short-term presynaptic plasticity. LTP induced in the dentate gyrus by high-frequency stimulation of the perforant path and followed for one hour in anesthetised mice was also comparable in wild-type and homozygous mutant mice. In contrast, investigation of LTP over several days in awake, freely-moving mice confirmed that homozygous *zif268* mutant mice exhibited early LTP in the dentate gyrus for at least one hour, but 24 and 48 hours after induction, late LTP was absent in contrast to wild-type littermates in which LTP was maintained for at least 48 hours after induction. *In situ* hybridization studies confirmed the complete absence of *zif268* mRNA in the mutant mice, while the mRNA was normally expressed in cortex and CA1 and induced in the dentate gyrus after LTP in wild-type mice. In addition, the constitutive and LTP-inducible expression of the *lacZ* gene in the mutant mice was comparable to that of *zif268* in wild-type mice, suggesting that signaling events upstream of *zif268* transcription were not affected in the mutants. Interestingly, in heterozygous mice who show a level of *zif268* expression approximately half that of wild-type mice, LTP also decayed to baseline within 24 hours, suggesting that half the complement of *zif268* is insufficient to achieve the levels of *zif268* activation required for the maintenance of the late phases of LTP in the dentate gyrus. In all, these studies have provided clear evidence that the *zif268* gene is required for the stabilization of late-phase LTP in the dentate gyrus.

Very few studies have examined the effect of *zif268* inactivation on LTP in other brain regions or in other forms of plasticity. *Zif268* is up-regulated after the induction of LTP in structures such as the insular cortex following stimulation of the basolateral amygdala (Jones et al., 1999) and the visual cortex following tetanic stimulation of the thalamocortical tract (Heynen and Bear, 2001), but not after induction of LTP in CA1 *in vivo* (French et al., 2001). It is yet unknown, however, whether *zif268* is required for the expression of LTP in these structures or whether there is a regional or cell-type specificity for the role of *zif268* in synaptic plasticity. In CA1, one study of the *in vitro* hippocampal slice preparation suggested that LTP followed for 45 minutes is not affected in *zif268* mutant mice, but reported LTP blockade in one experiment with stronger tetanic stimuli (Wei et al., 2000). However, this observation should be taken with caution because of the complete absence of post-tetanic or short-term potentiation in this experiment compared to what was observed with weaker tetani. Three other forms of plasticity associated with *zif268* expression have been examined in *zif268* mutant mice, but no clear phenotype was observed. First, despite a strong induction of *zif268* in the suprachiasmatic nucleus by photic stimuli capable of inducing a phase shift in the circadian rhythm, entrainment of phase shifting of the circadian system is not affected in *zif268* mutant mice (Kilduff et al., 1998). Similarly, the development of kindling by amygdala stimulation, which is associated with over-expression of *zif268*, as well as the associated axonal sprouting of mossy fibres, is not affected in mice carrying a null mutation of *zif268*, suggesting that neither constitutive nor seizure-induced *zif268* expression is required for kindling development and seizure-associated synaptic reorganization (Zheng et al., 1998). Finally, Mataga and colleagues (2001) examined experience-dependent plasticity in the mouse visual cortex using *zif268* mutant mice with the mutant allele generated by Lee and colleagues (1996). They found that *zif268* expression increases in the visual cortex with eye opening at approximately P14. During the critical period (P24–28) as well as in the adult, photo-stimulation after a period of dark adaptation increases *zif268* expression, suggesting activity-dependent regulation of *zif268* in visual cortex. However, they showed ocular dominance plasticity was not affected in *zif268* mutant mice, suggesting that *zif268* is not required in developmental plasticity in this system.

Thus, several forms of plasticity associated with *zif268* expression are not impaired in *zif268* mutant mice and the only clear phenotype in terms of neural plasticity to date is a complete abolilition of the late phases of LTP in the dentate gyrus when *zif268* is inactivated. It is worth noting, however, that the three studies just presented showing no obvious phenotype used the mice generated by Lee and colleagues (1996) with a mutant allele distinct from that of the mice used by Jones et al. (2001) in which the presence of truncated proteins can be excluded and where the mutation appears more stringent (Topilko et al., 1998).

4. Memory Consolidation Deficits Resulting from Inactivation of the *zif268* Gene

In the experiments in which we examined LTP in *zif268* mutant mice, we also examined learning and memory performance in the mice using a variety of behavioral tasks making use of single or repeated training, different types of reinforcement, and the processing of spatial or non-spatial information, in a attempt to determine how important *zif268* is for the formation of different memories (Jones et al., 2001). We first examined short-term spatial memory by testing spontaneous non-reinforced alternation in a T-maze, an innate behavior that relies upon spatial working memory and depends on the integrity of the hippocampus. We found, with a 30-sec or a 10-min delay, that *zif268* mutant mice have unaltered short-term spatial memory. In contrast, a learning impairment was observed in a spatial reference memory task in the open field water maze using a massed training protocol. This deficit in long-term spatial memory formation was confirmed in a probe trial conducted 48 hours after training. We next examined performance in three non-spatial tasks in which learning occurs after a single trial or a brief training episode. A long-term memory deficit (at 24 and 48 hours) was observed in a conditioned taste aversion task. This is an associative task in which the mice learn to associate a novel taste; it does not require the integrity of the hippocampus but is dependent on structures such as the basolateral amygdala and the insular cortex. A similar long-term deficit in memory formation (24 hours) was also observed in a social transmission of food preference task, an olfactory discrimination task in which mice show a preference for a novel food that has been previously smelled on the breath of a demonstrator mouse. Importantly, short-term memory (tested at 10 min) was not compromised in this task. Finally, we assessed the role of *zif268* in recognition memory in an object recognition task, a behavioral paradigm widely used in humans to probe declarative memory (Manns et al., 2003). This task is rapidly learned and does not require explicit reinforcement; in rodents it is based on the spontaneous preference for novelty and the ability to remember previously encountered objects (Ennanceur and Delacour, 1988; Clark et al., 2000). In the standard task, rodents are placed in a small arena and briefly exposed to two different objects that they can explore freely. Then, after a variable delay interval, one object is replaced by a novel object. Normal animals prefer to explore the novel rather than the familiar object, thus demonstrating they remember the two objects they had previously had experience with. If the memory of the familiar objects has faded, however, they will spend an equal amount of time exploring the two objects. Again, although short-term recognition memory was not affected in *zif268* mutant mice, they were not able to form a long-term memory for the objects they have explored 24 hours earlier (Fig. 10.1). These results were replicated and extended in a following experiment (Bozon et al., 2002) in which we used a more complex recognition task allowing us to assess memory for objects and memory for the spatial configuration of objects independently. In this protocol, three distinct objects were used instead of two during the training phase, and a cue card was placed above one of the sidewalls of the open field to aid the mapping of the location of each object

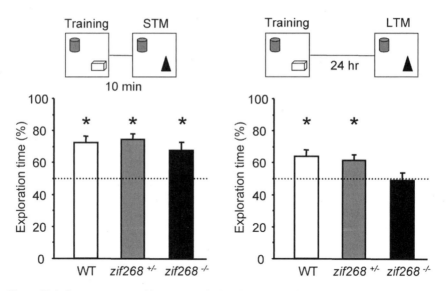

Figure 10.1. Long-term recognition memory is impaired in *zif268* mutant mice. Wild-type (WT), heterozygous ($zif268^{+/-}$) and homozygous ($zif268^{-/-}$) mutant mice were exposed to two objects and memory was tested 10 minutes or 24 hours later by replacing one object by a new one. The histograms represent the time spent exploring the novel object. Only $zif268^{-/-}$ mice had a long-term memory deficit (right). Short-term memory was intact (left). (Modified from Jones et al., 2001).

in space. The test consisted in either replacing one object by a new object, or by displacement of one familiar object to a new location in the arena. In this task, we found that *zif268* mutant mice were severely impaired in both object recognition memory and memory for the spatial location of objects (Fig. 10.2). Overall, these experiments provide evidence to suggest that *zif268* is required for the consolidation or expression of several types of long-term memories.

In all these memory tasks, we also examined the performance of *zif268* heterozygous mice, which carry half the complement of the gene, to examine whether there is a *zif268* gene dosage effect in the memory deficits. Surprisingly, we found the performance of heterozygous mice appeared to be labile. They were able to form a long-term memory as well as wild-type mice in the object recognition task (Figs 10.1 and 10.2); they showed a partial memory deficit in conditioned taste aversion and social transmission of food preference, as their performance fell halfway between that of wild-type and homozygous mice, and a retention deficit as severe as that of the homozygous mice in spatial navigation in the water maze and in the spatial version of the object recognition memory task (Jones et al., 2001; Bozon et al., 2002). These results indicate that there is not a simple relationship between gene dosage and behavioral performance and that, if a minimum threshold of activation of *zif268* is necessary for the consolidation of memories, this threshold may be different depending on task complexity or the type of memory at hand. They suggest that when *zif268* is not fully disrupted,

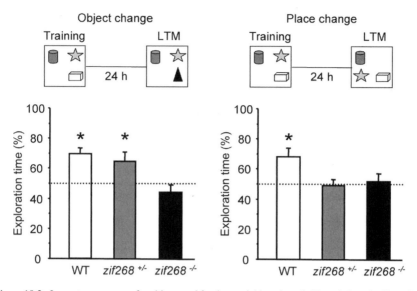

Figure 10.2. Long-term memory for objects and for the spatial location of objects is impaired in *zif268* mutant mice. The mice were exposed to three objects and memory was tested 24 hours later with either one familiar object replaced by a novel one (left) or one familiar object moved to a new location in the arena (right). The histograms represent the time spent exploring the novel or the displaced objects. *zif268*$^{-/-}$ homozygous mice were impaired in both object recognition memory and memory for the spatial location of objects, whereas only memory for the spatial location of objects was impaired in *zif268*$^{+/-}$ heterozygous mice. (Modified from Bozon et al., 2002).

memory deficits of increasing severity occur as the spatial demand of the task increases.

In 2004, two studies were published that examined the possible contribution of *zif268* to the formation of fear-associated memories (Lee at al., 2004; Malkani et al., 2004). These two groups used a *zif268* antisense oligonucleotide approach, a method that allowed the authors to directly assess the effect of interfering with Zif268 in an acute and spatially restricted manner in a specific brain structure. In the first experiment, Lee and colleagues (2004) infused *zif268* antisense oligonucleotides into the hippocampus 90 minutes prior to contextual fear conditioning in rats and found no impairment in long-term memory to the context associated with shock. It is unclear how much knock-down of Zif268 protein levels was induced in the hippocampus, but an analysis of protein content following retrieval, a procedure shown to increase Zif268 expression, showed a 66% reduction in Zif268 protein levels in area CA1, with no apparent reduction in the dentate gyrus. To the extent that a similar reduction in Zif268 protein levels was achieved when the antisense was injected prior to conditioning, these results suggest that Zif268 in area CA1 of the hippocampus is not required for the consolidation of contextual fear memory. However, it is possible that a 30–40% reduction in Zif268 content in CA1 was not sufficient to affect the consolidation of this type of memory. Using a similar approach, Malkani and colleagues (2004) infused *zif268* antisense oligonucleotides bilaterally into the amygdala prior to fear conditioning and found a dose-dependent impairment in the consolidation of contextual fear memory. In their experiment, biochemical analyses showed that there was a 11–25% reduction in Zif268 protein levels in the lateral nucleus of the amygdala. The results of these two studies suggest that a small reduction in Zif268 in the amygdala is sufficient to impair consolidation of contextual fear memory while a greater, but not complete reduction of Zif268 in area CA1 of the hippocampus has no adverse effect on memory associated with this task.

In summary, the findings to date show that the complete absence of *zif268* in null mutant mice spares short-term memory but results in a deficit in memory consolidation in many tasks such as object recognition memory, different types of spatial memory and associative olfactory and gustatory memories. Half the normal expression of *zif268* appears to affect the formation of these forms of memories much less, with the exception of a more severe impairment with increasing demand on the spatial component of the task. However, even a small amount of *zif268* inactivation in the amygdala is sufficient to severely impair consolidation of contextual fear memory, while it seems insufficient to impair this form of memory when restricted to the CA1 region of the hippocampus.

5. *Zif268* and the Reconsolidation of Memories

Recent research has revived interest in the possibility that previously consolidated memories, when recalled, become temporarily labile and may once again require stabilization necessary for their further long-term storage and availability for later recall (Misanin et al., 1968; Mactutus et al., 1979; Przybyslawski and Sara, 1997; Nader et al., 2000; Sara, 2000). This process, referred to as reconsolidation (for reviews see Nader, 2003; Alberini, 2005), evolved out of the conceptual framework proposed by Lewis (1979) who suggested that the memory trace can shift between two states: an inactive, stable, consolidated state corresponding to a stored memory and a transiently active state following both the initial encoding and the readout of the trace during retrieval. The implication is that when the memory is again converted into an active state following reactivation, a further storage process is required for the trace to remain in long-term memory and be available once again for recall. A central issue in the current literature is whether reconsolidation of recalled memories recruits the same mechanisms as those used during the initial consolidation process. The demonstration that well-consolidated memories, when reactivated, are vulnerable to systemic (Judge and Quartermain, 1982; Milekic and Alberini, 2002) or region-specific (Nader et al., 2000) protein synthesis inhibition in a similar manner as newly formed memories has provided direct evidence in favor of this hypothesis and has led to a model of cellular reconsolidation which posits that intracellular events necessary for the initial consolidation of memories are re-engaged after retrieval and required for later recall (Debiec et al., 2002; Myers and Davis, 2002). Although there is a substantial number of studies that have shown that inhibiting the synthesis of proteins with anisomycin blocks the reconsolidation process, there are few studies investigating the potential role of individual proteins and genes (for review see Alberini, 2005).

In our own studies (Bozon et al., 2003), we have used a recognition memory task in *zif268* mutant mice to test whether *zif268* is necessary for reconsolidation. However, in order to assess reconsolidation, consolidation must be shown to be unimpaired which, as described above, was not true for the *zif268* mutant mice. In an attempt to overcome this issue, we called on observations from previous experiments showing that, at least for spatial learning in the water-maze, the consolidation deficit can be overcome by extended and distributed training (Jones et al., 2001). We therefore attempted to overcome the consolidation deficit in recognition memory by extending the number of sessions of exploration of the objects and spacing the sessions over a wider interval of time, as was done with the water-maze training. When we tested retention 2 or 5 days after this extended and distributed training, we found that the deficit was completely overcome in the mutant mice, which showed as good a performance as that of wild-type littermates in preferentially exploring the novel object in the retention test (Fig. 10.3). With this protocol, we were able to assess the effect of retrieval interposed between training and retention. In a first series of experiments, memory for the objects was reactivated 24 hours after training by a brief re-exposure of the mice to the familiar objects, and retention was tested 10 minutes or 24 hours later. In this experiment, wild-type mice showed preferential exploration of the novel object both 10 minutes and 24 hours after reactivation. In contrast, *zif268* mutant mice were impaired when tested 24 hours after reactivation, while they were not 10 minutes after reactivation (Fig. 10.3). These results showing disrupted post-reactivation long-term memory and intact post-reactivation short-term memory provides evidence that *zif268* is required for reconsolidation of recognition memory.

To test the specificity of the effects of reactivation, we also examined post-reactivation performance when the context alone, or two completely different objects in the training context (pseudo-reactivation), or the two relevant 'target' objects in a completely different context, were presented during the reactivation session, and found in all cases that *zif268* mutant mice were not impaired 24 hours later (Fig. 10.4). Thus, *zif268*-dependent reconsolidation of recognition memory after retrieval requires the target memory to be actively reactivated. However, this cannot be achieved by contextual information alone in this task or by the target objects in an irrelevant context, but requires a match between the target items and the context within which they occurred.

In further experiments, we tested recognition memory 5 days after learning and examined the temporal constraints on the requirement for *zif268* in recon-solidation. To reactivate the memory the target objects were presented in the original training context, but this was done either 1 day or 4 days after learning. In both cases, post-reactivation performance of *zif268* mutant mice was impaired (Fig. 10.5). These results suggest that, within the limited time-scale that could be explored, there is no spontaneous recovery of the memory after reactivation when *zif268* is inactivated and the memory does not become immune to reconsolidation rapidly. Overall, these findings indicate that an emotionally neutral form of memory such as object recognition memory is subject to reconsolidation when

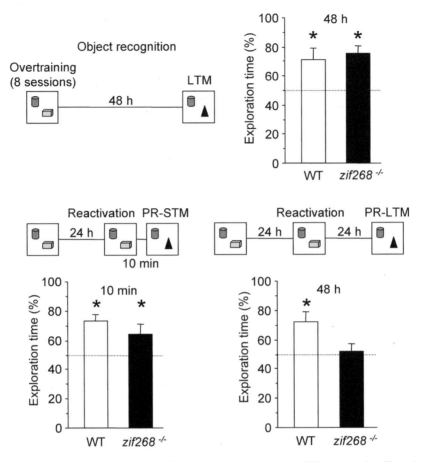

Figure 10.3. Reconsolidation of recognition memory is impaired in *zif268* mutant mice. The mice were exposed to eight briefly spaced training sesions (overtraining) and memory was tested two days later. In these conditions, *zif268*$^{-/-}$ mice had normal long-term recognition memory (top panel). When the *zif268*$^{-/-}$ mutant mice were briefly reexposed to the familiar objects 24 hours after training, post-reactivation short-term memory (PR-STM) was intact (bottom panel, left) but post-reactivation long-term memory (PR-LTM) was impaired (bottom panel, right). (Modified from Bozon et al., 2003).

recalled and they show that *zif268*-mediated transcriptional regulation in neurons is not required for retrieval, but is required for reconsolidation of recognition memory after retrieval (Bozon et al., 2003).

A role for *zif268* in reconsolidation was recently confirmed in another task, contextual fear conditioning, by Lee and colleagues (2004). In this task, reactivation of a consolidated fear memory by exposure to the context previously associated with shock is associated with an increase in the expression of *zif268* in several cortico-limbic structures (Hall et al., 2001; Thomas et al., 2002). As discussed above, Lee and colleagues (2004) injected *zif268* antisense oligonucleotides into the hippocampus before the learning of a context-shock association and found no effect on the expression of contextual fear memory. However, when

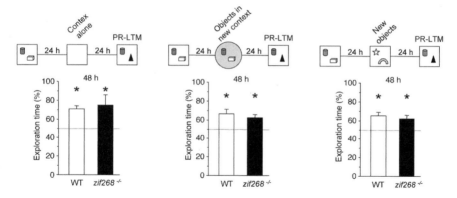

Figure 10.4. The impairement in post-reactivation reconsolidation of object recognition memory in *zif268* mutant mice is specific to reactivation of the target memory in the relevant context. Reexposure to either the context alone (left panel), the target objects in an entirely different context (middle panel), or two completely novel objects in the same context (right panel) had no effect on post-reactivation long-term recognition memory. (Modified from Bozon et al., 2003).

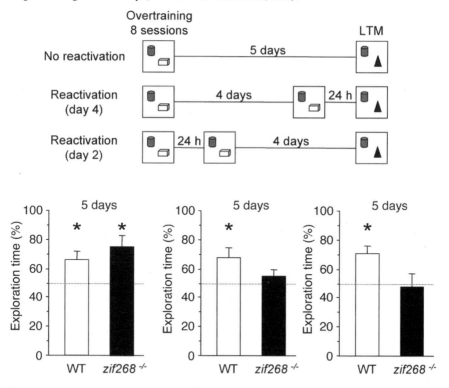

Figure 10.5. The impairement in reconsolidation of object recognition memory in *zif268* mutant mice is long-lasting and not temporally graded. With the overtraining paradigm, *zif268*[−/−] mice could form long-term recognition memory lasting 5 days (left panel). When reactivation was given 4 days after training and memory was tested 24 hours later (middle panel), or when reactivation was given 24 hours after training but memory was tested 4 days later (right panel), post-recativation long-term recognition memory was impaired in *zif268*[−/−] mutant mice. (Modified from Bozon et al., 2003).

they injected the antisense into the hippocampus just prior to reactivation of the memory, they found a profound impairment in post-reactivation long-term memory. They also reported that *zif268* antisense injection had no adverse effect on either retrieval *per se* or post-reactivation short-term memory. Although the inactivation of *zif268* in this study was not complete, it was sufficient to block reconsolidation of contextual fear memory and thus suggested an important role of hippocampal *zif268* activation for reconsolidation of this type of memory.

It is interesting to add that proteins of the MAPK/ERK cascade, a signaling pathway shown to be instrumental in triggering plasticity-dependent *zif268* expression in neurons (Davis et al., 2000), have also been shown to be implicated in memory reconsolidation. MAPK/ERK itself has been shown to be required for both consolidation and reconsolidation of object recognition memory (Kelly et al., 2003) and of fear memory (Duvarci et al., 2005), and the transcription factor CREB has been shown to be required for both consolidation and reconsolidation of fear memory (Kida et al., 2002).

6. Where to next?

To date, it is clearly apparent that the inducible-transcription factor *zif268* is necessary for the stabilization of late LTP in the dentate gyrus and that, for many types of memory, the absence of *zif268* prevents the normal consolidation process. This evidence therefore provides support for the hypothesis that rapid transcriptional regulation of *zif268* is one key molecular mechanism implicated in the stabilization of long-term memories. However, there is much information still to be gained to fully understand the role of *zif268* in this process. For example, *zif268* does not seem to be required in the hippocampus for the consolidation of contextual fear memory while it is for its reconsolidation (Lee et al., 2004), its activation is however required in the amygdala for consolidation of fear memory (Malkani et al., 2004). On the other hand *zif268* seems to be required both for consolidation and reconsolidation of recognition memory (Jones et al., 2001; Bozon et al., 2003). It beggars the question as to whether these differences reflect task-specificity, or process-specificity, or an interaction between both. Also, to what extent does this interaction depend on the type of memory and brain circuits involved in the laying down of these memories? Are there different levels and dynamics in the requirement for *zif268* transcriptional regulation, again depending on task complexity, type of memory, structures involved, and/or memory processes at hand? In an interesting recent study, Frankland and colleagues (2004) reported an increased expression of *zif268* in cortical regions such as the anterior cingulate, prelimbic, infralimbic and temporal cortices that was specific to remote (36 days) fear memory, and they showed that inactivation of the anterior cingulate cortex disrupted remote memories, but not recently acquired contextual fear memory. The early studies described here using methods to inactivate *zif268* can be used as scaffolding upon which to build new experiments to address issues concerning the requirement of *zif268* in different phases or time-dependent memory processes within different brain circuits. This will ultimately lead to a greater understanding of its role in acquisition, consolidation, retrieval and reconsolidation of information for different forms of memory.

Another important issue is that of functional compensation in the absence of *zif268*. As we have seen, the consolidation deficit in *zif268* null mice can be completely overcome by extended and distributed training in at least two paradigms: spatial learning in the water-maze and object recognition memory (Jones et al., 2001; Bozon et al., 2003). One core hypothesis is that *zif268* may be required only for *efficient* learning and consolidation, but, in the complete absence of *zif268*, other mechanisms may be recruited if the behavioral conditions are favorable. What are these mechanisms underlying the compensation, how do they come into play and why are they non-operative in standard training conditions? We have postulated elsewhere (Bozon et al., 2003) that other members of the *Egr* family might be good candidates. This is based on the similarity in their regional expression, the differential dynamics of regulation of *Egr*-family members after the induction of LTP (Williams et al., 1995) and, more fundamentally perhaps, the high homology in the zinc finger proteins they encode and therefore the likelihood that they would activate a similar set of downstream effector genes containing consensus EGR-binding motifs in their promoter regions (Swirnoff and Milbrandt, 1995). In this sense, the analysis of the consequences of inactivating other *Egr*-family members would provide valuable information for understanding the functional role of this family of transcription factors.

In the same way, at the cellular and molecular levels, are *zif268* and other EGR-family members similarly required for LTP in all brain structures, or is there a degree of regional or cell-type specificity? What other mechanisms of functional or structural plasticity do they participate in? What are the specific programs of *Egr*-dependent gene expression that are set into play in a finely tuned and co-ordinated manner with different types of neural activity or behavioral situations, and if the programs are specific, how are the instructions given and processed? All *Egr* target genes, including *zif268* itself, contain one or more EGR response elements in their promoter region, but they also harbor other regulatory sites that may contribute to or limit their expression. Based on the current understanding, it appears likely that *Egr* target genes are not all activated similarly in all circumstances as soon as one *Egr* transcription factor is expressed or overexpressed. Given the complexity of these regulatory mechanisms, the number of IEGs encoding inducible transcription factors that may come into play together with the *Egr* family members, and the rich cross-talk and feedback potential, research in these aspects of the functional role of transcriptional mechanisms for memory formation will certainly be a difficult task, but should benefit from the refinements in large-scale genomic and proteomic approaches.

Finally, the increasing evidence for an important role of *zif268*-mediated transcription in neural plasticity and cognitive functions may have implications in the context of the disease states of brain function. For example, studies have reported a reduced expression of *zif268* in the aging brain (Yau et al., 1996; Desjardins et al., 1997) as well as in mice models of Huntington's disease (Spektor et al., 2002) and Alzheimer's disease (Dickey et al., 2003). Taking a wider perspective, a more detailed knowledge of the mechanisms of transcriptional control of activity-dependent changes in genomic programs of neurons underlying enduring modification of neural networks required for the stability of memories may offer the prospect of developing strategies to intervene against neuronal malfunction and disease.

Acknowledgements

Supported by grants from the Human Frontier Science Programme (HFSP RGP0040/2002-C) to SL and from the French Ministry of Research ACI Neurosciences Intégratives et Computationnelles to SD.

References

Abraham WC, Dragunow M, Tate WP (1991). The role of immediate early genes in the stabilization of long-term potentiation. Mol Neurobiol 5:297-314.

Abraham WC, Mason SE, Demmer J, Williams JM, Richardson CL, Tate WP, Lawlor PA, Dragunow M (1993). Correlations between immediate early gene induction and the persistence of long-term potentiation. Neuroscience 56:717-727.

Alberini CM (2005). Mechanisms of memory stabilization: are consolidation and reconsolidation similar or distinct processes? Trends Neurosci 26:51-56.

Beckman AM, Wilce PA (1997). Egr transcription factors in the nervous system. Neurochem Int 31:477-510.

Bliss TVP, Lomo T (1973). Long-lasting potentiation of synaptic transmission in the dentate area of the anaesthetized rabbit following stimulation of the perforant path. J Physiol 232:331-356.

Bolhuis JJ, Zijlstra GO, den Boer-Visser AM, Van der Zee EA (2000). Localised neuronal activation in the zebra finch brain is related to the strength of song learning. Proc Natl Acad Sci USA 97:2282-2285.

Bozon B, Davis S, Laroche S (2002). Regulated transcription of the immediate-early gene Zif268: mechanisms and gene dosage-dependent function in synaptic plasticity and memory formation. Hippocampus 12:570-577.

Bozon B, Davis S, Laroche S (2003). A requirement for the immediate early gene zif268 in reconsolidation of recognition memory after retreival. Neuron 40:695-701.

Burmeister SS, Fernald RD (2005). Evolutionary conservation of the egr-1 immediate-early gene response in a teleost. J Comp Neurol 48:220-232.

Chandhuri A, Matsubara JA, Cynader MS (1995). Neuronal activity in primate visual cortex assessed by immunostaining for the transcription factor Zif268. Vision Neurosci 12:35-50.

Christy BA, Lau LF, Nathans D (1988). A gene activated in mouse 3T3 cells by serum growth factors encodes a protein with "zinc finger" sequences. Proc Natl Acad Sci USA 85:7857-7861.

Christy BA, Nathans D (1989). DNA binding site of the growth factor-inducible protein Zif268. Proc Natl Acad Sci USA 86:8737-8741.

Clark RE, Zola SM, Squire LR (2000). Impaired recognition memory in rats after damage to the hippocampus. J Neurosci 20:8853-8860.

Cole AJ, Saffan DW, Baraban JM, Worley PF (1989). Rapid increase if an immediate early gene messsenger RNA in hippocampal neurons by synaptic NMDA receptor activation. Nature 340:474-476.

Davis HP, Squire LR (1984). Protein synthesis and memory. A review. Psychol Bull 95:518-559.

Davis S, Vanhoutte P, Pages C, Caboche J, Laroche S (2000). The MAPK/ERK cascate targets both Elk-1 and cAMP response element-binding protein to control long-term potentiation-dependent gene expression in the dentate gyrus in vivo. J Neurosci 20:4563-4572.

Davis S, Bozon B, Laroche S (2003). How necessary is the activation of the immediate early gene zif268 in synaptic plasticity and learning? Behav Brain Res 142:17-30.

Debiec J, LeDoux JE, Nader K (2002). Cellular and systems reconsolidation in the hippocampus. Neuron 36:527-538.

Desjardins S, Mayo W, Vallee M, Hancock D, Le Moal M, Simon H, Abrous DN (1997). Effect of aging on the basal expression of c-Fos, c-Jun, and Egr-1 proteins in the hippocampus. Neurobiol Aging 18:37-44.

Dickey CA, Loring JF, Montgomery J, Gordon MN, Eastman PS, Morgan D (2003). Selectively reduced expression of synaptic plasticity-related genes in amyloid precursor protein + presenilin-1 transgenic mice. J Neurosci 23:5219-5226.

Dragunow M (1996). A role for immediate early transcription factors in learning and memory. Behav Genetics 20:293-299.

Duvarci S, Nader K, LeDoux JE (2005). Activation of extracellular signal-regulated kinase- mitogen-activated protein kinase cascade in the amygdala is required for memory reconsolidation of auditory fear conditioning. Eur J Neurosci 21:283-289.

Ennanceur A, Delacour J (1988). A new one-trial test for neurobiological studies of memory in rats: 1. Behavioural data. Behav Brain Res 100:85-92.

Frankland PW, Bontempi B, Talton LE, Kaczmarek L, Silva AJ (2004). The involvement of the anterior cingulate cortex in remote contextual fear memory. Science 304:881-883.

French PJ, O'Connor V, Jones MW, Davis S, Errington ML, Voss K, Truchet B, Wotjak C, Stean T, Doyere V, Maroun M, Laroche S, Bliss TVP (2001). Subfield-specific immediate early gene expression associated with hippocampal long-term potentiation in vivo. Eur J Neurosci 13:868-976.

Frey U, Morris RG (1997). Synaptic tagging and long-term potentiation. Nature 385:533-536.

Gashler A, Sukhatme VP (1995). Early growth-response protein 1 (Egr-1): prototype of a zinc-finger family of transcription factors. Prog Nucl Acid Res Mol Biol 50:191-224.

Grimm R, Tishmeyer W (1997). Complex patterns of immediate early gene induction in rat brain following brightness discrimination training and pseudotraining. Behav Brain Res 84:109-116.

Guzowski JF, Setlow B, Wagner EK, McGaugh JL (2001). Experience-dependent gene expression in the rat hippocampus after spatial learning: a comparison of the immediate-early genes Arc, c-fos, and zif268. J Neurosci 21:5089-5098.

Guzowski JF (2002). Insights into immediate-early gene function in hippocampal memory consolidation using antisense oligonucleotide and fluorescent imaging approaches. Hippocampus 12:86-104.

Hall J, Thomas KL, Everitt BJ (2000). Rapid and selective induction of BDNF expression expression in the hippocampus during contextual learning. Nat Neurosci 3:533-535.

Hall J, Thomas KL, Everitt BJ (2001). Cellular imaging of zif268 expression in the hippocampus and amygdala during contextual and cued fear memory retreival: selective activation of hippocampal CA1 neurons during the recall of contextual memories. J Neurosci 21:2186-2193.

Herms J, Zurmöhle U, Schlingensiepen R, Brysch W, Schlingensiepen KH (1994). Developmental expression of the transcription factor zif268 in rat brain. Neurosci Lett 165:171-174.

Heynen AJ, Bear MF (2001). Long-term potentiation of thalamocortical transmission in the adult visual cortex in vivo. J Neurosci 15:9801-9813.

Jarvis ED, Mello CV, Nottebohm F (1995). Associative learning and stimulus novelty influence in the song-induced expression of an immediate early gene in the canary forebrain. Learn Memory 2:62-80.

Jones MW, French PJ, Bliss TVP, Rosenblum K (1999). Molecular mechanisms of long-term potentiation in the insular cortex in vivo. J Neurosci 19:RC36.

Jones MW, Errington ML, French PJ, Fine A, Bliss TV, Garel S, Charnay P, Bozon B, Laroche S, Davis S (2001). A requirement for the immediate early gene Zif268 in the expression of late LTP and long-term memories. Nat Neurosci 4:289-296.

Judge ME, Quartermain D (1982). Characteristics of retrograde amnesia following reactivation of memory in mice. Physiol Behav 28:585-590.

Kelly A, Laroche S, Davis S (2003). Activation of mitogen-activated protein kinase/extracellular signal-regulated kinase in hippocampal circuitry is required for consolidation and reconsolidation of recognition memory. J Neurosci 23:5354-5360.

Kida S, Josselyyn SA, Pena de Ortiz S, Kogan JH, Chevere I, Masushige S, Silva AJ (2002). CREB required for the stability of new and reactivated memories. Nat Neurosci 5:248-355.

Kilduff TS, Vugrinic C, Lee SL, Milbrandt JD, Mikkelsen JD, O'Hara BF, Heller HC (1998). Characterization of the circadian system of NGFI-A and NGFI-A/NGFI-B deficient mice. J Biol Rhythms 13:347-357.

Knapska E, Kaczmarek L (2004). A gene for neuronal plasticity in the mammalian brain: Zif268/Egr-1/NGFI-A/Krox-24/TIS8/ZENK? Prog Neurobiol 74:183-211.

Krug M, Lossner, B, Ott T (1984). Anisomycin blocks the late phase of LTP in the dentate gyrus of freely moving rats. Brain Res Bull 13:39-42.

Lee SL, Sadovsky Y, Swirnoff AH, Polish JA, Goda P, Gavrinian G, Milbrandt J (1996). Luteininzing hormone deficiency and female infertility in mice lacking the transcription factor NGF1-A (Egr-1). Science 273:1219-1221.

Lee JL, Everitt BJ, Thomas KL (2004). Independent cellular processes for hippocampal memory consolidation and reconsolidation. Science 304:839-843.

Lemaire P, Revelant O, Bravo R, Charnay P (1988). Two mouse genes encoding potential transcription factors with identical DNA-binding domains are activated by growth factors in cultured cells. Proc Natl Acad Sci USA 85:4691-4695.

Lewis DJ (1979). Psychobiology of active and inactive memory. Psychol Bull 86:1054-1083.

Mack K, Day M, Milbrandt J, Gottlieb DI (1990). Localization of the NGF1-A protein in the rat brain. Mol Brain Res 8:177-180.

Mactus CF, Ricco DC, Ferek JM (1979). Retrograde amnesia for old (reactivated) memory:some anomalous characteristics. Science 204:1319-1320.

Malkani S, Rosen JB (2000). Specific induction of early growth response gene 1 in the lateral nucleus of the amygdalal following contextual fear conditioning in rats. Neuroscience 97:693-702.

Malkani S, Wallace KJ, Donley MP, Rosen JB (2004). An egr-1 (zif268) antisense oligodeoxynucleotide infused into the amygdala disrupts fear conditioning. Learn Memory 11:617-624.

Manns JR, Hopkins RO, Reed JM, Kitchener EG, Squire LR (2003). Recognition memory and the human hippocampus. Neuron 37:171-180.

Mataga N, Fujishima S, Condie BG, Hensch TK (2001). Experience-dependent plasticity of mouse visual cortex in the absence of the neuronal activity-dependent marker egr1/zif268. J Neurosci 21:9724-9732.

Meiri N, Rosenblum K (1998). Lateral ventricle injection of the protein synthesis inhibitor anisomycin impairs long-term memory in a spatial memory task. Brain Res 789:48-55.

Mello CV, Vicario DS, Clayton DF (1992). Song presentation induces gene expression in the sondbird forebrain. Proc Natl Acad Sci USA 89:6818-6822.

Milbrandt J (1987). A nerve growth factor-induced gene encodes a possible transcriptional regulatory factor. Science 238:797-799.

Milekic MH, Alberini CM (2002). Temporally graded requirement for protien synthsisi following memory reactivation. Neuron 36:521-525.

Misanin JR, Miller RR, Lewis DJ (1968). Retrograde amnesia produced by electroconvulsive shock after ractivation of consolidated memory trace. Science 160:554-555.

Myers KM, and Davis M (2002). System-level reconsolidation: reengagement of the hippocampus with memory reactivation. Neuron 36:340-343.

Nader K, Schafe GE, LeDoux JE (2000). Fear memories require protein synthesis in the amygdala for reconsolidation after retrieval. Nature 406:722-726.

Nader K (2003). Memory traces unbound. Trends Neurosci 26:65-72.

Nguyen PV, Abel T, Kandel ER (1994). Requirement of a critical period of transcription for induction of a late phase of LTP. Science 265:1104-1107.

Nikolaev E, Kaminska B, Tishmeyer W, Matthies H, Kaczmarek L (1992). Induction of expression of genes encoding transcription factors in the rat brain elicted by gehavioural traning. Brain Res Bull 28:479-484.

O'Donovan KJ, Tourtellotte WG, Milbrandt J, Baraban JM (1999). The EGR family of transcription-regulatory factors : progress at the interface of molecular and systems neuronscience. Trends Neurosci 22:167-173.

Okuno H, Miyashita Y (1996). Expression of the transcription factor Zif268 in the temoral cortex of monkeys during visual paired associate learning. Eur J Neurosci 8:2118-2128.

Otani S, Marshall CJ, Tate WP, Goddard GV, Abraham WC (1989). Maintenance of LTP in rat dentate gyrus requires protein synthesis but not mRNA synthesis immediately post-tetanization. Neuroscience 28:519-526.

Pinaud R (2004). Experience-dependent immediate early gene expression in the adult central nervous system: evidence from enriched-environment studies. Int J Neurosci 114:321-333.

Przybyslawski J, Sara S (1997). Reconsolidation of memory after its reactivation. Behav Brain Res 84:241-246.

Richardson CL, Tate WP, Mason SE, Lawlor PA, Dragunow M, Abraham WC (1992). Correlation between the induction of an immediate early gene, zif/268, and long-term potentiation in the dentate gyrus. Brain Res 580:147-154.

Sara S (2000). Retrieval and reconsolidation toward a neurobiology of remembering. Learn Memory 7:73-84.

Spektor BS, Miller DW, Hollingsworth ZR, Kaneko YA, Solano SM, Johnson JM, Penney JB Jr, Young AB, Luthi-Carter R (2002). Differential D1 and D2 receptor-mediated effects on immediate early gene induction in a transgenic mouse model of Huntington's disease. Mol Brain Res 102:118-128.

Sukhatme VP, Cao XM, Chang LC, Tsai-Morris CH, Stamenkovich D, Ferreira PC, Cohen DR, Edwards SA, Shows TB, Curran T (1988). A zinc finger-encoding gene coregulated wit hc-fos during growth and differentiation, and after cellular depolarization. Cell 53:37-43.

Swirnoff AH, Milbrandt J (1995). DNA-binding specificity of NGF1-A and related zinc finger transcription factors. Mol Cell Biol 15:2275-2287.

Thomas KL, Hall J, Everitt BJ (2002). Cellular imaging with zif268 expression in the rat nucleus accumbens and the frontal cortex further dissociated the neural pathways activated following the retrieval of contextual and cued memory. Eur J Neurosci 16:1789-1996.

Tishmeyer W, Grimm R (1999). Activation immediate early genes and memory formation. Cell Mol Life Sci 55:564-574.

Tokuyama W, Okuno H, Hashimoto T, Li YX, Miyashita Y (2002). Selective zif268 mRNA induction in the perirhinal cortex of macaque monkeys during formation of visual pair-association memory. J Neurochem 81:60-70.

Topilko P, Schneider-Maunoury S, Levy G, Trembleau A, Gourdji D, Driancourt MA, Rao CV, Charnay P (1998). Multiple pituitary and ovarian defcts in krox-24 (NGF1-A, Egr-1)-targeted mice. Mol Endocrinol 12:107-122.

Watson MA, Milbrandt J (1990). Expression of the nerve growth factor-regulated NGF1-A and NGF1-B genes in the developing rat. Development 110:173-183.

Wei F, Xu ZC, Qu Z, Milbrandt J, Zhuo M (2000). Role of EGR1 in hippocampal synaptic enhancement induced by tetanic stimulation and amputation. J Cell Biol 149:1325-1334.

Weitemier AZ, Ryabinin AE (2004). Subregion-specific differences in hippocampal activity between Delay and Trace fear conditioning: an immunohistochemical analysis. Brain Res 995:55-65.

Williams J, Dragunow M, Lawlor PA, Mason SE, Abraham WC, Leah J, Bravo R, Demmer J, Tate W (1995). Krox20 may play a key role in the stabilisation of long-term potentiation. Mol Brain Res 28:87-93.

Wisden W, Errington ML, Williams S, Dunnett SB, Waters C, Hitchcock D, Evans G, Bliss TV, Hunt SP (1990). Differential expression of immediate early genes in the hippocampus and spinal cord. Neuron 4:603-614.

Worley PF, Christy BA, Nakabeppu Y, Bhat RV, Cole AJ, Baraban JM (1991). Constitutive expression of zif268 in neocortex is regulated by synaptic activity. Proc Natl Acad Sci USA 88:5106-5110.

Worley PF, Bhat RV, Baraban JM, Erickson CA, McNaughton BL, Barnes CA (1993). Thresholds of synaptic activation of transcription factors in hippocampus: correlation with long-term enhancement. J Neurosci 13:4776-4786.

Yamagata K, Kaufmann WE, Lanahan A, Papapavlous M, Barnes CA, Andreasson KI, Worley PF (1994). Egr/Pilot Egr-1, a zinc finger transcription factor, is rapidly regulated by activity in brain neurons and colocalize with Egr-1/zif268. Learn Memory 1:140-152.

Yau JL, Olsson T, Morris RG, Noble J, Seckl JR (1996). Decreased NGFI-A gene expression in the hippocampus of cognitively impaired aged rats. Mol Brain Res 42:354-357.

Zheng D, Butler LS, McNamara JO (1998). Kindling and associated mossy fibre sprouting are not affected in mice deficient of NGFI-A/NGFI-B genes. Neuroscience 83:251-258.

PART III

Immediate Early Genes in Neurological Disorder: Clinical Implications

11

The Contribution of Immediate Early Genes to the Understanding of Brain Processing of Stressors

ANTONIO ARMARIO

Institut de Neurociències and Unitat de Fisiologia Animal (Facultat de Ciències),
Departament de Biologia Cellular, de Fisiologia i d'Inmunologia,
Universitat Autònoma de Barcelona, Spain

1. Introduction

1.1. Concept of Stress and Prototypical Physiological Changes

Exposure of animals to certain stimuli considered as stressful causes a wide range of behavioral and physiological changes that are orchestrated by the brain. Two of the prototypical physiological responses to stressful stimuli (stressors) are the activation of the hypothalamic-pituitary-adrenal (HPA) and sympatho-medullo-adrenal (SMA) axes. The activation of the HPA axis is initiated in the parvocellular region of the paraventricular nucleus of the hypothalamus (PVN), where projection neurons that target the pituitary-portal blood of the external median eminence are located. Upon stimulation, these neurons release hypothalamic factors (the main factors being the corticotropin-releasing factor—CRF—, and vasopressin) that stimulate the synthesis and release of adrenocorticotropin (ACTH) from the anterior pituitary. The main action of circulating ACTH is to stimulate the synthesis and secretion of glucocorticoids by the adrenal cortex. The activation of the SMA axis involves signals arising from different brain areas, including the PVN, and targeting sympathetic preganglionic nuclei located in the spinal cord, which in turn affect numerous physiological functions (e.g. activation of catecholamine-synthesizing chromaphin cells of the adrenal medulla). Typical consequences of sympathetic activation are increases in plasma levels of catecholamines (noradrenaline and adrenaline), blood pressure and skin conductance. The HPA and SMA axes are of great interest to investigators interested in the neurobiological basis of stress responses for

R. Pinaud, L.A. Tremere (Eds.), Immediate Early Genes in Sensory Processing, Cognitive Performance and Neurological Disorders, 199–221, ©2006 Springer Science + Business Media, LLC

several reasons: (a) they are activated by a wide range of stimulus considered as stressful; (b) the strength of their activation is proportional to the intensity of the stressful situation; and (c) they are implicated in most stress-associated pathologies (e.g., cardiovascular problems, immune suppression, drug addiction and affective disorders). Of particular importance is the fact that plasma levels of catecholamines (especially adrenaline) and ACTH (more so than glucocorticoids) are reasonably good markers of the intensity of stress experienced by animals (Martí and Armario, 1998; Márquez et al., 2002). In striking contrast, most of the physiological changes related to exposure to stressors, irrespective of their intensity, are of the same magnitude and, therefore, are not considered to be good stress markers (Armario et al., 1986).

According to the original definition proposed by Selye (1936), a stressor is any stimulus able to alter homeostasis, a term coined by Cannon (1929) to refer to the complex set of physiological mechanisms that allow organisms to adapt to environmental challenges (physical or biological) and to maintain certain critical physiological parameters within a narrow operational range (e.g., glycaemia, osmolality, core temperature). Recently, McEwen (2000) has introduced into the stress literature the use of the term allostasis to refer to the mechanisms used to stabilize critical variables despite environmental challenges. In addition, it is presently more accepted that the stress response is activated when challenges to homeostasis are strong (e.g., exhaustive exercise, infection, severe hemorrhage) and demand a special physiological response to allow survival. However, we also know that exposure of animals to certain stressful stimuli, such as the odor of a predator or to the predator itself, intra-specific fighting (with no injury) or restraint, cause emotional activation and a prototypical stress response, but do not have a direct impact on critical physiological variables. It is likely that emotional stressors elicit common physiological responses with systemic stressors because in nature the probability of physical demand (flight/fight) or injury is high under these situations. Consequently, there was probably an evolutionary convergence of the brain pathways activated by both systemic and emotional stressors into a response system.

1.2. Immediate Early Genes (IEG): Usefulness and Limitations in Stress Studies

Exposure to stressors is usually followed by dramatic behavioral and physiological changes. Therefore, information concerning the stimuli needs to be extensively processed within the brain and this involves electrophysiological activation of several important groups of neurons. The utility of IEGs is based on the evidence that their expression is linked with neuronal depolarization (Shen and Greenberg, 1990). This feature allows for the characterization of specific brain areas that are activated by different types of stressors. Most of the well-characterized IEGs, such as *c-fos* and *zif-268*, code for proteins that can act as transcription factors (reviewed in Herdegen and Leah, 1998). However, the proteins encoded by a few IEGs likely have a non-genomic and direct role in cellular physiology and are therefore referred to as effector IEGs. Among the effectors

IEGs, this review will focus on *arc* (activity-regulated cytoskeleton-associated protein) (Lyford et al., 1995; Steward et al., 1998; Steward and Worley, 2001, 2002; Pinaud et al., 2001). Before we proceed to discuss in detail the analysis of stress-induced brain IEG expression, we will consider some methodological aspects that can drastically affect the interpretation of the results obtained with this methodology.

1.2.1. Special Considerations   First, IEG expression is often measured with *in-situ* hybridization (that targets specific mRNAs encoded by IEGs) or immuno-cytochemical approaches (that targets the protein products generated by IEG expression). In particular with radioactive *in situ* hybridization (ISH) procedures, precise localization of mRNAs on film autoradiographies is sometimes difficult, particularly when complex anatomical regions are studied, as is the case of the bed nucleus of stria terminalis (BNST, intimately associated to the function of the amygdala). Anatomical resolution may be improved by using emulsion-dipped sections in radioactive ISH procedures, by the use of non-radioactive ISH or immunohistochemical (IHC) methods. IHC procedures rely on the specificity of the antibodies used and this may be particularly important in the case of *c-fos*, as some antibodies recognize epitopes that are shared among several members of family.

Second, is the activation of IEGs in response to stress restricted to neurons? Although glial cells express *c-fos* and other IEGs *in vivo* during brain injury and infection (e.g. Nitsch and Frotscher, 1992; Tchélingérian et al., 1997; Rubio and Martin-Clemente, 1999; Proescholdt et al., 2002), it is commonly assumed that brain *c-fos* expression is restricted to neurons after most common stressors. However, glial *c-fos* expression should not be ruled out without direct testing after exposure to some systemic stressors. For example, direct administration of hypertonic saline to the supraoptic nucleus of the hypothalamus (SON) caused *c-fos* induction in astrocytes, but not in magnocellular neurons, whereas the opposite was found after systemic administration of hypertonic saline (Ludwig et al., 1997). In addition, acute administration of glucose into the carotid artery was found to induce *c-fos* expression in both neurons and astrocytes of the arcuate hypothalamic nucleus (Guillod-Maximim et al., 2004).

Third, is IEG expression a reliable marker of neuronal depolarization? There is evidence that *in vivo* enhanced *c-fos* expression is only observed by the joint action of neuronal depolarization and synaptic inputs (Luckman et al., 1994). In addition, c-fos expression has been shown to be dissociated from electrophysiological activity in a number of preparations (reviewed in Sharp et al., 1993; Arckens, 2005; Pinaud, 2005). For example, studies in transgenic mice carrying mutations in different regulatory regions of the *c-fos* gene (Robertson et al., 1995) have revealed that kainate induced *c-fos* expression was abolished throughout the brain. In particular, mutations either in the serum response or the $Ca^{++}/cAMP$ elements, have supported the hypothesis for a requirement of simultaneous action of some intracellular pathways to enhance *c-fos* expression. In addition, *c-fos* is unlikely to be a reliable marker of neuronal activation in those neurons that are tonically active, as is the case of dopaminergic neurons of the medial basal hypothalamus that project to the median eminence thereby inhibiting the release of prolactin by the anterior pituitary (Hoffman et al., 1994). Thus, the lack of *c-fos* expression is not a proof of lack of neuronal activation. Since phosphorylation of the cAMP response element binding protein (CREB) is a rapid event linked to neuronal depolarization, IHC analysis directed at pCREB may be a good alternative to detect the activation of neuronal populations that lack the induction of *c-fos*. In accordance to this notion, Curtis et al. (2000) have found a greater number of pCREB positive cells, as compared to the number of c-Fos positive neurons in the central amygdala (CeA) after hypotension stress.

Forth, it is important to know the temporal dynamics of the response to particular stressors, especially considering that most studies use IEG expression at a single time point. Focusing on *c-fos*, although there are reports of significant increases in mRNA levels within 5–10 min after exposure to immobilization (Imaki et al., 1996; Umemoto et al., 1997) or alcohol administration (Ogilvie et al., 1998), consistent increases in expression levels are observed in the par-vocellular PVN 30 min after stimulus onset (Imaki et al., 1992, 1993, 1996; Cullinan et al., 1995; Kovacs and Sawchenko, 1996; Umemoto et al., 1997). Peak c-fos mRNA levels tend to occur 30 min after stimulus onset while maximum fos-like immunoreactivity (FLI) is often detected between 90–120 min after initial exposure to the stressor. It is commonly accepted that expression of *c-fos* reaches a maximum and then progressively declines to basal levels despite the persistency of the stimulus, a phenomenon that has also observed in response to stress (e.g. Imaki et al., 1992; Senba et al., 1994). However, previous reports have documented a distinct expression temporal pattern, with peaks in mRNA levels at 3–5 h after exposure to endotoxin (a component of the cell wall of gram-negative bacteria) or hypovolemia induced by the administration of polyethylene glycol (Lee et al., 1995; Rivest and Laflamme, 1995; Tanimura et al., 1998). It is important to note that the half-life of c-fos mRNA is about 10–15 min (Zangenehpour and Chaudhuri, 2002) and therefore sustained elevated levels involves continuous transcriptional activity. The results discussed above indicate that, in contrast to what is usually assumed, some neurons can maintain sustained c-fos transcriptional rates in response to certain stimuli.

Fifth, some IEGs are constitutively expressed, whereas others are not (at least to a significant degree), although it is sometimes difficult to define what is the actual basal expression of IEGs given the extreme sensitivity of some of them (e.g., c-fos) to minor stress. NGFI-A and c-jun are examples of IEGs showing high constitutive expression in a wide range of brain areas where c-fos expression is very low (Cullinan et al., 1995; Pinaud et al., 2002). It seems likely that those IEGs showing constitutive expression may be able to maintain a more sustained response to stress than those that do not. In practical terms, IEGs showing a more sustained response may be useful to map brain responses after more prolonged exposure to stress, although it is often more difficult to demonstrate significant changes against an already high background level observed in basal conditions.

Finally, a major issue is whether different IEGs are expressed by the same or by different neuronal populations in response to a particular stressor. Although this question has only been systematically approached in a few co-localization studies, it is commonly agreed that there is a substantial degree of co-localization across IEGs (Chan et al., 1993; Papa et al., 1993; Swank, 1994; Wang et al., 1997). Therefore, at least some IEGs appear to be co-expressed in certain cells types after stress and there is currently no evidence for the restricted expression of *c-fos*, or other IEGs, to any particular neuronal phenotype. Whether these rules apply to effector IEGs, such as *arc*, is currently unclear, as no studies have been conducted to directly address this question.

2. Stress-Induced c-fos Expression: Characterization of Brain Pathways Involved in the Brain's Stress Response

2.1. Categories of Stressors

The induction of *c-fos* after exposure to stressful situations has been extensively studied in the last decade and has greatly contributed to our knowledge of how the brain responds to and processes different stressful stimuli. This knowledge has been summarized in a series of recent reviews (Imaki et al., 1995; Herman and Cullinan, 1997; Kovacs, 1998; Sawchenko et al., 2000; Pacak and Palkovits, 2001; Herman et al., 2003) and therefore, we will focus here on the most critical aspects. Analysis of the overall picture yielded by these gene expression studies has led to the conclusion that there are two broad categories of stressors: systemic (physical, physiological, interoceptive) and processive (psychological, emotional, psychogenic). However, it should be taken into account that it is difficult to consider a stressor as purely emotional. Most stressors used in the field (e.g., forced swimming, electric shock, immobilization), despite having an important emotional component, also have a physical one. Stressors that may be considered basically emotional are exposure to novel environments or to conditioned stimuli after a classical Pavlovian conditioning procedure (Beck and Fibiger, 1995; Radulovic et al., 1998). On the other hand, some stressors having predominantly physical components (e.g., exhaustive exercise, hemorrhage, hypertonic saline administration, visceral pain, endotoxin and cytokines) are likely to induce discomfort in non-anesthetised animals and therefore could activate emotional circuits. It is noteworthy that most anesthetics are also able to induce *c-fos* expression in some brain areas and, therefore, may be considered as systemic stressors. Interestingly, the pattern of *c-fos* expression after ether treatment has the characteristics of both types of stressors (Emmert and Herman, 1999).

The expression of *c-fos* in response to systemic stressors has been extensively studied and details about brain areas activated are beyond the scope of the present review. Nevertheless, the overall conclusion is that each particular stressor only activates a rather restricted set of neuronal populations and brain circuits and, therefore, each one has its own signature (for reviews see Kovacs, 1998; Pacak and Palkovits, 2001). There are several possibilities through which stressors can reach brain areas critically involved in the response to stress, such as the PVN

(Sawchenko et al., 2000). First, circulating signals (e.g., plasma hypertonicity, LiCl) can directly reach neurons located in those areas devoid of blood brain barrier (circumventricular organs, CVO) that, in turn, may send stimulatory signals to the PVN. Second, signals generated in the viscera or in the cardiovascular system can inform the brain through afferent vegetative nerves (in most cases the vagus nerve), reaching the nucleus of the solitary tract (NST, that includes A2/C2 catecholaminergic regions). From this nucleus, information can be relayed to other catecholaminergic brainstem areas, such as the ventrolateral medulla (VLM, that includes A1/C1 catecholaminergic regions), the central nucleus of the amygdala (CeA) and the PVN. In striking contrast, emotional stressors are mainly processed in cortical and subcortical limbic areas, from which information is sent to effector areas such as the PVN, the mesencephalic periaqueductal gray (PGA) and preganglionic parasympathetic and sympathetic neurons located in the medulla and spinal cord. The difference between the two broad categories of stressors can be illustrated by the experiment conducted by Li et al. (1996), who studied the effects of unilateral knife cut of the brainstem (rostral to the NST and VLM) on c-fos induction in response to footshock or IL-1β treatment. After footshock, the c-fos response was similar in contralateral and ipsilateral sides of the PVN, whereas c-fos induction was clearly reduced in the ipsilateral NST. Conversely, c-fos expression after IL-1β, was similar in both sides of the NST, but was substantially reduced in the ipsilateral PVN. It appears, therefore, that the processing of information related to IL-1β takes place first at the level of brainstem (NST) and subsequently in other diencephalic areas such as the PVN, whereas the processing of footshock signals takes place primarily in telencephalic areas that, in turn, relay information to the PVN and brainstem areas, such as the NST.

2.2. Emotional Stressors

Exposure to different types of purely or predominantly emotional stressors, including painful stimuli, appears to elicit quite a similar pattern of c-fos expression (for review see Kovacs, 1998). Marked activation has been reported after exposure to novel environments, noise, restraint, forced swim or predator odor in all isocortical and allocortical areas, particularly in the medial prefrontal cortex (mPFC), that includes cingulate, prelimbic and infralimbic areas, the orbital cortex and the piriform cortex (Cullinan et al., 1995; Duncan et al., 1996; Bonaz and Rivest, 1998; Campeau and Watson, 1998; Li and Sawchewnko, 1998; Dielenberg et al., 2001; Day et al., 2004; Ons et al., 2004). Within limbic subcortical areas, activation in response to the above stressors has been consistently observed in the claustrum, dorsal endopiriform nucleus and lateral septum (LS). In contrast, low or absent c-fos induction has been usually reported in the medial septum and the hippocampal formation (HF, mainly formed by the dentate gyrus, CA regions and the subiculum) (Duncan et al., 1993; Imaki et al., 1993; Cullinan et al., 1995; Bonaz and Rivest, 1998). C-Fos expression has also been reported after emotional stressors (e.g. noise, restraint, forced swim or footshock) in the

BNST and amydala, although these results have not been reported in a consistent manner (see below). In particular in the BNST, controversies regarding which subregions are activated by stress (e.g. Cullinan et al., 1995; Duncan et al., 1996; Bonaz and Rivest, 1998; Campeau and Watson, 1998; Li and Sawchenko, 1998) are likely due to the anatomical complexity of this nucleus that poses technical difficulties to precisely delineate its different subdivisions in autoradiography films. Inconsistencies in the amygdala, however, particularly in the CeA, are more difficult to explain. ISH procedures have revealed consistent and higher c-fos induction in the medial amygdala (MeA), but low induction in the CeA after exposure to predominantly emotional stressors (Cullinan et al., 1995; Bonaz and Rivest, 1998; Campeau and Watson, 1998; Campeau et al., 1998; Ons et al., 2004). In contrast, using IHC procedures, some groups have reported clear increases in *c-fos* expression levels in the CeA (Honkaniemi, 1992; Bhatnagar et al., 1998; Martinez et al., 1998; Stampt and Herbert, 1999), while most other reports showed low or absent induction of this IEG (Arnold et al., 1992; Pezzone et al., 1992; Beck and Fibiger, 1995; Li and Sawchenko, 1998; Dayas et al., 2001; Dielenberg et al., 2001; Helfferich and Palkovits, 2003). Whether or not discrepancies measuring *c-fos* expression levels are due to antibody cross-reactivity is currently not known. However, there are consistent results regarding the activation of the CeA (specifically the lateral part, CeL) in response to a wide range of systemic stressors (Honkaniemi et al., 1992; Ericsson et al., 1994; Swank, 1994; Rivest and Laflamme, 1995; Rotllant et al., 2002; Nakagawa et al., 2003; Sullivan et al., 2003; Vallès et al., 2005). Ibotenic acid lesions of the MeA, but not the CeA, substantially reduced restraint-induced c-Fos expression in the PVN (Dayas et al., 1999), whereas lesions of the CeA dramatically reduced c-fos levels in PVN and SON, as well as in other brain areas, after IL-1β administration (Xu et al., 1999). These findings support the hypothesis that the CeA may be important for the control of behavioral and physiological responses to systemic stressors, whereas the MeA may be mainly involved in the response to emotional stressors. In addition, these data challenge the predominant view that the CeA is critical for the organization of most of the behavioral and physiological responses to emotionally-relevant situations.

Another important point is the controversial results regarding *c-fos* induction in the HF after exposure to emotional stressors (Cullinan et al., 1995; Kovacs, 1998; Pacak and Palkovits, 2001). In fact, we have observed a modest increase in c-fos levels after exposure to a novel environment. This increase was not observed after exposure of animals to a significantly more severe stressor (immobilization; Ons et al., 2004), and the same pattern was found measuring FLI (Rotllant et al., unpublished). It is unclear whether the inconsistent effects of emotional stressors on c-fos expression in the HF is a reflection of a lack of significant neuronal activation or a poor *c-fos* response to electrophysiological activation of neurons. In support of this latter possibility, it has been reported that NGFI-A induction is more sensitive than c-fos in response to electrical stimulation of entorhinal afferents that target the HF, whereby only high levels of stimulation trigger c-fos expression (Wisden et al., 1990; Worley et al., 1993).

The mPFC also appears to be a critical region for the processing of emotional stressors. There is evidence that lesions or inactivation of the mPFC increases ACTH response to emotional stressors but not to ether (Diorio et al., 1993; Figuereido et al., 2003; McDougall et al., 2004). In addition, although the latter two studies did not focus on the same brain areas, both observed increases in c-fos expression in the MeA and failed to find differences in BNST or LS (Figuereido et al., 2003; McDougall et al., 2004). These results strongly suggest that activation of the mPFC during emotional stress may have an inhibitory role on some of the behavioral and physiological responses associated to the stimulus. A role for other cortical areas is currently not known, however, Umegaki et al. (2003) observed induction of c-fos in the entorhinal cortex after immobilization, but not insulin-induced hypoglycemia. Accordingly, ibotenic acid lesions in this cortical area only reduced the ACTH response to the former stressor.

While exposure of animals to systemic stressors activates stressor-specific neuronal populations involved in the control of specific physiological responses, exposure to predominantly emotional stressors does not (apart from those areas directly involved in sensory processing). Therefore, emotional processing and coping (the particular cognitive and behavioral strategies used to face aversive situations) appear to involve a wide number of brain areas, most of them common to different stressors. Information from these brain areas (mPFC, amygdala-BNST, LS) may be conveyed to diencephalic and brainstem nuclei to trigger the set of behavioral and physiological responses to the stressor. Within this framework, a contribution of some brainstem nuclei to determine the magnitude of the response to emotional stress rather than to the elaboration of the response itself, has to be considered. Thus, although activation of areas such as the NST and the LC are likely to be in great part due to descending signals from telencephalic and diencephalic regions, there is some evidence that these areas may in turn activate the response to emotional stressors (Ziegler et al., 1999; Passerin et al., 2000; Kinzig et al., 2003). It is therefore possible that the intensity or duration of the response to emotional stressors may be influenced by a positive feedback loop between telencephalic and brainstem areas, and, consequently, pharmacological interventions in the latter areas may contribute to control of the response to stress. Fig. 11.1 summarizes the putative pathways involved in the processing of different categories of stressors.

3. Theories About Brain Processing of Stressors

There are two critical questions that have to be answered in order for a better understanding of the mechanisms underlying the brain processing of emotional stressful stimuli to emerge: (1) why in most brain areas the induction of *c-fos* appears to be similar regardless of the stressor intensity, while only in a few brain regions *c-fos* induction appears to be related to the intensity of the stimulus? (2) why clearly different emotional stressors induce *c-fos* in the same brain nuclei?

Surprisingly, only few attempts have been carried out in order to study the relationhip between the intensity of emotional stressors and the degree of brain

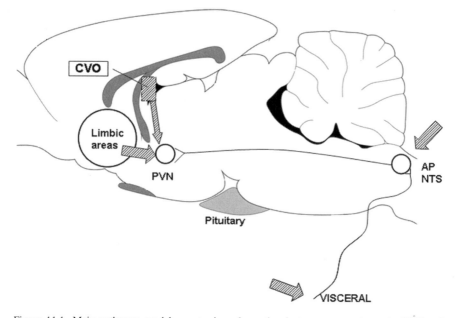

Figure 11.1. Main pathways used by systemic and emotional stressors to activate the PVN and to induce ACTH release. Systemic stressors can reach the brain through: (a) circumventricular organs (CVO), that include area postrema (AP) near the nucleus solitary tract (NST), that generates signals arriving at the PVN; or (b) peripheral signals from the viscera that activate afferent vegetative pathways integrated in the NST that directly project to the PVN. Emotional stressors are first processed in limbic cortical and subcortical areas and then signals are sent to the PVN through polysynaptic pathways than involve the bed nucleus of stria terminalis and other hypothalamic regions.

c-fos induction. To our knowledge, this problem has been studied by Campeau and colleagues (Campeau and Watson, 1998; Campeau et al., 1998) using exposure to different noise intensities or fear conditioning paradigms. In our lab we have compared *c-fos* induction by stressors known to differ in intensity (a novel environment, forced swimming and immobilization) (Ons et al., 2004). Although it is difficult to reach solid conclusions with only few studies available using clearly different approaches, it appears that cortical and thalamic areas are rather insensitive to the intensity of the stressors. Conversely, areas such as the LS, MeA, some regions of BNST, the PVN and the LC appear to discriminate between different intensity stressors.

The induction of *c-fos* in most brain areas may not be related to the intensity of stressors because (a) there are a limited number of neurons responding to stressors, (b) the induction of *c-fos* in individual neurons may be an all-or-none phenomenon, and (c) neurons may be so sensitive that most respond even to minor stimuli. One may argue that the evaluation of the number of c-Fos positive neurons would only indicate the number of activated neurons, but not the intensity of their activation. However, if this notion is correct, c-fos mRNA levels should reflect the intensity of activation, which is not the case: most brain areas show similar

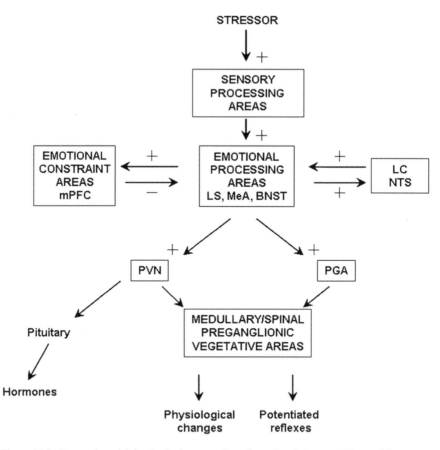

Figure 11.2. Proposed model for the brain processing of emotional stressors. The model assumes that sensory information about the stressor arrives at thalamic nuclei and subsequently at the sensory cortex, where information is conveyed to other isocortical areas, in addition to subcortical and cortical areas involved in emotional processing, such as the Lateral Septum (LS), Amygdala, particularly the medial amygdala (MeA) and the bed nucleus stria terminalis (BNST). Some of these areas (e.g., medial prefrontal cortex, mPFC) may play an important role in constraining emotional activation. Once processed, this information is sent to lower diencephalic and brainstem nuclei, such as the paraventricular nucleus of the hypothalamus (PVN) and the periaqueductal gray (PGA), that control lower level effector functions related to the endocrine system, the preganglionic medullary and spinal cord neurons of the vegetative nervous system and the motor reflexes. Some brainstem nuclei such as the locus coeruleus (LC) and the nucleus of the solitary tract (NST) may participate in positive feedback loops with cortical and subcortical limbic areas.

c-fos mRNA levels after exposure to two stressors strongly differing in intensity, such as exposure to a novel environment and immobilization (Ons et al., 2004). In those areas responding proportionally to the intensity of emotional stressors, gradual increases in both mRNA levels (Ons et al., 2004) and the number of c-Fos immunoreactive neurons (Rotllant et al., unpublished) have been observed. These data suggest that more neurons are recruited with severe stressors and that subsets of neurons having different thresholds exist, in order to respond to stimulatory

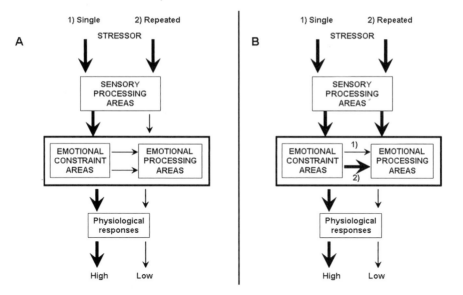

Figure 11.3. Alternative hypotheses to explain adaptation to a daily repeated stressor. The first hypothesis (A) proposes that repeated exposure to the same stressor is likely to reduce the magnitude of the activation of brain pathways involved in its processing at an early stage (i.e. sensory processing) and consequently the impact of the stressor throughout the brain. Inhibitory signals from emotional constraint areas are not changed. The second hypothesis (panel B) proposes that the reduced impact of repeated stress may be due to the superimposition of enhanced inhibitory synaptic inputs from critical (yet unknown) emotional constraint areas. Such inhibitory inputs may develop as a consequence of safety signals associated with previous experience of release from the stress situation. This second hypothesis appears to be more likely on the basis of available experimental data.

inputs. Particularly in the parvocellular PVN, in which we have the opportunity to indirectly evaluate the magnitude of activation of its neurons by measuring plasma levels of ACTH, some parallels exist between the magnitude of *c-fos* induction and the electrophysiological activity of its parvocellular neurons.

Two alternative explanations may account for the activation of brain areas regardless of the qualitatively aspects of emotional stressors. The first one is that the same neuronal population is activated by different types of stressors. This notion fits well with the concept of arousal and the non-specific activation of cortical areas through direct monoaminergic projections and non-stress-specific thalamic inputs (which are themselves also non-stress-specifically activated by different types of stimuli). However, some studies about cortical *c-fos* expression in response to sensory stimuli or stressors consistently indicate higher *c-fos* induction in deep (IV, V and VI) as compared to superficial layers (Schreiber et al., 1991; Morrow et al., 2000; Staiger et al., 2002), suggesting a relatively specific activation pattern. The second possibility is that a certain degree of selectivity may exist, however, it is not easily detectable by conventional histological methods of evaluation of neuronal activation. Thus, on the basis of different experimental approaches, including lesions, it is clear that functionally different neurons can be intermixed within a particular brain region, which suggests the existence of anatomically close, but functionally separate, circuits. Because each neuron can co-express at least one, and in some cases more, classical neurotransmitter, and several neuropeptides and neuromodulators, neurons involved in the processing of a particular stimulus could be different from those processing other stimuli regarding the repertoire of neurochemical agents they use to communicate to each other. Should this hypothesis be correct, further possibilities to better delineate brain circuits may be available.

It seems possible to test these two possibilities by exposing animals to two different stressors either separately or simultaneously. If all stressors activate the same neuronal population, those areas where c-fos induction is not sensitive to the intensity of stressors should exhibit the same number of c-Fos positive neurons in response to each individual stressor and after simultaneous exposure to both. Conversely, if different stressors activate different subsets of neurons, exposure to the two stressors would result in higher numbers c-Fos positive neurons as compared to those found after stimulation with each individual stressor. In those areas sensitive to the intensity of stressors, the same approach would result in a more ambiguous conclusion, because an increase in the mRNA signal or in the number of c-Fos positive neurons could be alternatively explained by the stronger impact of the simultaneous exposure to the two stressors. In addition, specificity in the neuronal population activated after emotional stressor may be more evident in higher level brain regions, as opposed to lower level areas, mainly involved in the low hierarchy control of the responses. For instance, specificity is expected to be greater in those areas involved in the evaluation of emotional consequences of stressors and the elaboration of appropriate strategies (cortical and subcortical limbic areas) as opposed to the PVN. Despite these caveats, the number of CRF-positive neurons displaying c-fos expression after exposure to a severe stressors such as immobilization is about 50% (Rotllant et al., unpublished), suggesting that different stressors may activate different subsets of parvocellular PVN neurons.

4. Adaptation to Chronic Repeated Stress and IEGs

When animals are repeatedly exposed to the same stressors on a daily basis, a reduced physiological response to the last exposure to the same (homotypic) stressor is sometimes observed. This reduced response is termed adaptation or habituation, the latter term being used because adaptation to chronic repeated stress, when observed, appears to follow the rules of habituation, as defined by Groves and Thompson (1970). Habituation is a non-associative form of learning and it is unlikely that adaptation to repeated stress may be so simple. In fact, there is evidence suggesting that conditioning may be involved (Riccio et al., 1991), although this is a matter that requires further investigation. Repeated exposure to predominantly emotional stressors has consistently been found to result in reduced adrenaline and ACTH response to the homotypic stressor, whereas the response to a novel (heterotypic) stressor is normal or enhanced

(Martí and Armario, 1998), demonstrating that only those variables sensitive to stress intensity are capable of reflecting adaptation.

Given the pivotal role of the PVN in the control of the HPA axis, attention has been paid to the *c-fos* response to repeated stress in this nucleus. Using ISH a consistent reduction of c-fos expression has been observed in the PVN after daily exposure of rats to noise (Campeau et al., 2002), restraint or immobilization (Umemoto et al., 1994a, b, 1997; Watanabe et al., 1994; Bonaz and Rivest, 1998) or male hamsters to social defeat (Kollack-Walker et al., 1999). C-Fos immunoreactivity has yield not so clear results: complete or strong adaptation has been found after repeated restraint (Chen and Herbert, 1995; Stampt and Herbert, 1999; Dumont et al., 2000; Viau and Sawchenko, 2002) whereas no significant reductions were found after repeated social defeat (Martinez et al., 1998) or footshock (Li and Sawchenko, 1998). It appears that c-fos mRNA levels, but not the overall numbers of c-Fos-positive neurons, may better reflect adaptation. In addition, the magnitude of adaptation may depend on the intensity, length of exposure and/or other characteristics of emotional stressors.

Lower activation of the PVN after repeated exposure to emotional stressors may be the reflection of lower stimulatory inputs arriving (directly or indirectly) at this nucleus from limbic areas. In accordance with this notion, reduced c-fos mRNA response has been observed in important brain areas, such as the mPFC, LS, PVN and LC, after repeated stress (Melia et al., 1994; Watanabe et al., 1994; Bonaz and Rivest, 1998; Campeau et al., 2002), with a normal response to heterotypic stressors (Melia et al., 1994; Watanabe et al., 1994). However, a widespread reduction of c-fos mRNA levels has not been observed with all repeated stress protocols. Thus, the pattern of adaptation of c-fos mRNA levels appears to be strongly dependent on the brain area studied, after 7 days of repeated social defeat in male hamsters (Kollack-Walker et al., 1999). Additional studies have evaluated FLI in response to chronic stress. After daily repeated restraint, marked adaptation of FLI was found in the PVN, LS, MeA and CeA, but not in lateral preoptic area or lateral hypothalamus (Chen and Herbert, 1995; Stampt and Herbert, 1999). After repeated social defeat in rats, no adaptation of FLI was observed in the PVN, MeA or some brainstem areas, such as PGA and raphe nuclei. However, adaptation was detected in these studies in the LS, CeA and LC (Martinez et al., 1998). Finally, Li and Sawchenko (1998), using a qualitative evaluation of FLI, were unable to observe reduction of FLI in most brain regions (including the PVN). As indicated above, it appears that adaptation of c-fos expression is less evident when evaluating FLI as opposed to c-fos mRNA levels although regional differences in the pattern of adaptation, depending on the intensity and nature of the stressor, are likely. Despite the paucity of data, some cortical and subcortical limbic regions that appear to be less activated by repeated exposure to the same stressor may be important for the control of the overall process of adaptation to the situation (e.g., LS and MeA). In addition, reduced activation of some hypothalamic and brainstem nuclei (e.g., PVN and LC) may be due to lower levels of stimulatory inputs arriving from those telencephalic areas.

Although repeated exposure to the same stressor results in reduced expression of c-fos and other IEGs of the fos family, it has been recently demonstrated that exposure to either chronic repeated restraint or chronic variable stress increased levels of ΔfosB, as measured 1 or 7 days after the last exposure to the stressors (Perrotti et al., 2004). This increase, however, was only observed in a subset of neurons that expressed c-fos during exposure to stress. For instance, ΔfosB increases were observed in the PFC, basolateral amygdala (BLA), nucleus accumbens, LS and LC, but not in BNST or PVN. The functional significance of a differential induction of ΔfosB in some brain areas is not known, but it may bear implications regarding susceptibility to addiction.

The mechanisms involved in the progressive reduction of c-fos (and other IEGs) expression, as well as the reduction of other prototypical responses (e.g., HPA activation) after repeated exposure to the same stressors are not known, however, two main possibilities may underlie these effects. First, as a consequence of repeated exposure to the same stressor, a progressively lower activation (desensitisation) of some neuronal populations strongly activated during the first exposure may occur. This possibility would fit well with the hypothesis that adaptation to repeated stress represents a habituation-like process. A second possibility is that repeated exposure to, and release from, the stressful situation may be elaborated in some critical areas (e.g., mPFC) as a safety signal, and translated into inhibitory inputs to those areas that initially responded to the stressors (limbic nuclei). That is, adaptation to repeated stress may be due to the experience-dependent progressive development of an inhibitory feedback onto the primary sites activated by the stressor.

In terms of IEGs, the second possibility appears to be more plausible. Adaptation of c-fos response after repeated exposure to the same stressor has been observed in those brain areas where c-fos induction is sensitive to the intensity of stressors (e.g. PVN) and in those where it is not (e.g. mPFC). When animals are repeated exposed to severe stressors (e.g. immobilization) an important degree of physiological activation persists that is greater than that observed after a first exposure to mild stressors (e.g. a novel environment). Therefore, if adaptation after repeated exposure to severe stressors was the consequence of a desensitization process that resulted in lower stimulatory inputs to all stress-sensitive neuronal populations, reduction of c-fos expression should not be observed in those brain areas that are maximally activated even by minor stressors. On the contrary, if adaptation was mainly due to the progressive superimposition of experience-dependent inhibitory processes (presumably, potentiated inhibitory GABAergic inputs), inhibitory inputs would target both those neurons which are sensitive and those which are insensitive to the intensity of stressors, with the consequent reduction of the magnitude of c-fos induction in the two types of neurons.

It is therefore necessary to characterize the phenotype of neurons showing FLI to demonstrate a putative increase in the number of GABAergic neurons activated in response to the homotypic stressors in repeatedly stressed versus naive rats. This second possibility also fits better with the evidence that an associative process may be involved in adaptation to repeated stress (Riccio et al., 1991). It also fits well with the results of Bodnoff et al. (1989), who found that the latency to eat in a novel environment was markedly reduced by repeated exposure to this environment, however, this effect was completely blocked by the administration of the benzodiazepine receptor antagonist flumazenil, suggesting that adaptation to the environment was likely to be the consequence of the release of an endogenous anxiolytic ligand that acted on GABA-A receptors.

5. A Comparison of the Response of Different IEGs to Stress

Using different IEGs to evaluate brain areas activated by stressors requires researchers to know their differential sensitivity to stressors and the temporal dynamics of their expression patterns. In this section we will discuss the response profiles of some IEGs and compare them to that of *c-fos*, with special focus on the PVN, where most of the studies have been done. Among the IEGs acting as transcription factor, we will discuss in this chapter the fos family (c-fos, fosB, Fra-1 and Fra-2), the jun family (c-jun, junB, junD), NGFI-A (also known as zif268, egr-1, krox-24 and zenk) and NGFI-B. As discussed previously, different IEGs appear to be induced in the same population of neurons, although each of them may have specific expression requirements. Therefore, a uniform pattern of activity is unlikely, even for genes of the same general family.

The effects of stress on the induction of other members of the fos family (Fra-1, Fra-2, fosB) has not been extensively studied. However, in the PVN, increases in Fra-2, but not Fra-1, have been observed after capsaicin administration (Honkaniemi et al., 1994), and increases in fosB have been repeatedly observed after restraint or immobilization (Senba et al., 1994; Umemoto et al., 1994a, 1997; Stampt and Herbert, 1999). In one of these studies a similar, but not identical, pattern of adaptation to repeated restraint was observed for fosB, as compared to *c-fos*, in several brain areas, including LS, PVN and NST (Stampt and Herbert, 1999).

Two of the most studied IEGs in stressful conditions are NGFI-A and NGFI-B, that belong to different families. Schreiber et al. (1991) described for the first time the induction of NGFI-A by emotional stressors and Cullinan et al. (1995) observed that, in contrast to *c-fos*, NGFI-A was constitutively expressed in isocortical and allocortical areas, HF, nucleus accumbens, some amygdaloid nuclei and cerebellum, with significantly lower basal expression in diencephalic and brainstem areas. Interestingly, although NGFI-A expression was increased in fewer areas compared to c-fos by 30 min restraint or swimming stressors, enhanced expression was observed in some areas where no *c-fos* induction was found: the dentate gyrus and CA3, BNST (lateral part) and the medial habenula (Cullinan et al., 1995).The induction of NGFI-A by stress in the hippocampus has been replicated (Olsson et al., 1997; Ueyama et al., 1999). These results are compatible with the finding that NGFI-A expression in the hippocampus has a lower threshold than c-fos after electrical stimulation of entorhinal cortical afferents to the hippocampus (Wisden et al., 1990; Worley et al., 1993), and demonstrate that, in some brain areas, *c-fos* is probably a rather insensitive marker of neuronal activation. Stress-induced NGFI-A expression has also been consistently observed in the PVN (Honkaniemi et al., 1994; Umemoto et al., 1994a, 1997; Watanabe et al., 1994; Cullinan et al., 1995). NGFI-A is therefore of interest for several reasons: first, enhanced expression of NGFI-A appears to be much more sustained than that of c-fos when a simulus persists for over 30 min (Zangenehpour and Chaudhuri, 2002) and therefore, may be more appropriate after longer exposure to stressors. Second, NGFI-A may be more sensitive than *c-fos* to neuronal activation. Finally, after repeated exposure to restraint, a marked reduction of the c-fos mRNA response to the homotypic stressor was observed in the PVN, whereas NGFI-A expression was maintained (Melia et al., 1994; Umemoto et al., 1994a, 1997; Watanabe et al., 1994), suggesting that the latter is resistant to adaptation to repeated stress. It is unclear whether there is a relationship between the more sustained NGFI-A response to stressors, its lack of adaptation and its constitutive expression.

NGFI-B appears to be constitutively expressed in telencephalic areas, but not in the hypothalamus, and its induction levels and time-course in the PVN after hemorrhage and IL-1β-treatment were similar to those of *c-fos* (Chan et al., 1993). Other studies have reported enhanced PVN NGFI-B transcription in response to different stressors (Honkaniemi et al., 1994; Umemoto et al., 1994a, 1997; Lee et al., 1995; Rivest and Laflamme, 1995; Imaki et al., 1996, 1998; Kovacs and Sawchenko, 1996; Lee and Rivier, 1998; Mansi et al., 1998; Ogilvie et al., 1998) and adaptation observed after repeated immobilization (Umemoto et al., 1994a, 1997). Widespread induction of NGFI-B has also been observed in telencephalic areas after exposure of animals to a novel environment (Hinks et al., 1996). Chan et al. (1993) reported that in hypothalamic and brainstem areas known to project to the PVN, NGFI-B induction by hemorrhage and IL-1β-treatment was smaller than that of *c-fos*. Thus, although there are few studies in areas other than the PVN, the results suggest that the pattern of expression of NGFI-B is qualitatively similar to that of *c-fos* and has not obvious advantages with regards to mapping of neuronal activity.

Regarding the jun family, *c-jun* is constitutively expressed in most brain areas (Cullinan et al., 1995), prominently in the PVN (Senba et al., 1994; Umemoto et al., 1994a; Cullinan et al., 1995), and it is rather insensitive to emotional stressors (novel environment, forced swim, restraint, immobilization) in the PVN (Senba et al., 1994; Umemoto et al., 1994a; Cullinan et al., 1995; Martinez et al., 1998), and in most other brain regions (Cullinan et al., 1995; Hinks et al., 1996; Nagahara et al., 1997). However, there is some evidence for the induction of *c-jun* in the HF in response to a novel environment or restraint (Papa et al., 1993; Lino de Oliveira et al., 1997), or in the LS and cortex in response to restraint (Ryabinin et al., 1995; Stampt and Herbert, 1999). The reasons for these discrepancies are unclear, although the induction of *c-jun* was, in any case, modest as compared to that of *c-fos*. Stress-induction of junD appears to be, if any, small, similar to that of c-jun, whereas junB expression is both more robust and consistent. Thus, immobilization or restraint procedures were found to increase junB mRNA levels in cortex (Melia et al., 1994; Ryabinin et al., 1995), HF (Ryabinin et al., 1995) and PVN (Senba et al., 1994; Umemoto et al., 1994a, 1997; Imaki et al., 1996), and induction of the junB protein has been reported in key areas such as LS, PVN and NST (Stampt and Herbert, 1999).

Assuming that all above mentioned IEGs are expressed in the same types of neurons and that their expression is the result of the convergent action of both membrane depolarization and neurotransmitter-associated intracellular signaling, there are several alternative hypothesis to explain the lower magnitude of adaptation of some IEGs as compared to others, and particularly the striking difference between *c-fos* and NGFI-A. First, NGFI-A may be much more sensitive than *c-fos* to depolarizing stimuli so that it would be maximally expressed despite a marked decrease in stimulatory inputs in repeatedly stressed rats. Second, *c-fos* may be more sensitive than NGFI-A to specific intracellular signals that are markedly decreased by repeated stress. Third, NGFI-A may be rather insensitive to repeated stress-induced superimposed inhibitory signals, that would, conversely, strongly reduce *c-fos* expression.

In recent years, much attention has been paid to effector IEGs, in particular to *arc*. Arc is an IEGs that codes for a protein that exhibits similarities to spectrin and is tightly associated to actin (for review see Steward and Worley, 2002). In addition, *arc* mRNA levels are transported to dendrites, specifically to active synapses, where the protein is locally synthesized (Steward et al., 1998; Steward and Worley, 2001; Moga et al., 2003). For these and other reasons, *arc* is considered to be involved in synaptic plasticity and learning (Steward and Worley, 2002). Because it is generally agreed that long-term effects of a single exposure to stress, and adaptation to repeated stress may involve synaptic changes, we became interested in the effects of stress on brain *arc* expression. We compared the responses of *arc* and *c-fos* to three different stressors differing both qualitatively and in severity (novel environment, forced swimming and immobilization) (Ons et al., 2004). Confirming previous reports, constitutive arc expression was observed in some telencephalic areas (e.g., cortex, HF, striatum), as well as in the reticular thalamic nucleus. After exposure to the three stressors, both arc and *c-fos* expression were enhanced in a wide range of areas. The pattern of *c-fos* expression was quite consistent with previous reports, but, importantly, *arc* expression was restricted to telencephalic areas. In addition, significant *arc* expression was found in areas where *c-fos* expression is low or less consistent, such as the accumbens shell or the BLA. Nevertheless, *arc* expression, like c-fos expression, was independent of the intensity of the stressor in most areas, but it was dependent on stressor intensity in the same brain areas where *c-fos* was (e.g., LS, accumbens shell, MeA). These results suggested that: (a) there appears to be two groups of brain areas that respond to stress: the first group encompasses areas where expression of *c-fos* and other IEGs, such as arc, are not related to the intensity of the stressor; the second group includes those brain regions that are sensitive to intensity; (b) the expression of arc, like that of NGFI-A, appears to be more sensitive than that of *c-fos* in some brain regions; (c) the restricted expression of *arc* in telencephalic areas after stress supports the hypothesis that this gene may be involved in synaptic plasticity associated to stress exposure.

6. Conclusions

The data discussed above indicate that the study of *c-fos* expression has greatly contributed to our knowledge of the brain processing of stressors. The use of IEGs other than *c-fos* (e.g., *zif268*, *arc*) reveals activation of brain areas not detected with *c-fos*, even though evaluation of other markers of neuronal activation (for example, phosphorylation of CREB) could be important in generating a complete picture of brain areas activated under stress. The expression of effector IEGs, such as *arc*, can provide researchers with additional important information concerning brain areas where synaptic plasticity associated to stress exposure may take place. However, much of the IEG response to stress remains to be characterized in order for a complete picture of brain processing of stressors emerges, and this includes: (a) a more complete characterization of the relationship between neuronal depolarization and IEG expression, (b) the description of the effects of neurotransmitters on stress-induced IEG expression (a topic not discussed in the present review), (c) the characterization of neuronal phehotypes activated during the stress response, and (d) the use of complementary approaches (e.g., lesions, tract-tracing studies, region-specific conditional expression or repression of particular IEGs in mutant mice) which should allow for directly testing hypotheses regarding the hierarchical control of the response to stress, an aspect poorly understood in this field, with the exception of the HPA axis. Finally, a more thorough characterization of the functional role of the protein products encoded by IEGs may shed light onto the precise adaptive values that result from their increased expression. This information will likely to reveal key aspects of the neural response to stress and will rival the critical importance of IEG expression in the mapping of brain activity.

Acknowledgments

This work was supported by grants from DGICYT (SAF2002-00623) and the ISCIII (G03/005, *Redes temáticas, Ministerio de Sanidad*). Thanks are given to Raúl Delgado for their help with figures and Roser Nadal for critical reading of the manuscript.

References

Arckens L (2005). The molecular biology of sensory map plasticity in adult mammals. In: Plasticity in the Visual System: From Genes to Circuits (Pinaud R, Tremere LA, De Weerd P, eds), pp. 181-203. New York: Springer-Verlag.

Armario A, López-Calderón S, Jolín T, Castellanos JM (1986). Sensitivity of anterior pituitary hormones to graded levels of psychological stress. Life Sci 39:471-475.

Arnold FJL, Bueno ML, Shiers H, Hancock DC, Evan GI, Herbert J (1992). Expression of c-fos in regions of the basal limbic forebrain following intracerebroventricular corticotropin-releasing factor in unstressed or stressed male rats. Neuroscience 51:377-390.

Beck CHM, Fibiger HC (1995). Conditioned fear-induced changes in behavior and in the expression of the immediate early gene c-fos: with and without diazepam pretreatment. J Neurosci 15:709-720.

Bhatnagar S, Dallman M (1998). Neuroanatomical basis for facilitation of hypothalamic-pituitary-adrenal responses to a novel stressor after chronic stress. Neuroscience 84:1025-1039.

Bodnoff SR, Suranyi-Cadotte BE, Quirion R, Meaney MJ (1989). Role of the central benzodiacepine receptor system in behavioral habituation to novelty. Behav Neurosci 103:209-212.

Bonaz B, Rivest S (1998). Effect of a chronic stress on CRF neuronal activity and expression of its type 1 receptor in the rat brain. Am J Physiol 275:R1438-R1449.

Campeau S, Dolan D, Akil H, Watson SJ (2002). c-fos mRNA induction in acute and chronic audiogenic stress: possible role of the orbitofrontal cortex in habituation. Stress 5:121-130.

Campeau S, Falls WA, Cullinan WE, Helmreich DL, Davis M, Watson SJ (1998). Elicitation and reduction of fear: behavioural and neuroendocrine indices and brain induction of the immediate-early gene c-fos. Neuroscience 78:1087-1104.

Campeau S, Watson SJ (1998). Neuroendocrine and behavioral responses and brain pattern of c-fos induction associated with audiogenic stress. J Neuroendocrinol 9:577-588.

Cannon W (1929). Organization for physiological homeostasis. Physiol Rev 9:399-431.

Chan RKW, Brown ER, Ericsson A, Kovacs KJ, Sawchenko PE (1993). A comparison of two immediate-early genes, c-fos and NGFI-B, as markers for functional activation in stress-related neuroendocrine circuits. J Neurosci 13:5126-5138.

Chen X, Herbert J (1995). Regional changes in c-fos expression in the basal forebrain and brainstem during adaptation to repeated stress. Correlations with cardiovascular, hypothermic and endocrine responses. Neuroscience 64:675-685.

Cullinan WE, Herman JP, Battaglia DF, Akil H, Watson SJ (1995). Pattern and time course of immediate early gene expression in rat brain following acute stress. Neuroscience 64:477-505.

Curtis AL, Bello NT, Connolly KR, Valentino RJ (2002). Corticotropin-releasing factor neurones of the central nucleus of the amygdala mediate locus coeruleus activation by cardiovascular stress. J Neuroendocrinol 14:667-682.

Day HEW, Masini CV, Campeau S (2004). The pattern of brain c-fos mRNA induced by a component of fox odor, 2,5-dihydro-2,4,5-trimethylthiazoline (TMT), in rats, suggests both systemic and processive stress characteristics. Brain Res 1025:139-151.

Dayas CV, Buller KM, Day TA (1999). Neuroendocrine responses to an emotional stressor: evidence for involvement of the medial but not the central amygdala, Eur J Neurosci 11:2312-2322.

Dayas CV, Buller KM, Day TA (2001). Medullary neurones regulate hypothalamic corticotropin-releasing factor cell responses to an emotional stressor. Neuroscience 105:707-719.

Dielenberg RA, Hunt GE, McGregor IS (2001). When a rat smells a cat: the distribution of Fos immunoreactivity in rat brain following exposure to a predator odor. Neuroscience 104:1085-1097.

Diorio D, Viau V, Menaye MJ (1993). The role of the medial prefrontal cortex (cyngulate gyrus) in the regulation of the hypothalamic-pituitary-adrenal responses to stress. J Neurosci 13:3839-3847.

Dumont EC, Kinkead R, Trottier J-F, Gosselin I, Drolet G (2000). Effect of chronic psychogenic stress exposure on enkephalin neuronal activity and expression in the rat hypothalamic paraventricular nucleus. J Neurochem 75:2200-2211.

Duncan GE, Johnson KB, Breese GR (1993). Topographic patterns of brain activity in response to swim stress: assessment by 2-deoxyglucose uptake and expression of Fos-like immunoreactivity. J Neurosci 13:3932-3943.

Emmert MH, Herman JP (1999). Differential forebrain c-fos m RNA induction by ether inhalation and novelty: evidence for distinctive stress pathways. Brain Res 845:60-67.

Ericsson A, Kovacs KJ, Sawchenko P (1994). A functional anatomical analysis of central pathways subserving the effects of interleukin-1 on stress-related neuroendocrine neurons. J Neurosci 14:897-913.

Figuereido HF, Bruestle A, Bodie B, Dolgas CM, Herman JP (2003). The medial prefrontal cortex differentially regulates stress-induced c-fos expression in the forebrain depending on type of stressor. Eur J Neurosci 18:2357-2364.

Groves PM, Thompson RF (1970). Habituation: a dual-process theory. Psychol Rev 77:419-450.

Guillod-Maximim E, Lorsignol A, Alquier T, Penicaud L (2004). Acute intracarotid glucose injection towards the brain induces specific c-fos activation in hypothalamic nuclei: involvement of astrocytes in cerebral glucose-sensing in rats. J Neuroendocrinol 16:464-471.

Helfferich F, Palkovits M (2003). Acute audiogenic stress-induced activation of CRH neurons in the hypothalamic paraventricular nucleus and catecholaminergic neurons in the medulla oblongata. Brain Res 975:1-9.

Herdegen T, Leah JD (1998). Inducible and constitutive transcription factors in the mammalian nervous system: control of gene expression by Jun, Fos and Krox, and CREB/ATF proteins. Brain Res Brain Res Rev 28:370-490.

Herman JP, Cullinan WE (1997). Neurocircuitry of stress: central control of the hypothalamo-pituitary-adrenocortical axis. TINS 20:78-84.

Herman JP, Figueiredo H, Mueller NK, Ulrich-Lai Y, Ostrander MM, Choi DC, Cullinan WE (2003). Central mechanisms of stress integration: hierarchical circuitry controlling hypothalamo-pituitary-adrenocortical responsiveness. Front Neuroendocrinol 24:151-180.

Hinks GL, Brown P, Field M, Poat JA, Hughes J (1996). The anxiolytics CI-988 and chlordiazepoxide fail to reduce immediate early gene mRNA stimulation following exposure to the rat elevated X-maze. Eur J Pharmacol 312:153-161.

Hoffman GE, Le W-W, Abbud R, Lee W-S, Smith MS (1994). Use of fos-related antigens (FRAs) as markers of neuronal activity: FRA changes in dopamine neurons during proestrus, pregnancy and lactation. Brain Res 654:207-215.

Honkaniemi J (1992). Colocalization of peptide- and tyrosine hydroxylase-like immunoreactivities with Fos-immunoreactive neurons in rat central amygdaloid nucleus after immobilization stress. Brain Res 598:107-113.

Honkaniemi J, Kainu T, Ceccatelli S, Rechardt L, Hokfelt T, Pelto-Huikko M (1992). Fos and jun in rat central amygdaloid nucleus and paraventricular nucleus after stress. NeuroReport 3:849-852.

Honkaniemi J, Kononen J, Kainu T, Pykonen I, Pelto-Huikko M (1994). Induction of multiple immediate early genes in rat hypothalamic paraventricular nucleus after stress. Mol Brain Res 25:234-241.

Imaki T, Naruse M, Harada S, Chikada N, Nakajima K, Yoshimoto T, Demura H (1998). Stress-induced changes of gene expression in the paraventricular nucleus are enhanced in spontaneously hypertensive rats. J Neuroendocrinol 10:635-643.

Imaki T, Shibasaki T, Chikada N, Harada S, Naruse M, Demura H (1996). Different expression of immediate-early genes in the rat paraventricular nucleus induced by stress: relation to corticotropin-releasing factor gene transcription. Endocrin J 43:629-638.

Imaki T, Shibasaki T, Demura H (1995). Regulation of gene expression in the central nervous system by stress: molecular pathways of stress responses. Endocrin J 42:121-130.

Imaki T, Shibasaki T, Hotta M, Demura H (1992). Early-induction of c-fos precedes increased expression of corticotropin-releasing factor messenger ribonucleic acid in the paraventricular nucleus after immobilization stress. Endocrinology 131:240-246.

Imaki T, Shibasaki T, Hotta M, Demura H (1993). Intracerebroventricular administration of corticotropin-releasing factor induces c-fos mRNA expression in brain regions related to stress responses: comparison with pattern of c-fos mRNA induction after stress. Brain Res 616:114-125.

Kinzig KP, D'Alessio DA, Herman JP, Sakai RR, Vahl TP, Figueiredo HF, Murphy EK, Seeley RJ (2003). CNS glucagon-like peptide-1 receptors mediate endocrine and anxiety responses to interoceptive and psychogenic stressors. J Neurosci 23:6163-6170.

Kollack-Walker S, Don C, Watson SJ, Akil H (1999). Differential expression of c-fos mRNA within neurocircuits of male hamsters exposed to acute and chronic defeat. J Neuroendocrinol 11:547-559.

Kovacs KJ (1998). c-fos as a transcription factor: a stressful (re)view from a functional map. Neurochem Int 33:287-297.

Kovacs KJ, Sawchenko PE (1996). Sequence of stress-induced alterations in indices of synaptic and transcriptional activation in parvocellular neurosecretory neurons. J Neurosci 16:262-273.

Lee S, Barbanel G, Rivier C (1995). Systemic endotoxin increases steady-state gene expression in the hypothalamic nitric oxide synthase: comparison with corticotropin-releasing factor and vasopressin gene transcripts. Brain Res 705:136-148.

Lee S, Rivier C (1998). Interaction between corticotropin-releasing factor and nitric oxide in mediating the response of the rat hypothalamus to immune and non-immune stimuli. Mol Brain Res 57: 54-62.

Li H-Y, Ericsson A, Sawchenko PE (1996). Distinct mechanisms underlie activation of hypothalamic neurosecretory neurons and their medullary catecholaminergic afferents in categorically different stress paradigms. Proc Natl Acad Sci USA 93:2359-2364.

Li H-Y, Sawchenko PE (1998). Hypothalamic effector neurons and extended circuitries activated in neurogenic stress: a comparison of footshock effects exerted acutely, chronically, and in animals with controlled glucocorticoid levels. J Comp Neurol 393:244-266.

Lino de Oliveira C, Guimaraes FS, Del Bel EA (1997). c-jun mRNA expression in the hippocampal formation induced by restraint stress. Brain Res 753:202-208.

Luckman SM, Dyball RE, Leng G (1994). Induction of c-fos expression in hypothalamic magnocellular neurons requires synaptic activation and not simply increased spike activity. J Neurosci 14:4825-4830.

Ludwig M, Johnstone LE, Neumann I, Landgraf R, Russell JA (1997). Direct hypertonic stimulation of the rat supraoptic nucleus increases c-fos expression in glial cells rather than magnocellular neurones. Cell Tissue Res 287:79-90.

Lyford GL, Yamagata K, Kaufmann WE, Barnes CA, Sanders LK, Copeland NG, Gilbert DJ, Jenkins NA, Lanahan AA, Worley PF (1995). Arc, a growth factor and activity-regulated gene, encodes a novel cytoskeleton-associated protein that is enriched in neuronal dendrites. Neuron 14:433-445.

Mansi JA, Rivest S, Drolet G (1998). Effect of immobilization stress on transcriptional activity of inducible immediate-early genes, corticotropin-releasing factor, its type 1 receptor, and enkephalin in the hypothalamus of borderline hypertensive rats. J Neurochem 70:1556-1566.

Márquez C, Belda X, Armario A (2002). Post-stress recovery of the pituitary-adrenal hormones and glucose, but not the response during exposure to the stressor, is a marker of stress intensity in highly stressful situations. Brain Res 926:181-185.

Martí O, Armario A (1988). Anterior pituitary response to stress: time-related changes and adaptation. Int J Devl Neurosci 16:241-260.

Martinez M, Phillips PJ, Herbert J (1998). Adaptation in patterns of c-fos expression in the rat brain associated with exposure to either single or repeated social stress in male rats. Eur J Neurosci 10:20-33.

McDougall SJ, Widdop RE, Lawrence AJ (2004). Medial prefrontal cortical integration of psychological stress in rats. Eur J Neurosci 20:2430-2440.

McEwen, BS (2000). The neurobiology of stress: from serendipity to clinical relevance. Brain Res 886:172-189.

Melia KR, Ryabinin AE, Schroeder R, Bloom FE, Wilson MC (1994). Induction and habituation of immediate early gene expression in rat brain by acute and repeated restraint stress. J Neurosci 14:5929-5938.

Moga DE, Calhoun ME, Chowdhury A, Worley P, Morrison JH, Shapiro ML (2003). Activity-regulated cytoskeletal-associated protein is localized to recently activated excitatory synapses. Neuroscience 125:7-11.

Morrow BA, Elsworth JD, Lee EJK, Roth RH (2000). Divergent effects of putative anxiolytics on stress-induced Fos expression in the mesoprefrontal system of the rat. Synapse 36:143-154.

Nagahara AH, Handa RJ (1997). Age-related changes in c-fos mRNA induction after open-field exposure in the rat brain. Neurobiol Aging 18:45-55.

Nakagawa T, Katsuya A, Tanimoto S, Yamamoto J, Yamauchi Y, Minami M, Satoh M (2003). Differential patterns of c-fos mRNA expression in the amygdaloid nuclei induced by chemical somatic and visceral noxious stimuli in rats. Neurosci Lett 344:197-200.

Nitsch R, Frotscher M (1992). Reduction of posttraumatic transneuronal "early gene" activation and dendritic atrophy by N-methyl-D-aspartate receptor antagonist MK-801. Proc Natl Acad Sci USA 89:5917-5200.

Ogilvie KM, Lee S, Rivier C (1998). Divergence in the expression of molecular markers of neuronal activation in the parvocellular paraventricular nucleus of the hypothalamus evoked by alcohol administration via different routes. J Neurosci 18:4344-4352.

Olsson T, Hakansson A, Seckl JR (1997). Ketanserin selectively blocks acute stress-induced changes in NGFI-A and mineralocorticoid receptor gene expression in hippocampal neurons. Neuroscience 76:441-448.

Ons S, Martí O, Armario A (2004). Stress-induced activation of the immediate early gene Arc (activity-regulated cytoskeleton-associated protein) is restricted to telencephalic areas in the rat brain: relationship to c-fos mRNA. J Neurochem 89:111-1118.

Pacak K, Palkovits M (2001). Stressor-specificity of central neuroendocrine responses: implications for stress-related disorders. Endocr Rev 22:502-548.

Papa M, Pellicano MP, Welz H, Sadile AG (1993). Distributed changes in c-Fos and c-Jun immunore-activity in the rat brain associated with arousal and habituation to novelty. Brain Res Bull 32:509-515.

Passerin AM, Cano G, Rabin BS, Delano BA, Napier JL, Sved AF (2000). Role of locus coeruleus in foot shock-evoked Fos expression in rat brain. Neuroscience 101:1071-1082.

Perrotti LI, Hadeishi Y, Ulery PG, Barrot M, Monteggia L, Duman RS, Nestler EJ (2004). Induction of ΔFosB in reward-related brain structures after chronic stress. J Neurosci 24:10594-10602.

Pezzone MA, Lee W-S, Hoffman GE, Rabin BS (1992). Induction of c-Fos immunoreactivity in the rat forebrain by conditioned and unconditioned aversive stimuli. Brain Res 597:41-50.

Pinaud R (2005). Critical calcium-regulated biochemical and gene expression programs involved in experience-dependent plasticity. In: Plasticity in the Visual System: From Genes to Circuits (Pinaud R, Tremere LA, De Weerd P, eds), pp. 153-180. New York: Springer-Verlag.

Pinaud R, Penner MR, Robertson HA, Currie RW (2001). Upregulation of the immediate early gene arc in the brains of rats exposed to environmental enrichment: implications for molecular plasticity. Brain Res Mol Brain Res 91:50-56.

Pinaud R, Tremere LA, Penner MR, Hess FF, Robertson HA, Currie RW (2002). Complexity of sensory environment drives the expression of candidate-plasticity gene, nerve growth factor induced-A. Neuroscience 112:573-582.

Proescholdt MG, Chakravarty S, Foster JA, Foti SB, Briley EM, Herkenham M (2002). Intracere-broventricular but not intraveous interleukin-1β induces widespread vascular-mediated leukocyte infiltration and immune signal mRNA expression followed by brain-wide glial activation. Neuroscience 112:731-749.

Radulovic J, Kammermeier J, Spiess J (1998). Relationship between Fos production and classical fear conditioning: effects of novelty, latent inhibition, and unconditioned stimulus preexposure. J Neurosci 18:7452-7461.

Riccio DC, MacArdy EA, Kissinger SC (1991). Associative processes in adaptation to repeated cold exposure in rats. Behav Neurosci 105:599-602.

Rivest S, Laflamme N (1995). Neuronal activity and neuropeptide gene transcription in the brains of immune-challenged rats. J Neuroendocrinol 7:501-525.

Robertson LM, Kerppola TM, Vendrell M, Luk D, Smeyne RJ, Bocchiaro C, Morgan JI, Curran T (1995). Regulation of c-fos expression in transgenic mice requires multiple interdependent transcription control elements. Neuron 14:241-252.

Rotllant D, Ons S, Carrasco J, Armario A (2002). Evidence that metyrapone can act as a stressor: effect on pituitary-adrenal hormones, plasma glucose and brain c-fos induction. Eur J Neurosci 16:693-700.

Rubio N, Martin-Clemente B (1999). Theiler's murine encephalomyelitis virus infection induces early expression of c-fos in astrocytes. Virology 268:21-29.

Ryabinin AE, Melia KR, Cole M, Bloom FE, Wilson MC (1995). Alcohol selectively attenuates stress-induced c-fos expression in rat hippocampus. J Neurosci 15:721-730.

Sawchenko PE, Li H-Y, Ericsson A (2000). Circuits and mechanisms governing hypothalamic responses to stress: a tale of two paradigms. Prog Brain Res 122:61-78.

Schreiber SS, Tocco G, Shors TJ, Thompson RF (1991). Activation of immediate early genes after acute stress. NeuroReport 2:17-20.

Selye H (1936). A syndrome produced by diverse nocuous agents. Nature 138:32.

Senba E, Umemoto S, Kawai Y, Noguchi K (1994). Differential expression of fos famili and jun family mRNAs in the rat hypothalamo-pituitary-adrenal axis after immobilization stress. Mol Brain Res 24:283-294.

Sharp FR, Sagar SM, Swanson RA (1993). Metabolic mapping with cellular resolution: c-fos vs. 2-deoxyglucose. Crit Rev Neurobiol 7:205-228.

Shen M, Greenberg ME (1990). The regulation and function of c-fos and other immediate early genes in the nervous system. Neuron 4:477-485.

Staiger JF, Masanneck C, Bisler S, Schleicher A, Zuschratter W, Zilles K (2002). Excitatory and in-hibitory neurons express c-Fos in barrel-related columns after exploration of a novel environment. Neuroscience 109:687-699.

Steward O, Worley PF (2001). Selective targeting of newly synthesized Arc mRNA to active synapses requires NMDA receptor activation. Neuron 30:227-240.

Steward O, Worley P (2002). Local synthesis of proteins at synaptic sites on dendrites: role in synaptic plasticity and memory consolidation? Neurobiol Learn Memory 78:508-527.

Steward O, Wallace CS, Lyford GL, Worley PF (1998). Synaptic activation causes the mRNA for the IEG Arc to localize selectively near activated postsynaptic sites on dendrites. Neuron 21:741-751.

Stampt JA, Herbert J (1999). Multiple immediate-early gene expression during physiological and endocrine adaptation to repeated stress. Neuroscience 94:1313-1322.

Sullivan GM, Arpegis J, Gorman JM, LeDoux JE (2003). Rodent doxapram model of panic: behavioural effects and c-Fos immunoreactivity in the amygdala. Biol Psychiatry 53:863-870.

Swank NW (1994). Coordinate regulation of Fos and Jun proteins in mouse brain by LiCl. NeuroReport 10:3685-3689.

Tanimura SM, Sanchez-Watts G, Watts AG (1998). Peptide gene activation, secretion, and steroid feedback during stimulation of rat neuroendocrine corticotropin-releasing hormone neurons. Endocrinology 139:3822-3829.

Tchélingérian J-L, Le Saux F, Pouzet B, Jacques C (1997). Widespread neuronal expression of c-Fos throughout the brain and local expression in glia following a hippocampal injury. Neurosci Lett 226:175-178.

Umegaki H, Zhu W, Nakamura A, Suzuki Y, Takada M, Endo H, Iguchi A (2003). Involvement of the entorhinal cortex in the stress response to immobilization but not to insulin-induced hypoglycaemia. J Neuroendocrinol 15:237-241.

Ueyama T, Ohya H, Yoshimura R, Senba E (1999). Effects of ethanol on the stress-induced expression of NGFI-A mRNA in the rat brain. Alcohol 18:171-176.

Umemoto S, Kawai Y, Senba E (1994a). Differential regulation of IEGs in the rat PVH in single and repeated stress models. NeuroReport 6:201-204.

Umemoto S, Kawai Y, Ueyama T, Senba E (1997). Chronic glucocorticoid administration as well as repeated stress affects the subsequent acute immobilization stress-induced expression of immediate early genes but not that of NGFI-A. Neuroscience 80:763-773.

Umemoto S, Noguchi K, Kawai Y, Senba E (1994b). Repeated stress reduces the subsequent stress-induced expression of Fos in rat brain. Neurosci Lett 167:101-104.

Vallés A, Martí O, Armario A (2005). Mapping the areas sensitive to long-term endotoxin tolerance in the rat brain: a c-fos mRNA study. J Neurochem 93:1177-1188.

Viau V, Sawchenko PE (2002). Hypophysiotropic neurons of the paraventricular nucleus respond in spatially, temporally and phenotypically differentiated manners to acute vs. repeated restraint stress. J Comp Neurol 445:293-307.

Wang K, Guldenaar SEF, McCabe JT (1997). Fos and Jun expression in rat supraoptic nucleus neurons after acute vs. repeated osmotic stimulation. Brain Res 746:117-125.

Watanabe Y, Stone E, McEwen BS (1994). Induction and habituation of c-fos and zif/268 by acute and repeated stressors. NeuroReport 5:1321-1324.

Wisden W, Errington ML, Williams S, Dunnett SB, Waters C, Hitchocock D, Evans G, Bliss TV, Hunt SP (1990). Differential expression of immediate early genes in the hippocampus and spinal cord. Neuron 4:603-614.

Worley PF, Bhat RV, Baraban JM, Erickson CA, McNaughton BL, Barnes CA (1993). Thresholds for synaptic activation of transcription factors in the hippocampus: correlation with long-term enhancement. J Neurosci 13:4776-4786.

Xu Y, Day TA, Buller KM (1999). The central amygdala modulates hypothalamic-pituitary-adrenal axis responses to systemic interleukin-1β administration. Neuroscience 94:175-183.

Zangenehpour S, Chaudhuri A (2002). Differential induction and decay curves of c-fos and zif268 revealed through dual activity maps. Mol Brain Res 109:221-225.

Ziegler DR, Cass WA, Herman JP (1999). Excitatory influence of the locus coeruleus in hypothalamic-pituitary-adrenocortical axis response to stress. J Neuroendocrinol 11:361-369.

12

Transcriptional Control of Nerve Cell Death, Survival and Repair

RACHEL CAMERON[a] and MIKE DRAGUNOW[a,b]

[a]*Department of Pharmacology and Clinical Pharmacology, The University of Auckland, Auckland, New Zealand*
[b]*National Research Centre for Growth and Development, The University of Auckland, Auckland, New Zealand*

1. Introduction

Many neurological disorders are caused by death and atrophy of nerve cells in different brain regions. This neurodegeneration is partly offset by regenerative processes presumably activated to maintain brain homeostasis. Over the past 10 years there has been a wealth of information gathered on the mechanisms responsible for these processes. Transcription factors are critically involved in both the neurodegenerative and neuroregenerative changes and a number of excellent reviews have been written on these topics (see Herdegen and Leah, 1998).

In the present manuscript we focus upon a number of interrelated basic leucine-zipper (bZIP) transcription factors, notably, c-Jun, activating transcription factor 3 (ATF-3), activating transcription factor 2 (ATF-2) and cyclic AMP response element binding protein (CREB). These transcription factors have many diverse functions but we will concentrate on their regulation of nerve cell death, survival and axonal repair.

2. c-Jun and Jun N-terminal Kinases

c-Jun is an inducible bZIP transcription factor that is a key component of the activator protein-1 complex (AP-1). c-Jun is interesting in that it regulates the expression of genes involved in seemingly opposite cellular functions and is a mediator of cell death (Ham et al., 1995) as well as cell proliferation (Shaulian and Karin, 2001), survival and differentiation (Dragunow et al., 2000). Activation of c-Jun-mediated gene expression occurs in response to phosphorylation of two amino-terminal serine residues (Ser63, Ser73) through a process that is

R. Pinaud, L.A. Tremere (Eds.), Immediate Early Genes in Sensory Processing, Cognitive Performance and Neurological Disorders, 223–242, ©2006 Springer Science + Business Media, LLC

principally mediated by the three c-Jun N-terminal kinases (JNKs); JNK1, JNK2 and JNK3 (Pulverer et al., 1991; Smeal et al., 1991). While JNK1 and JNK2 are present ubiquitously, JNK3 has a more restricted pattern of expression and is an important activator of c-Jun in neurons (Kumagae et al., 1999). Once activated by phosphorylation, c-Jun must dimerise with itself or another bZIP transcription factor in order to alter gene expression. Numerous possible partners exist including members of the Jun, Fos and activating transcription factor (ATF) family (Herdegen and Leah, 1998) and it is the composition of the c-Jun dimer that confers DNA-binding specificity. Thus, by associating with different partners, signals that initiate c-Jun expression can result in a wide variety of cellular responses.

There is much scientific interest in c-Jun because its expression is induced in injured neurons (detailed below) suggesting that c-Jun alters the expression of genes involved in either nerve cell death, or survival and repair. Over the years, it has become apparent that the actions of c-Jun are not restricted to one or other of these processes, and there is now strong evidence suggesting that c-Jun is an important mediator of both neuronal degeneration and regeneration.

3. c-Jun and Nerve Cell Death

A growing body of evidence suggests a key role for c-Jun in nerve cell death. *In vivo*, c-Jun is expressed in damaged neurons undergoing delayed cell death after hypoxic ischemic episodes (Dragunow et al., 1993) and in axotomised neurons (Dragunow, 1992). Furthermore, c-Jun induction has been observed *in vitro* in cultured cerebellar granule neurons undergoing apoptosis (Watson et al., 1998). Maintenance of cerebellar granule neurons in culture requires the presence of a depolarizing concentration of potassium, and depletion of this ion results in apoptotic cell death with a concomitant induction of c-Jun (Watson et al., 1998). In a similar manner, c-Jun expression is elevated in apoptotic sympathetic neurons deprived of nerve growth factor (NGF) (Estus et al., 1994; Ham et al., 1995). Cultured sympathetic neurons depend on NGF for survival and NGF-deprivation induces apoptosis that is accompanied by an increase in c-Jun expression. In this model, blocking the action of c-Jun by microinjection of neutralising c-Jun antibodies (Estus et al., 1994) or with a dominant negative c-Jun mutant (Ham et al., 1995) prevents apoptosis after NGF withdrawal, demonstrating a pro-apoptotic role for c-Jun in sympathetic neurons.

Several studies have demonstrated the importance of phosphorylation in the pro-apoptotic actions of c-Jun. Mutant mice where the codons for Ser63 and Ser73 are changed to non-phosphorylatable alanines (c-JunAA) (Behrens et al., 1999) are protected from neuronal apoptosis when treated with kainic acid, a toxin that is known to promote c-Jun induction and apoptosis of hippocampal neurons (Yang et al., 1997; Behrens et al., 1999). Furthermore, transfection of cerebellar granule neurons with a non-phosphorylatable c-Jun mutant offers protection from apoptosis induced by potassium depletion, whereas apoptosis is induced by transfection of a constitutively active c-Jun mutant where phosphoacceptor sites are replaced with aspartate residues (Watson et al., 1998). Evidence now suggests that the pro-apoptotic actions of c-Jun are specifically mediated by JNK3-dependent phosphorylation. Primary cultures of rat cortical neurons undergo apoptosis when treated with the environmental toxin sodium arsenite (Namgung and Xia, 2000). In this paradigm, apoptosis is accompanied by c-Jun phosphorylation that is mediated by selective activation of JNK3, but not JNK1 or JNK2 (Namgung and Xia, 2000). Moreover, JNK3 knock out mice are resistant to kainic acid-induced seizures and show reduced c-Jun phosphorylation, AP-1 activity and apoptosis of hippocampal neurons (Yang et al., 1997). In a similar manner, hippocampal neurons of JNK3-deficient mice subjected to hypoxic ischemic episodes show reduced phosphorylation of c-Jun and are protected from apoptosis (Kuan et al., 2003). Thus, JNK3-mediated phosphorylation is implicated in the pro-apoptotic actions of c-Jun in neurons.

4. c-Jun, Bim and the Mitochondrial Pathway of Apoptosis

Despite mounting evidence suggesting a role for phosphorylated c-Jun in neuronal apoptosis, the mechanisms mediating this effect remain to be determined. However, recent progress has identified the involvement of Bcl-2 interacting mediator of cell death (Bim), and activation of the intrinsic mitochondrial pathway of apoptosis in c-Jun-mediated cell death. Bim is a pro-apoptotic protein of the Bcl-2 family that promotes the release of mitochondrial cytochrome c and formation of the apoptosome, a complex comprised of cytochrome c, apoptotic protease-activating factor 1 (Apaf-1), ATP and pro-caspase 9. Assembly of the apoptosome initiates sequential activation of caspases 9 and 3, and ultimately results in cell death by a series of caspase 3-mediated cleavage events.

Several lines of evidence now suggest that Bim plays an important role in promoting c-Jun-mediated apoptosis. For example, elevated Bim expression has been detected in dying sympathetic neurons deprived of NGF (Whitfield et al., 2001), cerebellar granule neurons deprived of potassium (Harris and Johnson, 2001) and retinal ganglion cells after optic nerve transection (Napankangas et al., 2003), and in all of these paradigms, Bim expression correlates with the induction of c-Jun. Furthermore, over-expression of Bim alone is sufficient to induce apoptosis of both sympathetic neurons (Whitfield et al., 2001) and cerebellar granule cells (Harris and Johnson, 2001), and blocking Bim with antisense nucleotides delays apoptosis in sympathetic neurons after NGF withdrawal (Whitfield et al., 2001).

Bim expression in dying neurons is likely to result from activation of the JNK pathway because an inhibitor of JNK-mediated signalling, CEP-1347, reduces Bim induction in cerebellar granule cells (Harris and Johnson, 2001) and transfection of a dominant negative c-Jun mutant partially prevents Bim expression in NGF-depleted sympathetic neurons (Whitfield et al., 2001). Furthermore, recent studies have demonstrated that Bim can undergo JNK-mediated phosphorylation (Lei and Davis, 2003; Putcha et al., 2003), which, in fibroblasts, has been shown to free Bim from sequestration by cytoplasmic dyenin and myosin V motor complexes, allowing translocation to the mitochondria and release of cytochrome *c* (whether a similar mechanism operates in neurons remains to be determined) (Lei and Davis, 2003). It is probable that Bim associates with a second pro-apoptotic Bcl-2 family member, Bax, to initiate cytochrome *c* release and apoptosis, as the expression of c-Jun, Bim and Bax are temporally correlated in retinal ganglion cells after optic nerve transection (Napankangas et al., 2003) and over-expression of Bim does not kill cerebellar granule cells from Bax-deficient mice (Harris and Johnson, 2001). Thus, while it is tempting to speculate that injured neurons undergo apoptosis through JNK/c-Jun-mediated activation of Bim, it should be noted that JNK-independent Bim induction in potassium-deprived cerebellar granule cells has recently been reported (Shi et al., 2005). Therefore, while Bim induction may result from JNK/c-Jun pathway activation in some contexts, this is not always the case, and suggests the involvement of additional pro-apoptotic targets of the JNK/c-Jun pathway.

5. c-Jun, the Fas Ligand and Apoptosis

Studies using T cells of the immune system identified the Fas ligand (FasL) as an additional pro-apoptotic target of the JNK/c-Jun pathway (Faris et al., 1998) and induction of the FasL has now been observed in apoptotic neurons (Le-Niculescu et al., 1999). FasL induces apoptosis by binding to the cell surface Fas receptor and initiating a sequence of intracellular signalling events that culminate in caspase activation and apoptotic cell death. In cerebellar granule neurons deprived of potassium, apoptosis is accompanied by c-Jun phosphorylation and expression of FasL (Le-Niculescu et al., 1999; Ginham et al., 2001), and addition of a decoy that sequesters FasL inhibits apoptosis in these cells (Le-Niculescu et al., 1999). Interestingly, the FasL promoter contains an AP-1 binding site, allowing regulation of FasL gene expression by c-Jun dimers (Faris et al., 1998). There is evidence to suggest that c-Jun:ATF-2 heterodimers regulate FasL gene expression, as transfection of T cells with non-phosphorylatable mutants of c-Jun or ATF-2 prevent activation of the FasL promoter (Faris et al., 1998). Similarly, FasL induction can be prevented in neuronal PC12 cells with an inhibitor of p38 mitogen-activated protein kinase (p38 MAPK), which is the principal kinase that phosphorylates and activates ATF-2 (Le-Niculescu et al., 1999). Thus, it is possible that in some instances, c-Jun induces apoptosis by dimerizing with ATF-2 and inducing the expression of the FasL.

6. c-Jun, Apoptosis and the Neurofibrillar Pathology of Alzheimer's Disease

In addition to the cell death paradigms already discussed, recent developments have also identified a role for c-Jun in neurodegenerative disorders. Upregulation of neuronal c-Jun has been detected in post-mortem tissue from Parkinson's disease sufferers (Hunot et al., 2004) and c-Jun is also activated in a mouse model of Parkinson's disease where treatment with a neurotoxin, 1-methyl-4-phenyl-1,2,3,6-tetrahydropyridine (MPTP), induces selective degeneration of the dopaminergic neurons of the substantia nigra (Hunot et al., 2004). Since administration of a JNK inhibitor (SP600125) to MPTP-treated mice reduces c-Jun phosphorylation and prevents apoptosis of the dopaminergic substantia nigra neurons, it is possible that c-Jun plays a role in promoting neurodegeneration in Parkinson's disease (Wang et al., 2004).

c-Jun and JNKs may also be involved in Alzheimer's disease (AD), which is characterized by amyloid plaques and tau-positive tangles. We have detected c-Jun in tangle-bearing neurons in AD (MacGibbon et al., 1997) and Anderson et al. (Anderson et al., 1994) detected c-Jun co-localised with tau in AD brain suggesting that c-Jun expression occurs in tau-positive tangles. Furthermore, a number of protein kinases that can phosphorylate c-Jun are also activated in tau-positive, tangle-bearing neurons in AD brain, including JNKs, extracellular signal-regulated kinases (ERKs), and p38-MAPK (Hensley et al., 1999; Zhu et al., 2001, 2003; Pei et al., 2002; Helbecque et al., 2003; Swatton et al., 2004). Tangle-bearing neurons and other cells in AD brain also show an upregulation of c-Jun and evidence of apoptotic death as detected by DNA fragmentation (with TUNEL) and activation of caspases 3, 6 and 9, which are effectors of apoptosis in many cell types including neurons (Dragunow et al., 1993; Su et al., 1994; Dragunow and Preston, 1995; Anderson et al., 1996; Sheng et al., 1998; Stadelmann et al., 1999; Rohn et al., 2001, 2002; Guo et al., 2004). Thus, tau-positive tangle-bearing neurons in AD brain express activated kinases, c-Jun and undergo apoptosis.

Hyperphosphorylation, accumulation and aggregation of tau are hallmarks of the neurofibrillary lesions of AD, and this aggregation correlates with the cognitive decline in AD highlighting its importance in disease pathogenesis. Evidence from AD brains suggest that this tau pathology may occur because of a reduction of levels of protein phosphatase 2A (PP2A) (Gong et al., 1995; Vogelsberg-Ragaglia et al., 2001) a phosphatase that dephosphorylates tau. Okadaic acid, a drug which inhibits PP2A activity models important aspects of tau neurofibrillar lesions in dissociated cultures and in organotypic slice cultures, including abnormal phosphorylation and accumulation of tau, and activation of protein kinases (Pei et al., 2003). Furthermore, neuron-specific PP2A inhibition in transgenic mice leads to JNK, ERK and tau phosphorylation, and c-Jun induction (Kins et al., 2003), similar to that observed in AD. We have found, using cell cultures, that okadaic acid causes phosphorylation of c-Jun and ATF-2 in neuronal cells and causes apoptosis (Walton et al., 1998; Woodgate et al., 1999; Dragunow et al., 2000). Thus, the loss of PP2A activity in AD brain may be responsible for changes in stress kinases, transcription factors (c-Jun), tau phosphorylation and aggregation, and may cause the neurofibrillar tangle (NFT) pathology and the associated apoptotic nerve cell death.

Additionally, there is decreased proteasome activity in AD brain (Keller et al., 2000) which might itself be a result of tau aggregation (Keck et al., 2003). Pharmacological inhibition of proteasome function with MG132 stabilizes okadaic acid-induced tau inclusions *in vitro* (Goldbaum et al., 2003). This suggests that the proteasome clears tau inclusions, which is supported by a recent study (Oddo et al., 2004). The proteasome normally clears misfolded/aberrant proteins and it has recently gained prominence in a number of protein misfolding disorders including AD. Proteasomal inhibition also leads to JNK activation, c-Jun phosphorylation and alterations in neurofilaments characteristic of AD in neuronal cells (Masaki et al., 2000).

These combined results suggest that blockade of PP2A and the proteasome may model important aspects of the neurofibrillar pathology of AD and both also produce c-Jun induction, phosphorylation and apoptosis. c-Jun might therefore, contribute to AD by a playing a role in the development of neurofibrillar pathology and/or it may be a down-stream inducer of death in tangle-bearing neurons in AD brain.

7. c-Jun and Neuronal Regeneration

While there is clear evidence demonstrating the ability of c-Jun to promote neuronal apoptosis and neurodegeneration, paradoxically, there is equally convincing data demonstrating a role for c-Jun in nerve cell survival and repair. Interestingly, transfection of neuronal PC12 cells with c-Jun promotes neurite outgrowth without inducing cell death (Dragunow et al., 2000). In this model, rather than promoting apoptosis, c-Jun transfection confers resistance against apoptosis induced by the toxin okadaic acid, suggesting an anti-apoptotic action of c-Jun in this paradigm (Dragunow et al., 2000). Similar neurite outgrowth has also been achieved in PC12 cells by microinjection of a constitutively active mimic of phosphorylated c-Jun (Leppa et al., 1998). Therefore, these data demonstrate that the function of c-Jun in neuronal cells is not exclusively pro-apoptotic and that c-Jun expression also mediates signals involved in neuronal growth and regeneration.

Evidence implicating c-Jun in neuronal regeneration has also been provided by studies of axonal re-growth after axotomy. Different populations of neurons retain varying capacities for repair after injury. In simple terms, axons of peripheral neurons regenerate relatively easily, while those in the central nervous system display a poor regenerative response. Interestingly, the ability of a neuron to repair itself is correlated with its ability to express c-Jun. For example, injury to a peripheral branch of a primary sensory neuron is associated with a robust regenerative response and a strong induction of c-Jun (Jenkins et al., 1993). Conversely, injury of a central process of a primary sensory neuron is associated with a poor regenerative response and little induction of c-Jun (Jenkins et al., 1993). Similar patterns of c-Jun expression and regeneration have also been observed after peripheral or central axotomy of the dorsal root ganglia neurons of adult rats, where the strength of the regenerative response correlates with the induction of c-Jun, implicating c-Jun in the regenerative process (Broude et al., 1997).

Transgenic mice that do not express c-Jun in the neurons and glia of the central nervous system have been used to determine the role of c-Jun in axonal repair and regeneration (Raivich et al., 2004). Transection of the facial nerve in these mice is associated with reduced motor neuron reconnection to the whisker pad compared with control mice expressing c-Jun. This observation implies that c-Jun is required for axonal regeneration after axotomy; however, it should be noted that partial axonal regeneration still occurs in the c-Jun mutant mice, although at a greatly reduced rate (Raivich et al., 2004). These results indicate that the expression of c-Jun improves the efficiency of axonal repair but is not an absolute requirement for successful regeneration. Conversely, axonal growth during development of the c-Jun mutant mice was apparently unaltered, as the mutants displayed typical brain morphology and normal outcomes in assessments of locomotor function, hippocampal-dependent learning and memory, and amygdala-dependent fear conditioning tasks (Raivich et al., 2004). These observations suggest that axogenesis during brain development can take place in the absence of c-Jun and that c-Jun improves the efficiency of adult nerve cell repair but is not essential for regeneration to proceed.

The role of c-Jun in axonal regeneration has been further investigated using transgenic mice that selectively over-express c-Jun in Purkinje cells of the cerebellum (Carulli et al., 2002). Purkinje cells have a very limited capacity for axonal re-growth after injury and therefore provide a valuable model to study the effects of over-expression of c-Jun on nerve cell regeneration. Interestingly, Purkinje cells from c-Jun over-expressing transgenic mice develop normally with no morphological evidence of increased axonal sprouting or cell death despite the association of c-Jun with both of these processes (Carulli et al., 2002). Furthermore, axotomy of Purkinje cells from c-Jun transgenic mice did not result in an increased regenerative potential (Carulli et al., 2002). This lack of effect of c-Jun over-expression on axonal repair may suggest that c-Jun does not mediate axonal regeneration in these cells; however, as Purkinje cells are renowned for their poor rate of recovery after injury, it is possible that over-expression of c-Jun alone is insufficient to mount a regenerative response in this cell type, and that Purkinje cells may lack the necessary transcription factor partners to undertake a regeneration program. Thus, the lack of effect in this model does not exclude c-Jun from the nerve cell regeneration process but suggests that c-Jun cannot act alone to induce axogenesis in these cells.

8. ATF-3 and c-Jun-Mediated Regeneration

The data reviewed here suggests that the induction of c-Jun improves the efficiency of the regenerative response to nerve cell damage. Thus, it is important to consider the mechanisms that operate to induce c-Jun-facilitated axogenesis. As c-Jun is a transcription factor and a key member of the AP-1 complex, it is probable that the regenerative process involves the expression of growth and repair-related genes. As previously mentioned, c-Jun must dimerize in order to bind to DNA and alter gene expression, and increasing evidence implicates ATF-3 as the dimer partner that mediates the reparative actions of c-Jun.

ATF-3 is a bZIP transcription factor that belongs to the ATF/CREB family and like c-Jun, ATF-3 alters gene transcription acting as a dimer with other bZIP transcription factors. As a homodimer ATF-3 represses transcription, whereas heterodimers of ATF-3 and c-Jun activate transcription (Liang et al., 1996). In the nervous system, ATF-3 is not normally present (Tsujino et al., 2000), yet its expression is rapidly induced in response to a number of stressful stimuli such as ischemia and seizures. The mechanisms governing the induction of ATF-3 remain to be determined, yet the ATF-3 gene contains several regulatory transcription factor binding sites including AP-1 and CRE sites, indicating a role for c-Jun in the regulation of ATF-3 expression (Liang et al., 1996).

The functions of ATF-3 are incompletely understood, although a role in cellular survival seems probable. In rats, over-expression of ATF-3 protects hippocampal neurons from kainic acid-induced apoptosis (Francis et al., 2004). Similarly, over-expression of ATF-3 in human umbilical vein endothelial cells (HUVECs) confers protection from tumor necrosis factor α-induced apoptosis (Kawauchi et al., 2002). In the nervous system, the expression of ATF-3 has been observed in axotomised neurons (Takeda et al., 2000; Tsujino et al., 2000; Nakagomi et al., 2003) and evidence now suggests a key role for ATF-3 in the regenerative process. Transection of adult rat hypoglossal motor neurons leads to a strong induction of ATF-3 and c-Jun that is maintained for 4 weeks while nerve cell regeneration occurs (Nakagomi et al., 2003). Furthermore, axotomy of spinal motor neurons in adult rats is followed by co-localized induction of ATF-3 and c-Jun (Tsujino et al., 2000). As almost all motor neurons survive after axotomy, this suggests the possibility that ATF-3 and c-Jun are involved in the survival and repair response (Tsujino et al., 2000). Similarly, axotomy of sensory dorsal root ganglion neurons in adult rats is also associated with the induction of ATF-3 and c-Jun, and almost all c-Jun positive neurons express ATF-3 (Tsujino et al., 2000). However, some neurons express ATF-3 in the absence of c-Jun and some of these cells die (Tsujino et al., 2000). Thus, it is possible that c-Jun:ATF-3 dimers confer protection, but in the absence of c-Jun, ATF-3 alone cannot protect cells from death (Tsujino et al., 2000).

Further evidence of a role for ATF-3 and c-Jun in the regenerative response to nerve cell injury has been provided by studies of optic nerve crush in adult rats. Optic nerve crush is a model of central nervous system injury and is typically associated with an initial regenerative phase followed by degeneration of the retinal ganglion cells. ATF-3 is not expressed in uninjured retinas however, after optic nerve crush, ATF-3 induction occurs in the nuclei of injured retinal ganglion cells (Takeda et al., 2000). The expression of ATF-3 is transient and terminates prior to the onset of retinal ganglion cell loss (Takeda et al., 2000). c-Jun expression is also induced in this model; however, the temporal characteristics are different, with c-Jun expression occurring earlier and being maintained for longer than ATF-3 (Takeda et al., 2000). Thus, it is tempting to speculate that early expression of ATF-3:c-Jun dimers occurs during the regenerative phase of the response, and that later loss of functional ATF-3:c-Jun dimers signals the end of the regenerative phase and leads to nerve cell degeneration by a c-Jun-dependent mechanism.

Lindwall et al. (2004) have recently provided further evidence demonstrating the importance of c-Jun and ATF-3 in nerve cell regeneration. Using axotomised neurons from adult rats, these authors demonstrated activation of JNK, c-Jun and the induction of ATF-3 in the nuclei of axotomised neurons (Lindwall et al., 2004). Retrograde tracing showed that phosphorylated c-Jun and ATF-3 were present in regenerating neurons and inhibition of JNK almost completely prevented c-Jun activation and ATF-3 induction, and significantly reduced axonal re-growth (Lindwall et al., 2004). Furthermore, transfection of PC12 cells with c-Jun promotes neurite outgrowth that can be significantly increased on co-expression of ATF-3 (Pearson et al., 2003). Collectively, these studies provide evidence of the importance of c-Jun and ATF-3 in the neuronal regenerative response.

Additional confirmation of the importance of c-Jun:ATF-3 heterodimers in nerve cell regeneration has been provided by an elegant series of cell culture experiments conducted by Nakagomi et al. (2003). These authors used PC12 cells infected with an adenoviral vector to express the catalytic domain of the JNK activator MEKK1. Infection with MEKK1 alone resulted in activation of JNK and c-Jun-mediated PC12 cell death. Concurrent over-expression of ATF-3 in these cells promoted survival and led to the extension of neurite outgrowths. These results suggest that in the absence of ATF-3, c-Jun promotes cell death, yet when ATF-3 is present, a survival and repair response is initiated. Support for this hypothesis was provided using superior cervical ganglion cells undergoing NGF withdrawal-induced apoptosis that is know to occur through a c-Jun mediated process (Estus et al., 1994; Ham et al., 1995). Over-expression of ATF-3 protected these cells from apoptosis induced by NGF depletion and instead promoted neurite extension (Nakagomi et al., 2003). In this study, the protective effect of c-Jun:ATF-3 heterodimers was demonstrated to rely partly on the induction of an anti-apoptotic heat shock protein, Hsp27, and subsequent activation of the Akt (PKB) survival pathway. However, the molecular mechanisms mediating the neurite elongation response remain to be elucidated, although induction of target genes such as growth associated protein 43 (GAP43) (Tsuzuki et al., 2002) and cell adhesion integrin molecules (Carulli et al., 2004) seems probable.

The phosphorylation state of Jun may also regulate its functional effects. In our own laboratory, when c-Jun transfection induced PC12 cell neuritogenesis and survival, most transfected cells were phosphorylated on Ser73, but not on Ser63, and a mutant c-Jun construct that cannot be phosphorylated on Ser73 (mJun73) could not drive neuritogenesis, whereas a Ser63 mutant (mJun63) could (Dragunow et al., 2000). Others have found an association of Ser73 phospho-rylation of c-Jun with cell survival (Huang et al., 2000; Schwarz et al., 2002) and after axotomy (Herdegen et al., 1998). In contrast, during amylin-induced apoptosis we found that Ser63, but not Ser73, was phosphorylated on c-Jun and this was related to the induction of apoptosis (Zhang et al., 2002, 2003). Similar results have more recently been obtained by Li et al. (Li et al., 2004). They found that nitric oxide-induced apoptosis in SH-SY5Y human neuroblastoma cells was associated with Ser63, but not Ser73 phosphorylation of c-Jun. Furthermore, they showed that a mutant Ser63 construct, but not a mutant Ser73 construct inhibited nitric oxide-induced apoptosis (Li et al., 2004). These combined results suggest

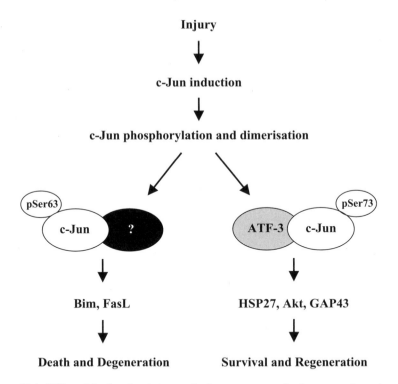

Figure 12.1. Differential phosphorylation and dimer partner selection may determine the c-Jun-mediated response to injury. In the proposed scheme, phosphorylation of c-Jun at Ser63 and dimerization (possibly with ATF-2) induces cell death and degeneration by the induction of Bim and FasL. In contrast, phosphorylation of c-Jun at Ser73 and dimerization with ATF-3 promotes cell survival and regeneration.

that the divergent functions of c-Jun on cell survival/axon growth and apoptosis, are mediated at least partly by a combination of dimer partner selection and differential phosphorylation of serine residues (Fig. 12.1).

9. ATF-2

A further bZIP transcription factor whose activation has been associated with cell survival and death, is the constitutively expressed activating transcription factor-2 (ATF-2). Like c-Jun, ATF-2 undergoes MAPK-mediated phosphorylation (typically by p38 MAPK although JNK can also phosphorylate ATF-2) and forms dimers in order to alter gene expression (Herdegen and Leah, 1998). As mentioned above, previous studies have shown that the dimerization of ATF-2 with c-Jun can drive the expression of FasL and this might be related to apoptotic cell death. Dimerization of c-Jun with the transcription factor ATF-2 may drive c-Jun towards apoptosis. We have found that ATF-2 is phosphorylated in neurons and pancreatic islet cells during apoptosis (Walton et al., 1998; Zhang et al., 2002) and blocking ATF-2 phosphorylation by inhibition of the upstream kinase p38

MAPK prevents islet cell apoptosis (Zhang et al., 2005). Thus, ATF-2 may signal for apoptosis both in neuronal cells and in islets (where it may be involved in the development of diabetes).

In contrast to these results and conclusions, studies of the expression of ATF-2 in normal and diseased human brain suggest that ATF-2 may be lost in neurons undergoing the death process (Pearson et al., 2005). Pearson et al. (2005) found that neurons, but not astrocytes, in the normal adult human brain expressed ATF-2 at very high levels under basal conditions. Regions of high expression included the cerebellum, cerebral cortex, brain stem, pigmented cells of the substantia nigra and locus coeruleus, and the hippocampus. In the brains of neurologically diseased cases they found regional losses of ATF-2 that reflected the patterns of neuronal death. In Parkinson's disease brain there was a dramatic loss of ATF-2 in pigmented substantia nigra neurons. In Alzheimer's disease there was a dramatic loss of ATF-2 in neurons in the hippocampus and in Huntington's disease there was a dramatic loss of ATF-2 in neurons in the caudate-putamen. This pattern suggests that ATF-2 might be involved in neuronal survival or at least may be lost in dying neurons. Paradoxically, Pearson et al. (2005) also found that cells in the sub-ependymal layer of the caudate nucleus of Huntington's disease brains expressed ATF-2 at high levels and triple-label studies showed that these were mainly GFAP- and PCNA-positive, although a small proportion were also MAP-2- and PCNA-positive. These cells are thought to be undergoing gliogenesis or neurogenesis (Curtis et al., 2003). This suggests that ATF-2 is expressed in cells undergoing gliogenesis or neurogenesis in Huntington's disease human brain. Whether ATF-2 is involved in neurogenesis or in the survival of the newly formed neurons and astrocytes is presently unclear.

10. CREB and Cell Survival

A number of years ago we first found that the transcription factor CREB was activated by phosphorylation in neurons that survived hypoxic-ischemic mediated cell death (Walton et al., 1996). We then showed that over-expressing wild-type CREB in neuronal cells promoted their resistance to the toxin okadaic acid (Walton et al., 1999a). Many reports have replicated and extended our studies and CREB is now recognized as a major transcriptional regulator of neuronal survival (reviewed in Walton and Dragunow, 2000 and Dragunow, 2004). There is also evidence that this CREB survival signaling cassette may be lost in some neurodegenerative disorders (reviewed in Dragunow, 2004).

11. CREB and Neuroprotective Drug Development

CREB also mediates the survival-promoting effects of various neurotrophins, for example, NGF and BDNF (Riccio et al., 1999; Dragunow, 2004), neurotransmitters such as glutamate (Mabuchi et al., 2001; Hardingham et al., 2002) as well as ischemic preconditioning (Hara et al., 2003; Lee et al., 2004). Recently, the phosphodiesterase 4 inhibitor rolipram has been shown to mediate neuroprotective effects through CREB (Gong et al., 2004). CREB may mediate both the neurotoxic and neuroprotective effects of NMDA receptor stimulation (Hardingham et al., 2002) and this may be related to the duration of CREB phosphorylation. Prolonged CREB phosphorylation and neuroprotection is produced by synaptically-driven NMDA receptor activation, whereas neurotoxic NMDA receptor stimulation with 50 μM NMDA elicits only brief CREB phosphorylation (Lee et al., 2005). Calcineurin activation by NMDA appears to be responsible for the de-phosphorylation of CREB during neurotoxic NMDA receptor stimulation (Lee et al., 2005). In the brain the threshold for activation of CREB by synaptically-driven stimulation may be higher as we found that only strong burst of LTP-inducing stimulation, which also lead to a near permanent form of LTP, drove CREB phosphorylation in dentate granule cells (Abraham et al., 2002). Understanding how drugs, neurotransmitters and neurotrophins promote cell survival through CREB activation will lead to the identification of targets that may facilitate the development of novel neuroprotective molecules. We recently proposed that the cell survival promoting effects of the neurotransmitter acetylcholine (Yan et al., 1995) might be due to activation of CREB (Dragunow, 2004), based upon our studies demonstrating a novel pathway of CREB phosphorylation in human neuroblastoma cells following activation of muscarinic acetylcholine receptors (Dragunow and Henderson, 1999; Greenwood and Dragunow, 2002).

12. Muscarinic Receptor Signalling and Neuroprotection

Historically acetylcholine has been targeted in the treatment of AD (using acetylcholinesterase inhibitors), not for its ability to promote neuronal survival, but for its permissive effects on learning and memory. AD is caused by dysfunction and death of nerve cells in the brain. One of the earliest symptoms of AD is memory impairment (perhaps due to the early loss of acetylcholine neurons), and treatments aim to reduce memory loss in patients. The M1 muscarinic receptor has been a focus for drug development to improve memory processes in AD, given its vital role in memory formation (Messer, 2002). Additionally, the M1 receptor might also mediate neuroprotection (Lindenboim et al., 1995; Murga et al., 1998; Leloup et al., 2000), but because it is depleted in AD brain (Rodriguez-Puertas et al., 1997) this may ultimately prove to be an ineffective therapeutic target for established AD patients. Activation of the M3 muscarinic receptor has also been shown to be neuroprotective (Yan et al., 1995; Budd et al., 2003; Greenwood and Dragunow, in preparation), but in contrast to the loss of the M1 muscarinic receptor the M3 receptor is preserved in AD brain (Flynn et al., 1995; Rodriguez-Puertas et al., 1997), indicating that it may be a better target for the development of neuroprotective AD treatments. However, the molecular basis of M3 muscarinic receptor-mediated neuroprotection is unclear.

13. Transcription Factors and M3 Muscarinic Receptor Signalling

A number of years ago, we found that activation of muscarinic receptors in rodent brain leads to the induction of transcription factors (EGR1, Nur77, Fos, JunB) in neurons in the neocortex, basal ganglia and limbic system including the hippocampus (Hughes and Dragunow, 1993, 1994, 1995; MacGibbon et al., 1995; Dragunow et al., 1997). These transcription factors have many functions, including in memory formation (Abraham et al., 1994; Dragunow, 1996; Dragunow et al., 1996; Walton et al., 1999b; Greenwood et al., 2004). To dissect out the receptor sub-types and signaling pathways involved in this gene activation, we embarked on *in vitro* studies using SK-N-SH human neuroblastoma cells (Dragunow and Henderson, 1999; Greenwood and Dragunow, 2002), which express muscarinic receptors and are widely used as models of human neuronal cells. We found that activation of muscarinic receptors with carbachol leads to induction of EGR1 (similar to our *in vivo* results; Hughes and Dragunow, 1994), and also to CREB phosphorylation (Dragunow and Henderson, 1999; Greenwood and Dragunow, 2002). Induction of EGR1, but not CREB, requires the phosphorylation of the MAP kinases ERK1/2. Rosenblum et al. (2000) also found that carbachol activates ERK in primary cortical neurons (Rosenblum et al., 2000). More recently, we found that the M3 muscarinic receptor antagonist 4-DAMP (but not M1 or M2 antagonists) inhibits carbachol-mediated activation of CREB, ERK and EGR1 (Greenwood and Dragunow, in preparation). This result indicates that M3 muscarinic acetylcholine receptor activation drives the ERK, CREB and EGR1 pathways in neuronal cells. We and others (Yan et al., 1995; Budd et al., 2003; De Sarno et al., 2003) have also shown that activation of the M3 receptor promotes cell survival.

14. Signalling Pathways Involved in M3 Muscarinic Receptor-Mediated Cell Protection

The signal transduction cascade linking M3 receptor stimulation to cell survival is unclear. M3-mediated neuroprotection (Budd et al., 2003), like M3-mediated CREB activation (Greenwood and Dragunow, 2002), does *not* seem to occur via the ERK pathway, suggesting that CREB rather than EGR1 (which is downstream of ERK) is involved, although more studies are required to determine the effects of MEK blockade on native versus ectopically expressed M3 receptors. Activation of CREB also promotes survival of human neuroblastoma cells following serum-deprivation (Ciani et al., 2002) and promotes survival of mouse neuroblastoma cells (Walton et al., 1999a). Furthermore, the anti-apoptotic effect of M3 receptor stimulation requires new gene expression, indicating that transcription factors are involved (Budd et al., 2004). M3-mediated neuroprotection may involve signaling through protein kinase C (PKC). Teber et al. (2004) showed that the PKC activator phorbol 12-myristate 13-acetate (PMA) induces, whereas the PKCα blocker bisindolylmaleimide I inhibits, carbachol-mediated gene induction in human neuroblastoma SH-SY5Y cells. PMA also induces the anti-apoptotic protein Bcl-2 in SH-SY5Y cells (Itano et al., 1996) and enhances the viability of these cells (Zeidman et al., 1999). Carbachol also induces Bcl-2 expression in SH-SY5Y cells and this induction is inhibited by the M3 antagonist 4-DAMP, but not by M1 or M2 antagonists (Itano et al., 1996). These results suggest that PKC may mediate the survival effects of M3 receptor stimulation. In contrast, Budd et al. (2003) found that activation of phospholipase C, which would activate PKC, is not involved in cell protection in M3 receptor-transfected CHO cells. Thus, the role of PKC and ERK in M3-mediated cell survival is unclear and requires further investigation.

Overall, it is clear that synaptically- or pharmacologically-driven CREB phosphorylation can promote nerve cell survival. Because CREB may also be involved in the survival of tumor cells, including gliomas (Abramovitch et al., 2004; Iwadate et al., 2004) care will need to be exercised in the development of drugs to promote neuroprotection through the CREB signalling cassette.

15. Conclusions

The data reviewed here demonstrate the paradoxical involvement of the bZIP transcription factors in the mechanisms mediating nerve cell death as well as those promoting survival and repair after injury. Precisely, how these transcription factors regulate such diverse cellular processes awaits clarification but differential phosphorylation and dimer partner availability are likely to be key factors. For example, c-Jun-mediated regeneration is associated with selective phosphorylation of Ser73 and dimerization with ATF-3, whereas Ser63 phosphorylation and dimerization with ATF-2 can (at least in some cases) result in c-Jun-mediated cell death. Similarly, the duration of CREB phosphorylation appears to be related to its function, since prolonged phosphorylation is associated with neuroprotection, whereas dephosphorylation is associated with neurotoxicity. Whether these observations represent fundamental conserved responses to injury or are dependent on the insult and cell-type remains to be determined. However, these transcription factors and the signaling mechanisms they control represent potential targets for the design of new therapies to prevent nerve cell death and promote survival and repair after injury.

Acknowledgments

Supported by grants from the Health Research Council of New Zealand and the National Research Centre for Growth and Development.

References

Abraham WC, Logan B, Greenwood JM, Dragunow M (2002). Induction and experience-dependent consolidation of stable long-term potentiation lasting months in the hippocampus. J Neurosci 22:9626-9634.

Abraham WC, Christie BR, Logan B, Lawlor P, Dragunow M (1994). Immediate early gene expression associated with the persistence of heterosynaptic long-term depression in the hippocampus. Proc Natl Acad Sci USA 91:10049-10053.

Abramovitch R, Tavor E, Jacob-Hirsch J, Zeira E, Amariglio N, Pappo O, Rechavi G, Galun E, Honigman A (2004). A pivotal role of cyclic AMP-responsive element binding protein in tumor progression. Cancer Res 64:1338-1346.

Anderson AJ, Cummings BJ, Cotman CW (1994). Increased immunoreactivity for Jun- and Fos-related proteins in Alzheimer's disease: association with pathology. Exp Neurol 125:286-295.

Anderson AJ, Su JH, Cotman CW (1996). DNA damage and apoptosis in Alzheimer's disease: colocalization with c-Jun immunoreactivity, relationship to brain area, and effect of postmortem delay. J Neurosci 16:1710-1719.

Behrens A, Sibilia M, Wagner EF (1999). Amino-terminal phosphorylation of c-Jun regulates stress-induced apoptosis and cellular proliferation. Nat Genet 21:326-329.

Broude E, McAtee M, Kelley MS, Bregman BS (1997). c-Jun expression in adult rat dorsal root ganglion neurons: differential response after central or peripheral axotomy. Exp Neurol 148:367-377.

Budd DC, Spragg EJ, Ridd K, Tobin AB (2004). Signalling of the M3-muscarinic receptor to the anti-apoptotic pathway. Biochem J 381:43-49.

Budd DC, McDonald J, Emsley N, Cain K, Tobin AB (2003). The C-terminal tail of the M3-muscarinic receptor possesses anti-apoptotic properties. J Biol Chem 278:19565-19573.

Carulli D, Buffo A, Strata P (2004). Reparative mechanisms in the cerebellar cortex. Prog Neurobiol 72:373-398.

Carulli D, Buffo A, Botta C, Altruda F, Strata P (2002). Regenerative and survival capabilities of Purkinje cells overexpressing c-Jun. Eur J Neurosci 16:105-118.

Ciani E, Guidi S, Della Valle G, Perini G, Bartesaghi R, Contestabile A (2002). Nitric oxide protects neuroblastoma cells from apoptosis induced by serum deprivation through cAMP-response element-binding protein (CREB) activation. J Biol Chem 277:49896-49902.

Curtis MA, Penney EB, Pearson AG, van Roon-Mom WM, Butterworth NJ, Dragunow M, Connor B, Faull RL (2003). Increased cell proliferation and neurogenesis in the adult human Huntington's disease brain. Proc Natl Acad Sci USA 100:9023-9027.

De Sarno P, Shestopal SA, King TD, Zmijewska A, Song L, Jope RS (2003). Muscarinic receptor activation protects cells from apoptotic effects of DNA damage, oxidative stress, and mitochondrial inhibition. J Biol Chem 278:11086-11093.

Dragunow M (1992). Axotomized medial septal-diagonal band neurons express Jun-like immunoreactivity. Brain Res Mol Brain Res 15:141-144.

Dragunow M (1996). A role for immediate-early transcription factors in learning and memory. Behav Genet 26:293-299.

Dragunow M (2004). CREB and neurodegeneration. Front Biosci 9:100-103.

Dragunow M, Preston K (1995). The role of inducible transcription factors in apoptotic nerve cell death. Brain Research—Brain Research Reviews 21:1-28.

Dragunow M, Henderson C (1999). An in vitro human model system to investigate muscarinic receptor mediated induction of the CREB and Krox 24 memory-related transcription factors. Keynote Symposium on Molecular Mechanism's of Alzheimer's Disease, Taos, New Mexico.

Dragunow M, Abraham W, Hughes P (1996). Activation of NMDA and muscarinic receptors induces nur-77 mRNA in hippocampal neurons. Brain Res Mol Brain Res 36:349-356.

Dragunow M, Young D, Hughes P, MacGibbon G, Lawlor P, Singleton K, Sirimanne E, Beilharz E, Gluckman P (1993). Is c-Jun involved in nerve cell death following status epilepticus and hypoxic-ischaemic brain injury? Brain Res Mol Brain Res 18:347-352.

Dragunow M, MacGibbon GA, Lawlor P, Butterworth N, Connor B, Henderson C, Walton M, Woodgate A, Hughes P, Faull RL (1997). Apoptosis, neurotrophic factors and neurodegeneration. Rev Neurosci 8:223-265.

Dragunow M, Xu R, Walton M, Woodgate A, Lawlor P, MacGibbon GA, Young D, Gibbons H, Lipski J, Muravlev A, Pearson A, During M (2000). c-Jun promotes neurite outgrowth and survival in PC12 cells. Brain Research Molecular Brain Research 83:20-33.

Estus S, Zaks WJ, Freeman RS, Gruda M, Bravo R, Johnson EM, Jr. (1994). Altered gene expression in neurons during programmed cell death: identification of c-jun as necessary for neuronal apoptosis. Journal of Cell Biology 127:1717-1727.

Faris M, Kokot N, Latinis K, Kasibhatla S, Green DR, Koretzky GA, Nel A (1998). The c-Jun N-terminal kinase cascade plays a role in stress-induced apoptosis in Jurkat cells by up-regulating Fas ligand expression. Journal of Immunology 160:134-144.

Flynn DD, Ferrari-DiLeo G, Mash DC, Levey AI (1995). Differential regulation of molecular subtypes of muscarinic receptors in Alzheimer's disease. J Neurochem 64:1888-1891.

Francis JS, Dragunow M, During MJ (2004). Over expression of ATF-3 protects rat hippocampal neurons from in vivo injection of kainic acid. Brain Res Mol Brain Res 124:199-203.

Ginham R, Harrison DC, Facci L, Skaper S, Philpott KL (2001). Upregulation of death pathway molecules in rat cerebellar granule neurons undergoing apoptosis. Neurosci Lett 302:113-116.

Goldbaum O, Oppermann M, Handschuh M, Dabir D, Zhang B, Forman MS, Trojanowski JQ, Lee VM, Richter-Landsberg C (2003). Proteasome inhibition stabilizes tau inclusions in oligodendroglial cells that occur after treatment with okadaic acid. J Neurosci 23:8872-8880.

Gong B, Vitolo OV, Trinchese F, Liu S, Shelanski M, Arancio O (2004). Persistent improvement in synaptic and cognitive functions in an Alzheimer mouse model after rolipram treatment. J Clin Invest 114:1624-1634.

Gong CX, Shaikh S, Wang JZ, Zaidi T, Grundke-Iqbal I, Iqbal K (1995). Phosphatase activity toward abnormally phosphorylated tau: decrease in Alzheimer disease brain. J Neurochem 65:732-738.

Greenwood JM, Dragunow M (2002). Muscarinic receptor-mediated phosphorylation of cyclic AMP response element binding protein in human neuroblastoma cells. J Neurochem 82:389-397.

Greenwood JM, Curtis P, Logan B, Lawlor P, Dragunow M (2004). Immediate-early genes. In: From Messengers to Molecules: Memories are made of These (Reidel G, Platt B, eds), pp. 506-513. New York: Kluwer Academic/Plenum Publishers.

Guo H, Albrecht S, Bourdeau M, Petzke T, Bergeron C, LeBlanc AC (2004). Active caspase-6 and caspase-6-cleaved tau in neuropil threads, neuritic plaques, and neurofibrillary tangles of Alzheimer's disease. Am J Pathol 165:523-531.

Ham J, Babij C, Whitfield J, Pfarr CM, Lallemand D, Yaniv M, Rubin LL (1995). A c-Jun dominant negative mutant protects sympathetic neurons against programmed cell death. Neuron 14:927-939.

Hara T, Hamada J, Yano S, Morioka M, Kai Y, Ushio Y (2003). CREB is required for acquisition of ischemic tolerance in gerbil hippocampal CA1 region. J Neurochem 86:805-814.

Hardingham GE, Fukunaga Y, Bading H (2002). Extrasynaptic NMDARs oppose synaptic NMDARs by triggering CREB shut-off and cell death pathways. Nat Neurosci 5:405-414.

Harris CA, Johnson EM, Jr. (2001). BH3-only Bcl-2 family members are coordinately regulated by the JNK pathway and require Bax to induce apoptosis in neurons. J Biol Chem 276:37754-37760.

Helbecque N, Abderrahamani A, Meylan L, Riederer B, Mooser V, Miklossy J, Delplanque J, Boutin P, Nicod P, Haefliger JA, Cottel D, Amouyel P, Froguel P, Waeber G (2003). Islet-brain1/C-Jun N-terminal kinase interacting protein-1 (IB1/JIP-1) promoter variant is associated with Alzheimer's disease. Mol Psychiatry 8:363, 413-422.

Hensley K, Floyd RA, Zheng NY, Nael R, Robinson KA, Nguyen X, Pye QN, Stewart CA, Geddes J, Markesbery WR, Patel E, Johnson GV, Bing G (1999). p38 kinase is activated in the Alzheimer's disease brain. J Neurochem 72:2053-2058.

Herdegen T, Leah JD (1998). Inducible and constitutive transcription factors in the mammalian nervous system: control of gene expression by Jun, Fos and Krox, and CREB/ATF proteins. Brain Research—Brain Research Reviews 28:370-490.

Herdegen T, Claret FX, Kallunki T, Martin-Villalba A, Winter C, Hunter T, Karin M (1998). Lasting N-terminal phosphorylation of c-Jun and activation of c-Jun N-terminal kinases after neuronal injury. J Neurosci 18:5124-5135.

Huang Y, Hutter D, Liu Y, Wang X, Sheikh MS, Chan AM, Holbrook NJ (2000). Transforming growth factor-beta 1 suppresses serum deprivation-induced death of A549 cells through differential effects on c-Jun and JNK activities. J Biol Chem 275:18234-18242.

Hughes P, Dragunow M (1993). Muscarinic receptor-mediated induction of Fos protein in rat brain. Neurosci Lett 150:122-126.

Hughes P, Dragunow M (1994). Activation of pirenzepine-sensitive muscarinic receptors induces a specific pattern of immediate-early gene expression in rat brain neurons. Brain Res Mol Brain Res 24:166-178.

Hughes P, Dragunow M (1995). Induction of immediate-early genes and the control of neurotransmitter-regulated gene expression within the nervous system. Pharmacological Reviews 47:133-178.

Hunot S, Vila M, Teismann P, Davis RJ, Hirsch EC, Przedborski S, Rakic P, Flavell RA (2004). JNK-mediated induction of cyclooxygenase 2 is required for neurodegeneration in a mouse model of Parkinson's disease. Proc Natl Acad Sci USA 101:665-670.

Itano Y, Ito A, Uehara T, Nomura Y (1996). Regulation of Bcl-2 protein expression in human neuroblastoma SH-SY5Y cells: positive and negative effects of protein kinases C and A, respectively. J Neurochem 67:131-137.

Iwadate Y, Sakaida T, Hiwasa T, Nagai Y, Ishikura H, Takiguchi M, Yamaura A (2004). Molecular classification and survival prediction in human gliomas based on proteome analysis. Cancer Res 64:2496-2501.

Jenkins R, McMahon SB, Bond AB, Hunt SP (1993). Expression of c-Jun as a response to dorsal root and peripheral nerve section in damaged and adjacent intact primary sensory neurons in the rat. Eur J Neurosci 5:751-759.

Kawauchi J, Zhang C, Nobori K, Hashimoto Y, Adachi MT, Noda A, Sunamori M, Kitajima S (2002). Transcriptional repressor activating transcription factor 3 protects human umbilical vein endothelial cells from tumor necrosis factor-alpha- induced apoptosis through down-regulation of p53 transcription. J Biol Chem 277:39025-39034.

Keck S, Nitsch R, Grune T, Ullrich O (2003). Proteasome inhibition by paired helical filament-tau in brains of patients with Alzheimer's disease. J Neurochem 85:115-122.

Keller JN, Hanni KB, Markesbery WR (2000). Impaired proteasome function in Alzheimer's disease. J Neurochem 75:436-439.

Kins S, Kurosinski P, Nitsch RM, Gotz J (2003). Activation of the ERK and JNK signaling pathways caused by neuron-specific inhibition of PP2A in transgenic mice. Am J Pathol 163:833-843.

Kuan CY, Whitmarsh AJ, Yang DD, Liao G, Schloemer AJ, Dong C, Bao J, Banasiak KJ, Haddad GG, Flavell RA, Davis RJ, Rakic P (2003). A critical role of neural-specific JNK3 for ischemic apoptosis. Proc Natl Acad Sci USA 100:15184-15189.

Kumagae Y, Zhang Y, Kim OJ, Miller CA (1999). Human c-Jun N-terminal kinase expression and activation in the nervous system. Brain Res Mol Brain Res 67:10-17.

Lee B, Butcher GQ, Hoyt KR, Impey S, Obrietan K (2005). Activity-dependent neuroprotection and cAMP response element-binding protein (CREB): kinase coupling, stimulus intensity, and temporal regulation of CREB phosphorylation at serine 133. J Neurosci 25:1137-1148.

Lee HT, Chang YC, Wang LY, Wang ST, Huang CC, Ho CJ (2004). cAMP response element-binding protein activation in ligation preconditioning in neonatal brain. Ann Neurol 56:611-623.

Lei K, Davis RJ (2003). JNK phosphorylation of Bim-related members of the Bcl2 family induces Bax-dependent apoptosis. Proc Natl Acad Sci USA 100:2432-2437.

Leloup C, Michaelson DM, Fisher A, Hartmann T, Beyreuther K, Stein R (2000). M1 muscarinic receptors block caspase activation by phosphoinositide 3-kinase- and MAPK/ERK-independent pathways. Cell Death Differ 7:825-833.

Le-Niculescu H, Bonfoco E, Kasuya Y, Claret FX, Green DR, Karin M (1999). Withdrawal of survival factors results in activation of the JNK pathway in neuronal cells leading to Fas ligand induction and cell death. Mol Cell Biol 19:751-763.

Leppa S, Saffrich R, Ansorge W, Bohmann D (1998). Differential regulation of c-Jun by ERK and JNK during PC12 cell differentiation. Embo J 17:4404-4413.

Li L, Feng Z, Porter AG (2004). JNK-dependent phosphorylation of c-Jun on serine 63 mediates nitric oxide-induced apoptosis of neuroblastoma cells. J Biol Chem 279:4058-4065.

Liang G, Wolfgang CD, Chen BP, Chen TH, Hai T (1996). ATF3 gene. Genomic organization, promoter, and regulation. Journal of Biological Chemistry 271:1695-1701.

Lindenboim L, Pinkas-Kramarski R, Sokolovsky M, Stein R (1995). Activation of muscarinic receptors inhibits apoptosis in PC12M1 cells. J Neurochem 64:2491-2499.

Lindwall C, Dahlin L, Lundborg G, Kanje M (2004). Inhibition of c-Jun phosphorylation reduces axonal outgrowth of adult rat nodose ganglia and dorsal root ganglia sensory neurons. Mol Cell Neurosci 27:267-279.

Mabuchi T, Kitagawa K, Kuwabara K, Takasawa K, Ohtsuki T, Xia Z, Storm D, Yanagihara T, Hori M, Matsumoto M (2001). Phosphorylation of cAMP response element-binding protein in hippocampal neurons as a protective response after exposure to glutamate in vitro and ischemia in vivo. J Neurosci 21:9204-9213.

MacGibbon GA, Lawlor PA, Hughes P, Young D, Dragunow M (1995). Differential expression of inducible transcription factors in basal ganglia neurons. Brain Res Mol Brain Res 34:294-302.

MacGibbon GA, Lawlor PA, Walton M, Sirimanne E, Faull RL, Synek B, Mee E, Connor B, Dragunow M (1997). Expression of Fos, Jun, and Krox family proteins in Alzheimer's disease. Experimental Neurology 147:316-332.

Masaki R, Saito T, Yamada K, Ohtani-Kaneko R (2000). Accumulation of phosphorylated neurofilaments and increase in apoptosis-specific protein and phosphorylated c-Jun induced by proteasome inhibitors. J Neurosci Res 62:75-83.

Messer WS, Jr. (2002). The utility of muscarinic agonists in the treatment of Alzheimer's disease. J Mol Neurosci 19:187-193.

Murga C, Laguinge L, Wetzker R, Cuadrado A, Gutkind JS (1998). Activation of Akt/protein kinase B by G protein-coupled receptors. A role for alpha and beta gamma subunits of heterotrimeric G proteins acting through phosphatidylinositol-3-OH kinasegamma. J Biol Chem 273:19080-19085.

Nakagomi S, Suzuki Y, Namikawa K, Kiryu-Seo S, Kiyama H (2003). Expression of the activating transcription factor 3 prevents c-Jun N-terminal kinase-induced neuronal death by promoting heat shock protein 27 expression and Akt activation. J Neurosci 23:5187-5196.

Namgung U, Xia Z (2000). Arsenite-induced apoptosis in cortical neurons is mediated by c-Jun N-terminal protein kinase 3 and p38 mitogen-activated protein kinase. Journal of Neuroscience 20:6442-6451.

Napankangas U, Lindqvist N, Lindholm D, Hallbook F (2003). Rat retinal ganglion cells upregulate the pro-apoptotic BH3-only protein Bim after optic nerve transection. Brain Res Mol Brain Res 120:30-37.

Oddo S, Billings L, Kesslak JP, Cribbs DH, LaFerla FM (2004). Abeta immunotherapy leads to clearance of early, but not late, hyperphosphorylated tau aggregates via the proteasome. Neuron 43:321-332.

Pearson AG, Curtis MA, Waldvogel HJ, Faull RL, Dragunow M (2005). Activating transcription factor 2 expression in the adult human brain: association with both neurodegeneration and neurogenesis. Neuroscience 133:437-451.

Pearson AG, Gray CW, Pearson JF, Greenwood JM, During MJ, Dragunow M (2003). ATF3 enhances c-Jun-mediated neurite sprouting. Brain Res Mol Brain Res 120:38-45.

Pei JJ, Braak H, An WL, Winblad B, Cowburn RF, Iqbal K, Grundke-Iqbal I (2002). Up-regulation of mitogen-activated protein kinases ERK1/2 and MEK1/2 is associated with the progression of neurofibrillary degeneration in Alzheimer's disease. Brain Res Mol Brain Res 109:45-55.

Pei JJ, Gong CX, An WL, Winblad B, Cowburn RF, Grundke-Iqbal I, Iqbal K (2003). Okadaic-acid-induced inhibition of protein phosphatase 2A produces activation of mitogen-activated protein kinases ERK1/2, MEK1/2, and p70 S6, similar to that in Alzheimer's disease. Am J Pathol 163:845-858.

Pulverer BJ, Kyriakis JM, Avruch J, Nikolakaki E, Woodgett JR (1991). Phosphorylation of c-jun mediated by MAP kinases. Nature 353:670-674.

Putcha GV, Le S, Frank S, Besirli CG, Clark K, Chu B, Alix S, Youle RJ, LaMarche A, Maroney AC, Johnson EM, Jr. (2003). JNK-mediated BIM phosphorylation potentiates BAX-dependent apoptosis. Neuron 38:899-914.

Raivich G, Bohatschek M, Da Costa C, Iwata O, Galiano M, Hristova M, Nateri AS, Makwana M, Riera-Sans L, Wolfer DP, Lipp HP, Aguzzi A, Wagner EF, Behrens A (2004). The AP-1 transcription factor c-Jun is required for efficient axonal regeneration. Neuron 43:57-67.

Riccio A, Ahn S, Davenport CM, Blendy JA, Ginty DD (1999). Mediation by a CREB family transcription factor of NGF-dependent survival of sympathetic neurons. Science 286:2358-2361.

Rodriguez-Puertas R, Pascual J, Vilaro T, Pazos A (1997). Autoradiographic distribution of M1, M2, M3, and M4 muscarinic receptor subtypes in Alzheimer's disease. Synapse 26:341-350.

Rohn TT, Rissman RA, Davis MC, Kim YE, Cotman CW, Head E (2002). Caspase-9 activation and caspase cleavage of tau in the Alzheimer's disease brain. Neurobiol Dis 11:341-354.

Rohn TT, Head E, Su JH, Anderson AJ, Bahr BA, Cotman CW, Cribbs DH (2001). Correlation between caspase activation and neurofibrillary tangle formation in Alzheimer's disease. Am J Pathol 158:189-198.

Rosenblum K, Futter M, Jones M, Hulme EC, Bliss TV (2000). ERKI/II regulation by the muscarinic acetylcholine receptors in neurons. J Neurosci 20:977-985.

Schwarz CS, Seyfried J, Evert BO, Klockgether T, Wullner U (2002). Bcl-2 up-regulates ha-ras mRNA expression and induces c-Jun phosphorylation at Ser73 via an ERK-dependent pathway in PC 12 cells. Neuroreport 13:2439-2442.

Shaulian E, Karin M (2001). AP-1 in cell proliferation and survival. Oncogene 20:2390-2400.

Sheng JG, Mrak RE, Griffin WS (1998). Progressive neuronal DNA damage associated with neurofibrillary tangle formation in Alzheimer disease. J Neuropathol Exp Neurol 57:323-328.

Shi L, Gong S, Yuan Z, Ma C, Liu Y, Wang C, Li W, Pi R, Huang S, Chen R, Han Y, Mao Z, Li M (2005). Activity deprivation-dependent induction of the proapoptotic BH3-only protein Bim is independent of JNK/c-Jun activation during apoptosis in cerebellar granule neurons. Neurosci Lett 375:7-12.

Smeal T, Binetruy B, Mercola DA, Birrer M, Karin M (1991). Oncogenic and transcriptional cooperation with Ha-Ras requires phosphorylation of c-Jun on serines 63 and 73. Nature 354:494-496.

Stadelmann C, Deckwerth TL, Srinivasan A, Bancher C, Bruck W, Jellinger K, Lassmann H (1999). Activation of caspase-3 in single neurons and autophagic granules of granulovacuolar degeneration in Alzheimer's disease. Evidence for apoptotic cell death. Am J Pathol 155:1459-1466.

Su JH, Anderson AJ, Cummings BJ, Cotman CW (1994). Immunohistochemical evidence for apoptosis in Alzheimer's disease. Neuroreport 5:2529-2533.

Swatton JE, Sellers LA, Faull RL, Holland A, Iritani S, Bahn S (2004). Increased MAP kinase activity in Alzheimer's and Down syndrome but not in schizophrenia human brain. Eur J Neurosci 19:2711-2719.

Takeda M, Kato H, Takamiya A, Yoshida A, Kiyama H (2000). Injury-specific expression of activating transcription factor-3 in retinal ganglion cells and its colocalized expression with phosphorylated c-Jun. Invest Ophthalmol Vis Sci 41:2412-2421.

Teber I, Kohling R, Speckmann EJ, Barnekow A, Kremerskothen J (2004). Muscarinic acetyl-choline receptor stimulation induces expression of the activity-regulated cytoskeleton-associated gene (ARC). Brain Res Mol Brain Res 121:131-136.

Tsujino H, Kondo E, Fukuoka T, Dai Y, Tokunaga A, Miki K, Yonenobu K, Ochi T, Noguchi K (2000). Activating transcription factor 3 (ATF3) induction by axotomy in sensory and motoneurons: A novel neuronal marker of nerve injury. Mol Cell Neurosci 15:170-182.

Tsuzuki K, Noguchi K, Mohri D, Yasuno H, Umemoto M, Shimobayashi C, Fukazawa K, Sakagami M (2002). Expression of activating transcription factor 3 and growth-associated protein 43 in the rat geniculate ganglion neurons after chorda tympani injury. Acta Otolaryngol 122:161-167.

Vogelsberg-Ragaglia V, Schuck T, Trojanowski JQ, Lee VM (2001). PP2A mRNA expression is quantitatively decreased in Alzheimer's disease hippocampus. Exp Neurol 168:402-412.

Walton M, Sirimanne E, Williams C, Gluckman P, Dragunow M (1996). The role of the cyclic AMP-responsive element binding protein (CREB) in hypoxic-ischemic brain damage and repair. Brain Res Mol Brain Res 43:21-29.

Walton M, Woodgate AM, Sirimanne E, Gluckman P, Dragunow M (1998). ATF-2 phosphorylation in apoptotic neuronal death. Brain Res Mol Brain Res 63:198-204.

Walton M, Woodgate AM, Muravlev A, Xu R, During MJ, Dragunow M (1999a). CREB phosphory-lation promotes nerve cell survival. Journal of Neurochemistry 73:1836-1842.

Walton M, Henderson C, Mason-Parker S, Lawlor P, Abraham WC, Bilkey D, Dragunow M (1999b). Immediate early gene transcription and synaptic modulation. J Neurosci Res 58:96-106.

Walton MR, Dragunow I (2000). Is CREB a key to neuronal survival? Trends Neurosci 23:48-53.

Wang W, Shi L, Xie Y, Ma C, Li W, Su X, Huang S, Chen R, Zhu Z, Mao Z, Han Y, Li M (2004). SP600125, a new JNK inhibitor, protects dopaminergic neurons in the MPTP model of Parkinson's disease. Neurosci Res 48:195-202.

Watson A, Eilers A, Lallemand D, Kyriakis J, Rubin LL, Ham J (1998). Phosphorylation of c-Jun is necessary for apoptosis induced by survival signal withdrawal in cerebellar granule neurons. Journal of Neuroscience 18:751-762.

Whitfield J, Neame SJ, Paquet L, Bernard O, Ham J (2001). Dominant-negative c-Jun promotes neuronal survival by reducing BIM expression and inhibiting mitochondrial cytochrome c release. Neuron 29:629-643.

Woodgate A, Walton M, MacGibbon GA, Dragunow M (1999). Inducible transcription factor expression in a cell culture model of apoptosis. Brain Research Molecular Brain Research 66:211-216.

Yan GM, Lin SZ, Irwin RP, Paul SM (1995). Activation of muscarinic cholinergic receptors blocks apoptosis of cultured cerebellar granule neurons. Mol Pharmacol 47:248-257.

Yang DD, Kuan CY, Whitmarsh AJ, Rincon M, Zheng TS, Davis RJ, Rakic P, Flavell RA (1997). Absence of excitotoxicity-induced apoptosis in the hippocampus of mice lacking the Jnk3 gene. Nature 389:865-870.

Zeidman R, Pettersson L, Sailaja PR, Truedsson E, Fagerstrom S, Pahlman S, Larsson C (1999). Novel and classical protein kinase C isoforms have different functions in proliferation, survival and differentiation of neuroblastoma cells. Int J Cancer 81:494-501.

Zhang S, Liu J, Dragunow M, Cooper GJ (2003). Fibrillogenic amylin evokes islet beta-cell apoptosis through linked activation of a caspase cascade and JNK1. J Biol Chem 278:52810-52819.

Zhang S, Liu J, MacGibbon G, Dragunow M, Cooper GJ (2002). Increased expression and activation of c-Jun contributes to human amylin-induced apoptosis in pancreatic islet beta-cells. J Mol Biol 324:271-285.

Zhang S, Liu H, Liu J, Tse C, Dragunow M, Cooper GJS (2005). Activation of ATF-2 by p38MAPK/JNK1 in apoptosis induced by human amylin in pancreatic islet beta-cells. Diabetalogica, submitted.

Zhu X, Ogawa O, Wang Y, Perry G, Smith MA (2003). JKK1, an upstream activator of JNK/SAPK, is activated in Alzheimer's disease. J Neurochem 85:87-93.

Zhu X, Raina AK, Rottkamp CA, Aliev G, Perry G, Boux H, Smith MA (2001). Activation and redistribution of c-jun N-terminal kinase/stress activated protein kinase in degenerating neurons in Alzheimer's disease. J Neurochem 76:435-441.

13

Immediate Early Genes, Inducible Transcription Factors and Stress Kinases in Alzheimer's Disease

ISIDRO FERRER, GABRIEL SANTPERE and BERTA PUIG

Institut de Neuropatologia, Servei Anatomia Patològica, Hospital Universitari de Bellvitge, Universitat de Barcelona, Spain

1. Introduction

Alzheimer's disease (AD) is a progressive neurological disease mainly manifested by aphasia, apraxia, agnosia, memory loss, cognitive impairment, and dementia. Histological studies in the AD brain show the presence of β-amyloid (βA) depositions in the form of senile plaques and amyloid angiopathy, hyper-phosphorylated tau deposits in neurons consisting of neurofibrillary tangles (NFTs), which are composed of paired helical filaments (PHFs), dystrophic neurites surrounding β-amyloid deposits in senile plaques, and neuropil threads (reviewed in Duyckaerts and Dickson, 2003). β-amyloid deposition results, in part, from the abnormal cleavage of the β-amyloid precursor protein (βAPP or APP) by γ-secretases. Tau hyper-phosphorylation appears to be, in part, the consequence of an imbalance between kinases and phosphatases. The disease first involves the entorhinal cortex, then the limbic structures and, finally, all the neocortex. The process is accompanied by neuronal loss and advancing cerebral atrophy (Braak and Braak, 1999).

The majority of patients suffer from sporadic AD forms, but a number of familial cases with early onset are associated with mutations in the APP gene (*PAD* gene) or with mutations in the genes encoding Presenilin 1 and Presenilin 2 (*PS1* and *PS2*, respectively), which participate in γ-secretase activity. Apolipoprotein E (ApoE) also regulates AD as the expression of the ε4-allele increases the risk of developing late onset AD (Herreman et al., 2000; Zhang et al., 2000; Bertram and Tanzi, 2003). Whatever its origin, oxidative stress plays a significant role in

R. Pinaud, L.A. Tremere (Eds.), Immediate Early Genes in Sensory Processing, Cognitive Performance and Neurological Disorders, 243–260, ©2006 Springer Science + Business Media, LLC

the pathogenesis of AD (Smith et al., 1991, 1994; Papolla et al., 1992, 2005; Markesbery and Carney, 1999; Aksenov et al., 2001; Perry et al., 2003).

Even though AD is a human disease, several transgenic animals, including mice, *C. elegans* and flies, have been produced to mimic distinct aspects of AD (van Leuven, 2000; McGowan et al., 2003). Common APP transgenic mice leading to β-amyloid brain deposition are those bearing the double Swedish mutation K670N/M671L (Tg2576) and those expressing the Indiana mutation V717F (PDAPP). Double *APP/PS1* mutants further augment β-amyloid deposition when compared with transgenics expressing the corresponding single mutations. Furthermore, tau hyper-phosphorylation in mice has been re-created in transgenic mice over-expressing short and long isoforms of human tau, and in mice carrying human pathogenic mutations in the *tau* gene (Götz, 2001; Hutton et al., 2001; Lewis and Dickson, 2003). Interestingly, mice expressing a repressible human tau under a constitutively active promoter (CamKII) develop age-related NFTs, neuronal loss and behavioral impairment. After the suppression of transgenic tau memory recovered and neurons number stabilized but NFTs continued to accumulate (Santacruz et al., 2005).

2. Immediate Early Genes and the Transcription Factor AP-1

Proto-oncogenes of the *fos* and *jun* families encode the proteins Fos and Jun. Fos:Jun heterodimers and Jun:Jun homodimers are the major components of the transcription factor activator protein 1 (AP-1), the basic region of which binds with high affinity to the consensus sequence TGAC/GTCA in DNA (Halazonetis et al., 1988; Nakabeppu et al., 1988; Rauscher et al., 1988; Curran and Franza, 1989). The composition of dimers alters DNA binding capacity, and post-translational modifications of Fos and Jun, including phosphorylation at different sites, affect binding capacity and transactivation potential (Hunter and Karin, 1992; Karin, 1994, 1995).

Fos and Jun participate in a varied number of metabolic scenarios including those controlling cell development and cell death. A primary interest on the possible role of Fos and Jun in neurodegenerative diseases was focused on the possible relationship of these factors and neuronal degeneration and death. The role of Fos and Jun has also been examined in a large number of experimental models *in vivo* and *in vitro* (Munujos et al., 1993; Soriano et al., 1995; Ferrer et al., 1996b, 1997a, b, 2000; Pozas et al., 1997; Sanz et al., 1997).

2.1. AP-1 Binding Sites in the Promoters of AAP and APP-Related Genes

The *PAD* gene has two AP-1 consensus sequences (Salbaum et al., 1989; Quitschke and Golgaber, 1992). Therefore, it is feasible that members of the Fos and Jun family regulate the transcription of *PAD*. Activation of the APP gene in a human glial cell line induces components of AP-1, increases protein DNA-binding to AP-1 sequences within the *PAD* promoter, and suggests that AP-1 binding is likely to be composed of Jun:Jun homodimers (Trejo et al., 1994). In addition, the carboxy-terminal APP intracellular domain binds to the mouse

JNK-interacting protein-1 (JIP1) and its human homologue islet-brain1 (IB1), and scaffolds APP with JNK (Matsuda et al., 2001; Scheinfeld et al., 2002; Helbecque et al., 2003). Studies in *Drosophila* APP and JIP analogues further strengthen the role of JIPs in the regulation of APP by JNK (Taru et al., 2002). Compounds that modulate APP adaptor protein interactions such as JIP1 may inhibit β-amyloid generation by specifically targeting the substrate (King and Scott Turner, 2004). Together these observations point to Jun as a pivotal transcription factor in the regulation of APP production.

Interestingly, binding sites for AP-1 and NFκB are also present in the *ApoE* promoter (Lahiri, 2004). Thus, members of the AP-1 family may also regulate the transcription of a determining factor in the pathogenesis of sporadic AD.

2.2. c-Fos and c-Jun in AD

Some years ago, *c-fos* was examined as a possible locus for early-onset familial AD (EOFAD) on the basis of its localization on chromosome 14q24.3 (Bonnycastle et al., 1993; Rogaev et al., 1993; Cruts et al., 1994). Subsequently, several studies ruled out the possibility that *c-fos* was in the chromosome 14 EOFAD locus (Morris and St Clair, 1994).

However, c-Fos has also been implicated in other aspects of AD. c-Fos immunoreactivity is increased in certain areas of the hippocampus in patients with AD (Zhang et al., 1992; Anderson et al., 1994; MacGibbon et al., 1997; Marcus et al., 1998). Yet whether senile plaques and NFTs are stained or not with anti-Fos antibodies is still controversial (Zhang et al., 1992). Increased c-Fos expression has been associated with β-amyloid-induced toxicity in a mouse hippocampal cell line (Guillardon et al., 1996), but these observations have not been replicated. Together, these data provide only weak support for a role of c-Fos in the pathogenesis of AD.

Early immunohistochemical studies showed c-Jun immunoreactivity in some neurons bearing NFTs in AD, and in astrocytes in AD and control cases (Anderson et al., 1994; Marcus et al., 1998). These findings, together with observations with the method of *in situ* end-labeling of nuclear DNA fragmentation, suggested that strong c-Jun immunoreactivity precedes apoptosis in AD (Anderson et al., 1996). Increased c-Jun mRNA has also been demonstrated in the CA1 region of the hippocampus in AD, but no direct evidence exists of a connection between c-Jun mRNA expression and cell death (MacGibbon et al., 1997).

Interestingly, the staining characteristics of antibodies to certain transcription factors vary when different antibodies directed to the same protein are used (MacGibbon et al., 1997). This is an important point as certain commercial c-Jun antibodies recognize autophagy-related proteins and caspase-3-cleaved products in addition to c-Jun (Hammond et al., 1998; Pozas et al., 1999; Ferrer et al., 2002; Ribera et al., 2002). Therefore, studies using the antibody c-Jun/AP-1 (N), or antibodies directed against the same c-Jun amino acid sequence, may demonstrate c-Jun as well as other unrelated products cross-reacting with c-Jun. In our experience, the c-Jun/AP-1 (N) antibody decorates the cytoplasm of many glial cells in the cerebral cortex, some of them surrounding senile plaques, but very few, if any, neurons (Ferrer et al., 1996a). Astrocytes and microglial cells stained with the c-Jun/AP-1 (N) antibody are not decorated with other anti-c-Jun antibodies directed to other protein domains, thus supporting the idea that other products rather than c-Jun are recognized with this antibody in AD.

In contrast, antibodies to phospho c-Jun^{Ser63} stain the nuclei of individual neurons and glial cells in both control and AD brains (unpublished observations). There is no evidence that this immunostaining, the localization of which is consistent with that expected for a transcription factor, is associated with cell death.

More recent studies have shown two interesting aspects. On the one hand, c-Jun appears to be a contributory factor to amyloid-induced neuronal apoptosis but is not necessary for amyloid-beta-induced c-jun induction (Kihiko et al., 1999). On the other, Tg-APP (Sw, V717F)/B6 mice expressing amyloid precursor protein show progressive cognitive impairment and anxiety together with reduced expression of c-Fos and absence of plaque deposition (Lee et al., 2004).

3. Stress Kinase Pathways

The family of mitogen-activated protein kinases (MAPKs) is composed of several members, including extracellular signal-regulated kinases (ERKs), and stress-activated protein kinases (SAPKs), c-Jun N-terminal kinases (JNKs) and p38 kinases (Cobb and Goldsmith, 1995; Robinson and Cobb, 1997; Mielke and Herdegen, 2000). The JNKs are activated by phosphorylation through JNK kinases (JNKKs), also called MAP kinase kinases (MKKs) (Minden and Karin, 1997; Yang et al., 1997; Ip and Davis, 1998). The p38 kinases are activated by dual phosphorylation mediated by MKKs (Dérijard et al., 1995; Raingaud et al., 1995, 1996; Enslen et al., 1998). Some MKKs activate JNK and p38 equally (Lin et al., 1995; Yang et al., 1997).

Active, phosphorylated JNKs (SAPK/JNK-P or JNK-P) phosphorylate c-Jun at Ser63 and Ser73 (Kallunki et al., 1994, 1996; Gupta et al., 1996; Minden and Karin, 1997). Active, phosphorylated p38 (p38-P) activates several substrates through phosphorylation, including activating transcription factor-2 (ATF-2) at Thr69 and Thr71, but also calcium/cAMP response element binding protein (CREB) and Elk-1 (Hazzalin et al., 1996; Tan et al., 1996; Shaywitz and Greenberg, 1999). ATF-2 and Elk-1 are also phosphorylated by JNKs (van Dam et al., 1993; Cavigelli et al., 1995; Gupta et al., 1995; Whitmarsh and Davis, 1996).

Interest on (MAPKs) and Alzheimer's disease is not limited to their possible role in gene transcription, but also it is due to the fact that tau, which is phosphorylated in AD, is a substrate of MAPKs, and therefore, MAPKs may play a dual pivotal role in tau phosphorylation and cell fate.

3.1. SAPK/JNK and p38 Activation in Neurofibrillary Tangles and Dystrophic Neurites of Senile Plaques in AD

Stress-activated protein kinases SAPK/JNK and p-38 have the capacity to phosphorylate tau at specific sites *in vitro* (Goedert et al., 1997; Reynolds et al., 1997a, b; Jenkins et al., 2000; Reynolds et al., 2000; Anderton et al., 2001). Increased expression of active SAPK/JNK and p38 (SAPK/JNK-P Tyr183/185 and p-38-P Thr180/Tyr182) has been observed in association with abnormal tau deposits in AD, including NFTs and dystrophic neurites of senile plaques (Zhu et al., 2000, 2001a; Atzori et al., 2001; Ferrer et al., 2001; Pei et al., 2001; Hoozemans et al., 2004; Swatton et al., 2004). Both active SAPK/JNK and p38 expression appear at early stages of AD, and increase with disease progression (Pei et al., 2001; Sun et al., 2003). Furthermore, JNK kinase 1, an upstream activator of JNK/SAPK, is activated in AD (Zhu et al., 2003a). Similarly, MKK6, one of the upstream activators of p38, is activated in AD and expressed in neurons with neurofibrillary tangles, dystrophic neurites and neuropil threads (Zhu et al., 2001b). The percentage of neurons bearing phosphorylated tau that co-localize active SAPK/JNK-P or p38-P is variable and represents between 30 and 70% of the total number of NFTs (Ferrer et al., 2001). This is an interesting point as it suggests kinase sequestration by tau aggregates in NFTs.

Oxidative stress is likely a promoting factor of SAPK/JNK and p38 activation, and this activation in the vicinity of β-amyloid deposits is associated with tau phosphorylation in dystrophic neurites. It is feasible that β-amyloid-induced oxidative stress triggers SAPK/JNK and p38 activation, which in turn phosphorylates tau in dystrophic neurites (Zhu et al., 2003b). Evidence for such a mechanism may also be inferred from observations in the brains of patients with AD treated with β-amyloid peptides in an attempt to reduce β-amyloid loading. The number of amyloid plaques is dramatically decreased in the cerebral cortex of β-amyloid-immunized AD patients (Nicoll et al., 2003; Ferrer et al., 2004). Interestingly, SAPK/JNK-P and p38-P, together with SOD-1 immunoreactivities, are decreased in areas with reduced β-amyloid burden. Moreover, the number of tau-immunoreactive aberrant neurites is significantly reduced in the same areas, but not the number of neurofibrillary tangles and neuropil threads (Nicoll et al., 2003; Ferrer et al., 2004). These observations demonstrate that curbing β-amyloid is associated with decreased oxidative stress response, decreased stress kinase activation, and decreased tau hyper-phosphorylation in neurites surrounding β-amyloid plaques in AD following β-amyloid immunization (Ferrer et al., 2004).

3.2. Expression of Active Stress Kinases SAPK/JNK and p38 in Other Tauopathies

Active, SAPK/JNK-P and p38-P are also expressed in neurons and glial cells with abnormal phospho-tau deposition in tauopathies, including progressive supranuclear palsy, corticobasal degeneration and Pick's disease (Atzori et al., 2001; Ferrer et al., 2001). Activation of the p38 pathway, as revealed by phospho-MKK6 and p38-P immunohistochemistry, has been corroborated in PiD and PSP (Hartzler et al., 2002). Increased expression of SAPK/JNK-P and p38-P is also found in grains, tangles, neurons with pre-tangles and tau-containing oligodendrocytes and astrocytes in argyrophilic grain disease (Ferrer et al., 2003a). Finally, stress kinases are also involved in tau phosphorylation in frontotemporal dementia and parkisonism syndromes linked to chromosome 17 (FTDP-17 syndromes) due to mutations in the *tau* gene. SAPK/JNK-P and p38-P are expressed in tau-bearing neurons and glial cells in several familial tauopathies, including those associated with delN296 and P301L mutations in the *tau* gene (Atzori et al., 2001; Ferrer et al., 2003b, c).

These findings show that SAPK/JNK and p38 activation is closely associated with tau hyper-phosphorylation in neurons and glial cells in AD and other tauopathies. Yet the mechanisms that induce SAPK/JNK and p38 activation in neurons, which in turn is associated with tau phosphorylation in NFTs and neuropil threads, are not clearly understood.

A recent study has shown cross-reactivity of p38-P and phosphorylated tau proteins in sarkosyl-insoluble fractions in AD, thus suggesting that such cross-reactivity is responsible for the p38-P immunolabeling of tau-positive inclusions (Sahara et al., 2004). This amazing observation weakens the putative role of active p38 in tau phosphorylation in AD and other tauopathies.

However, studies in Pick's disease have shown no cross-reactivity between p38-P and phospho-tau in sarkosyl-insoluble fractions (Puig et al., 2004). In addition, p38-P immunoprecipitated from sarkosyl-insoluble fractions, enriched in abnormal fibrils and phospho-tau, has the capacity to phosphorylate its specific substrate ATF-2 as well as recombinant tau (Puig et al., 2004). Moreover, SAPK/JNK- and p38-immunoprecipitated sarkosyl-insoluble fractions in AD cases, which are enriched in paired helical filaments, have the capacity to phosphorylate several substrates, including myelin basic protein and c-Jun at Ser63 and ATF-2 at Thr71 (Ferrer et al., 2005). In addition, these immunoprecipitated fractions have the capacity to phosphorylate recombinant tau (Ferrer et al., 2005). These combined observations support the alternative hypothesis suggesting a role for tau aggregates in the perpetuation of tau hyper-phosphorylation in neurofibrillary tangles and in Pick bodies in AD and Pick's disease, respectively.

3.3. SAPK/JNK and p38 Activation, and Tau Phosphorylation of Neurites Surrounding Aβ Plaques in Single APP and Double APP/PS1 Transgenic Mice

Active SAPK/JNK and p38 (SAPK/JNK-P Tyr183/185 and p-38-P Thr180/ Tyr182), and tau hyper-phophorylation in neurites surrounding Aβ plaques, have been observed in double mutants for APP (Swedish mutation, Tg2576) and PS-1 (P264L) (Savage et al., 2002). Along the same lines, enhanced expression of SAPK/JNK-P and p38-P surrounding β-amyloid deposits, together with increased phosphorylated neurofilaments and tau hyper-phosphorylation in neurites adjacent to β-amyloid deposits, has been found in Tg2576 mice (Otth et al., 2003; Puig et al., 2004). This is accompanied by increased expression levels of phosphorylated SAPK/JNK and p38, as seen in Western blots of total brain homogenates (Puig et al., 2004). These observations support a primary role for stress kinases in tau hyper-phosphorylation in neurites surrounding β-amyloid plaques in transgenic mice.

3.4. SAPK/JNK and p38 Activation in Tau Transgenic Mice, and in Mice with the Double Swedish and Tau (G272V, P301L and R406W) Mutation

Transgenic mice expressing the triple mutation G272V, P301L and R406W in the tau gene have high levels of tau in the cerebral cortex and tau hyperphosphorylation at specific sites (Lim et al., 2001). These mice show a weak focal increase in SAPK/JNK-P and p38-P immunoreactivity in neocortical neurons (Ferrer et al., 2005).

Mice with the double Swedish (Tg2576) and tau (G272V, P301L and R406W) mutation exhibit increased SAPK-P and p38-P immunoreactivity in relation with β-amyloid plaques and in scattered cortical neurons (Perez et al., 2005).

3.5. SAPK/JNK, p38 Kinase, and β-Amyloid-Induced Cell Death in vitro

Several pieces of evidence implicate the JNK/c-Jun cascade and p38 activation in β-amyloid-induced cell death in vitro (Bozyczko-Coyne et al., 2001; Daniels et al., 2001; Morishima et al., 2001; Jang and Surth, 2002; Wei et al., 2002; Fogarti et al., 2003). Introduction of β-amyloid peptides to primary culture cortical neurons induces JNK activation and cell death (Shoji et al., 2000). Similar results have been obtained in PC12 cells and in sympathetic neurons subjected to β-amyloid in which JNK activation precedes cell death. Introduction of a selective blocker of JNK protects PC12 and sympathetic neurons from dying, supporting JNK as an activator of β-amyloid-induced cell death (Troy et al., 2001). In the same line, dominant-negative c-Jun mutants protect cells from dying following β-amyloid exposure (Kihiko et al., 1999; Troy et al., 2001).

In spite of these achievements, the link between β-amyloid, and JNK and p38 activation, is barely known, although several pieces of evidence point to oxidative stress as a triggering factor. Neuroblastoma (Nα2) cells exposed to β-amyloid show oxidative stress, increased p38 phosphorylation in a dose-dependent manner, and reduced cell viability (Daniels et al., 2001). Recent studies in lymphocytes have also shown that JNK activation is probably mediated by an oxidative stress mechanism involving H_2O_2, NFκB, p53, c-Jun and caspase-3 (Vélez-Pardo et al., 2002). This is further supported in neuronally-differentiated SK-N-BE cells exposed to Aβ (25-35), Aβ (1-40) and Aβ (1-42) peptides (Tamagno et al., 2003). β-amyloid peptides generate 4-hydroxynonenal and H_2O_2, and activate JNK and p38. Antioxidants such as α-tocopherol and N-acetylcysteine prevent activation of JNK and p38, and prevent apoptosis (Tamagno et al., 2003). Direct JNK inhibitors also block β-amyloid-induced cell death in this paradigm (Tamagno et al., 2003).

In addition to β-amyloid peptides, APP may produce cell death via dimerization and activation of JNK and apoptosis signal-regulating kinase 1 (ASK-1) under appropriate conditions (Hashimoto et al., 2003). PC12 cells expressing the APP Swedish double mutation suffer, after oxidative stress, JNK-associated caspase-dependent apoptosis (Marques et al., 2003). Finally, V642I-APP-induced cytotoxicity in primary neurons involves JNK and p38. The V642I-APP-induced cytotoxicity is suppressed by secretase inhibitors, suggesting that mutant APP and not the resulting $A\beta$ peptide is directly neurocytotoxic (Niikura et al., 2004).

3.6. SAPK/JNK, p38 Kinase, and Cell Death in AD

Together, the data obtained from *in vitro* models indicate that APP and β-amyloid may produce apoptotic cell death mediated by JNK and p38 kinases in several paradigms. This information has permitted the suggestion of a scenario in which JNK and p38 kinases may also mediate β-amyloid-induced cell death in AD (Okazawa and Estus, 2002). However, the relationship between SAPK/JNK and p38, and cell death, is not so clear in AD and murine models of AD.

Double-labeling immunohistochemical studies have stressed the lack of co-localization of SAPK/JNK-P and p38-P, and cleaved caspase-3 and cell death in AD and other tauopathies (Selznick et al., 1999; Atzori et al., 2001; Ferrer et al., 2001, 2003b, 2004; Zhu et al., 2003b; Ferrer, 2004). Therefore, it seems feasible that stress-activated kinases participate in tau hyper-phosphorylation but do not mediate cell death in AD (Zhu et al., 2003b).

Increased SAP/JNK-P and p38-P immunoractivity is observed around β-amyloid deposits in Tg2576 mice (Ferrer, 2004; Puig et al., 2004). Yet no caspase-3 activation has been found in association with amyloid plaques in single APP and in double APP/PS1 transgenic mice (Selznick et al., 1999; Savage et al., 2002; Puig et al., 2004). This feature is consistent with the lack of neuronal loss in PDAPP, Tg2576 and Tg2576/(P264L)PS1 mice (Irizarry et al., 1997a, b; Savage et al., 2002).

Interestingly, increased SAPK/JNK and p38 expression has also been found in the brains of mice in which the APPsw was expressed under the control of the neuron-specific enolase promoter (Hwang et al., 2004). Increased caspase-3 and TUNEL-stained nuclei were observed in these mice (Hwang et al., 2004), thus suggesting variations depend on the transgene. Moreover, JNK is activated in neurons bearing β-amyloid immunoreactivity in aged transgenic mice over-expressing mutant (M146L) *PS1* which develop diffuse β-amyloid deposits (Shoji et al., 2000).

Cell death and neuron loss occur in very old double transgenic mice expressing the APP Swedish mutation and a triple tau mutation (V337M, G272V or R406W). Yet there is no evidence that SAPK/JNK-P and p38-P play a role in this process of dying (unpublished observations).

4. Expression of Other Transcription Factors Regulated by Stress Kinases *in vitro* and in AD

The role of other transcription factors regulated by JNK and p38 is less known. Total CREB levels appear unchanged but phosphorylation of CREB is decreased in AD, thus suggesting that cAMP signaling is impaired in AD and may contribute to the pathophysiology of the disease (Yamamoto-Sasaki et al., 1999). ATF-2 immunohistochemistry and western blotting have revealed subtle modifications in AD which are not conclusive (Yamada et al., 1997).

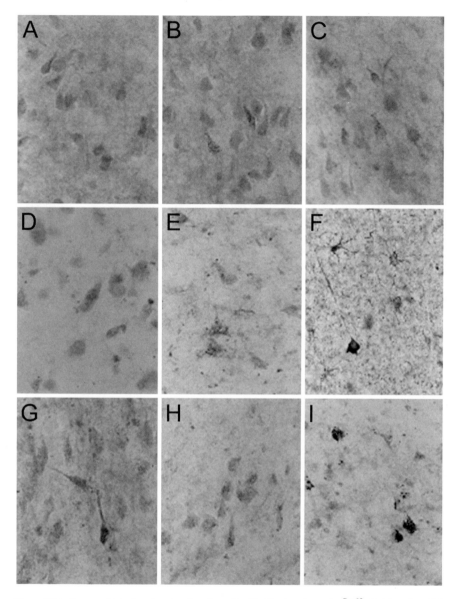

Figure 13.1. Immunohistochemistry to phospho-c-Fos (A-C), phospho-c-Jun^{Ser63} (D, E), c-Jun-AP1 (N) (F), phospho-CREB^{Ser12} (G) and phospho-Elk-1Ser^{383} (H, I) in control (A, C, H) and AD (B, C, E, F, G, I). C-Fos-P is found in small granules in neurons with granulovacuolar degeneration in AD. The antibody c-Jun-AP1 (N) decorates a few neurons and many astrocytes in AD. The antibody to phospho-Elk-1Ser^{383} immunostains granules of neurons with granulovacuolar degeneration.

Personal studies in Tg2576 mice have shown no phosphorylation CREB, ATF-2, c-Jun and Elk-1, as revealed with antibodies to phospho-CREB^{Ser12}, ATF-2^{Thr71}, c-Jun^{Ser63} and Elk-1^{Ser383} even at very old ages (unpublished observations).

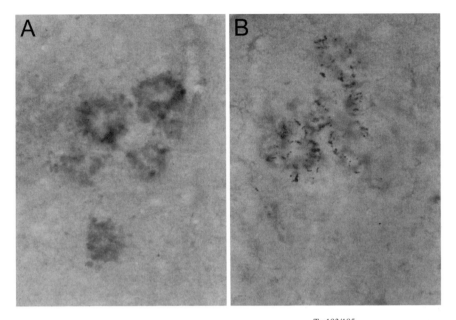

Figure   13.2. Immunohistochemistry   to   phospho-SAPK7JNK[Tyr183/185]   and   phospho-p38[Thr180/Tyr182] immunoreactivity surrounding amyloid plaques in Tg2576 mice expressing the APP Swedish mutation.

Finally, members of the Ets1/2 transcription factor family regulate *PS1* transcription. ER81 activates, whereas factor Elk1 represses, *PS1* transcription (Pastorcic et al., 2003). Whether this feature has a functional implication in AD is not known.

5. β-Amyloid and NFκB Expression in AD

The NFκB family is composed of several inducible transcription factors which regulate inflammatory responses and which control stress-induced apoptosis. Activation of NFκB results in the translocation of NFκB from the cytoplasm to the nucleus and activation of specific target genes (123). Several lines of evidence have implicated NFκB in AD (Bales et al., 2000). NFκB is activated in primary neurons by β-amyloid peptides (Kaltschmidt et al., 1997). Furthermore, the non-Aβ component of AD amyloid also generates reactive oxygen species and activates NFκB in cortical neurons of rat brain primary cultures (Tanaka et al., 2002). Western blots have shown increased NFκB expression levels in AD (Kitamura et al., 1997), and immunohistochemical studies have shown increased NFκB expression in association with amyloid deposits in diffuse and neuritic plaques (Kaltschmidt et al., 1997; Ferrer et al., 1998), and in the nuclei of both cortical neurons (Kaltschmidt et al., 1997; Ferrer et al., 1998) and cholinergic neurons in patients with AD (Boissiere et al., 1997). Recent studies have shown that the NFκB precursor, p105, and the NFκB inhibitor, IκBγ, are both elevated in AD (Huang et al., 2005). Brain IκB has been reported as being expressed in a distribution that corresponds to neurofibrillary pathology in AD, suggesting that disruption of the auto-regulatory mechanism of NFκB may play a role in the pathogenesis of AD (Yoshiyama et al., 2001). NFκB has also been implicated in cell death and survival in AD. β-amyloid-induced neurotoxicity and apoptosis are preceded by decreased NFκB activation in cultured fetal rat cortical neurons (Bales et al., 2000). Treatment of these cell cultures with anti-sense oligonucleotides to IκBα mRNA are neuroprotective (Bales et al., 2006). In contrast to neurons, exposure of cultured astrocytes to Aβ peptides is accompanied by increased NFκB activity (Bales et al., 2006). Additional support for the hypothesis that NFκB decreases neuronal vulnerability is provided by observations in *PS1* transgenic mice. Activation of NFκBp50 is impaired in transgenic mice expressing the human M146L *PS-1* gene mutation, which is associated with increased susceptibility of neurons to apoptosis (Kassed et al., 2003).

6. Conclusions

Considering all these observations, it is clear that JNK/SAPK and p38 kinases play crucial roles in tau phosphorylation in AD and related transgenic models. Whether these signal pathways also promote cell death in diseased brains is a matter for further study, as observations *in vitro* do not match observations in AD and related transgenic models. Difficulties in the interpretation of human post-mortem brain material largely result from the short life of Fos and Jun in cell death paradigms, which may minimize the possibility of detecting Fos and Jun activation at the precise time preceding cell death in human neurodegenerative disorders. In contrast to SAPK/JNK-P and p38-P, which are probably sequestered by phospho-tau aggregates, no similar situation occurs with Fos and Jun.

Additional studies have shown a possible link between p38 activation and microglial responses to activation stimuli, thus suggesting that activation of p38 by β-amyloid may participate in the activation of microglia and the inflammatory response in AD (Pocock and Liddle, 2001).

Controlling p38 and SAPK expression could be beneficial in AD disease progression and, for this reason, stress kinases have been seen as putative therapeutic targets in AD (Dalrimple, 2002; Zhu et al., 2002; Johnson and Bailey, 2003).

Acknowledgements

This work was supported in part by the EU project Brain Net II, and by FIS grants P1020004, P1030032 and G03-167. We wish to thank T. Yohannan for editorial assistance.

References

Aksenov MY, Aksenova MV, Butterfield DA, Geddes JW, Markesbery WR (2001). Protein oxidation in the brain in Alzheimer's disease. Neuroscience 103:373-383.

Anderson AJ, Cummings BJ, Cotman CW (1994). Increased immunoreactivity for Jun- and Fos-related proteins in Alzheimer's disease: association with pathology. Exp Neurol 125:286-295.

Anderson AJ, Su JH, Cotman CW (1996). DNA damage and apoptosis in Alzheimer's disease: co-localization with c-Jun immunoreactivity, relationship to brain area, and effect of postmortem delay. J Neurosci 16:1710-1719.

Anderton BH, Betts J, Blackstock WP, Brion JP, Chapman S, Connell J, Dayanandan R, Gallo JM, Gibb G, Hanger DP, Hutton M, Kardalinou E, Leroy K, Lovestone S, Mack T, Reynolds CH, van Slegtenhorst M (2001). Sites of phosphorylation in *tau* and factors affecting their regulation. Biochem Soc Symp 67:73-80.

Atzori C, Ghetti B, Piva R, Srinivasan AN, Zolo P, Delisle MB, Mirra SS, Migheli A (2001). Activation of the JNK/p38 pathway occurs in diseases characterized by *tau* protein pathology and is related to *tau* phosphorylation but not to apoptosis. J Neuropathol Exp Neurol 60:1190-1197.

Bales KR, Du Y, Holtzman D, Cordell B, Paul SM (2000). Neuroinflammation and Alzheimer's disease: critical role for cytokine/Aβ induced glial activation, NFκB, and apolipoprotein E. Neurobiol Aging 21:427-432.

Bertram L, Tanzi R (2003). Genetics of Alzheimer's disease. In: Neurodegeneration: The Molecular Pathology of Dementia and Movement Disorders (Dickson D, ed.), pp. 40-46. Basel: ISN Neuropath Press.

Boissiere F, Hunnot S, Faucheux B, Duyckaerts C, Hauw JJ, Agid Y, Hirsch EC (1997). Nuclear translocation of NFκB in cholinergic neurons of patients with Alzheimer's disease. NeuroReport 8:2849-2852.

Bonnycastle LL, Yu CE, Wijsman EM, Orr HT, Patterson D, Clancy KP, Goddard KA, Alonso ME, Nemens E, White JA (1993). The *c-fos* gene and early-onset familial Alzheimer's disease. Neurosci Lett 160:33-36.

Bozyczko-Coyne D, O'Kane TM, Wu ZL, Dobrzanski P, Murphy S, Vaught JL, Scott RW (2001). CEP-1347/KT-7515, an inhibitor of SAPK/JNK pathway activity, promotes survival and blocks multiple events associated with Aβ-induced cortical neuron apoptosis. J Neurochem 77:849-863.

Braak H, Braak E (1999). Temporal sequence of Alzheimer's disease-related pathology. In: Cerebral Cortex, vol. 14: Neurodegenerative and Age-Related Changes in Structure and Function of Cerebral Cortex (Peters A, Morrison JH, eds.), pp. 475-512. New York, Boston, Dordrecht, London, Moscow: Kluver Academic Press.

Cavigelli M, Dolfi F, Claret FX, Karin M (1995). Induction of cFos expression through JNK-mediated TCF-ELK-1 phosphorylation. EMBO J 14:5957-5964.

Cobb MH, Goldsmith EJ (1995). How MAP kinases are regulated. J Biol Chem 270:14843-14846.

Cruts M, Backhovens H, Martin JJ, van Broeckhoven C (1994). Genetic analysis of the cellular oncogene fos in patients with chromosome 14 encoded Alzheimer's disease. Neurosci Lett 174:97-100.

Curran T, Franza BR (1989). Fos and Jun: the AP-1 connection. Cell 55:395-397.

Dalrimple SA (2002). p38 mitogen activated protein kinase as a therapeutic target for Alzheimer's disease. J Mol Neurosci 19:295-299.

Daniels WM, Hendricks J, Salie R, Taljaard JJ (2001). The role of MAP-kinase superfamily in β-amyloid toxicity. Metab Brain Dis 16:175-185.

Dérijard B, Raingeaud J, Barrett T, Wu IH, Han J, Ulevitch RJ, Davis RJ (1995). independent human MAP kinase signal-transduction pathways defined by MEK and MKK isoforms. Science 267:682-685.

Duyckaerts C, Dickson DW (2003). Neuropathology of Alzheimer's disease. In: Neurodegeneration: The Molecular Pathology of Dementia and Movement Disorders (Dickson D, ed.), pp. 47-68. Basel: ISN Neuropath Press.

Enslen H, Raingeaud J, Davis RJ (1998). Selective activation pf p38 mitogen-activated protein MAP kinase isoforms by the MAP kinase kinases MKK3 and MKK6. J Biol Chem 273:1741-1748.

Ferrer I (2004). Stress kinases involved in tau phosphorylation in Alzheimer's disease, tauopathies and APP transgenic mice. Neurotox Res 6:469-475.

Ferrer I, Ballabriga J, Pozas E (1997a). Transient forebrain ischemia in the adult gerbil is associated with a complex c-Jun response. Neuroreport 8:2483-2487.

Ferrer I, Barrachina M, Tolnay M, Rey MJ, Vidal N, Carmona M, Blanco R, Puig B (2003a). Phosphorylated protein kinases associated with neuronal and glial *tau* deposits in argyrophilic grain disease. Brain Pathol 13:62-78.

Ferrer I, Blanco R, Carmona M, Puig B (2001). Phosphorylated mitogen-activated protein kinase (MAPK/ERK-P), protein kinase of 38 kDa (p38-P), stress-activated protein kinase (SAPK/JNK-P), and calcium/calmodulin-dependent kinase II (CaM kinase II) are differentially expressed in *tau* deposits in neurons and glial cells in tauopathies. J Neural Transm 108:1397-1415.

Ferrer I, Boada-Rovira M, Sanchez-Guerra ML, Rey MJ, Costa-Jussa F (2004). Neuropathology and pathogenesis of encephalitis following amyloid-β immunization in Alzheimer's disease. Brain Pathol 14:11-20.

Ferrer I, Gómez-Isla T, Puig B, Freixes M, Ribé E, Dalfó E, Avila J (2005). Current advances on different kinases involved in *tau* phosphorylation, and implications in Alzheimer's disease and tauopathies. Curr Alzheimer Res 2:3-18.

Ferrer I, Hernandez I, Puig B, Rey MJ, Ezquerra M, Tolosa E, Boada M (2003b). Ubiquitin-negative mini-Pick-like bodies in the dentate gyrus in P301L tauopathy. J Alzheimer's Dis 5:445-454.

Ferrer I, Lopez E, Blanco R, Rivera R, Krupinski J, Marti E (2000a). Differential c-Fos and caspase expression following kainic acid excitotoxicity. Acta Neuropathol 99:245-256.

Ferrer I, Martí E, López E, Tortosa A (1998). NFκB immunoreactivity is observed in association with βA4 diffuse plaques in patients with Alzheimer's disease. Neuropathol Appl Neurobiol 24:271-277.

Ferrer I, Olive M, Ribera J, Planas AM (1996a). Naturally-occurring (programmed) and radiation-induced apoptosis are associated with selective c-Jun expression in the developing brain. Eur J Neurosci 8:1286-1298.

Ferrer I, Pastor P, Rey MJ, Muñoz E, Puig B, Pastor E, Oliva R, Tolosa E (2003c). Tau phosphorylation and kinase activation in familial tauopathy linked to delN296 mutation. Neuropathol Appl Neurobiol 29:23-34.

Ferrer I, Planas AM, Pozas E (1997b). Rdaiation-induced apoptosis in developing rats and kainic acid-induced excitotoxicity in adult rats are associated with distinctive morphological and biochemical c-Jun/AP-1 (N) expression. Neuroscience 80:449-458.

Ferrer I, Pozas E, Planas AM (2000b). c-Jun/AP-1 (N) expression and apoptosis. Neuroscience 96:447-448.

Ferrer I, Seguí J, Planas AM (1996b). Amyloid deposition is associated with c-Jun expression in Alzheimer's disease and amyloid angiopathy. Neuropathol Appl Neurobiol 22:521-526.

Fogarty MP, Downer EJ, Campbell V (2003). A role for c-Jun N-terminal kinase 1 (JNK-1), but not JNK2, in the β-amyloid-mediated stabilization of protein p53 and induction of the apoptotic cascade in cultured cortical neurons. Biochem J 371:789-798.

Goedert M, Hasegawa M, Jakes R, Lawler S, Cuenda A, Cohen P (1997). Phosphorylation of microtubule-associated protein *tau* by stress-activated protein kinases. FEBS Lett 409:57-62.

Götz J (2001). *Tau* and transgenic animal models. Brain Res Rev 35:266-286.

Guillardon F, Skutella T, Uhlmann E, Holsboer F, Zimmermann M, Behl C (1996). Activation of c-Fos contributes to β-amyloid peptide-induced neurotoxicity. Brain Res 706:169-172.

Gupta S, Barrett, Whitmarsh AJ, Cavanagh J, Sluss HK, Dérijard B, Davis RJ (1996). Selective interaction of JNK protein kinase isoforms with transcription factors. EMBO J 15:2760-2770.

Gupta S, Campbell D, Dérijard B, Davis RJ (1995). Transcription factor ATF-2 regulation by JNK signal transduction pathway. Science 267:389-393.

Halazonetis TD, Georgopoulos K, Greenberg ME, Leder P (1988). c-Jun dimerizes with itself and with c-Fos, forming complexes of different DNA binding affinities. Cell 55:917-924.

Hammond ES, Brunet CL, Johnson GD, Parkhill J, Miller AE, Brady G, Gregory CD, Grand RJA (1998). Homology between a human apoptosis specific protein and the product of APG5, a gene involved in autophagy. FEBS Lett 425:391-395.

Hartzler AW, Zhu X, Siedlak SL, Castellani RJ, Avila J, Perry G, Smith MA (2002). The p38 pathway is activated in Pick disease and progressive supranuclear palsy: a mechanistic link between mitogenic pathways, oxidative stress and *tau*. Neurobiol Aging 23:855-859.

Hashimoto Y, Niikura T, Chiba T, Tsukamoto E, Kadowaki H, Nishitoh H, Yamagishi Y, Ishizaka M, Yamada M, Nawa M, Terashita K, Aiso S, Ichijo H, Nishimoto I (2003). The cytoplasmic domain of Alzheimer's amyloid-β protein precursor causes sustained apoptosis signal-regulating kinase 1/c-Jun NH2-terminal kinase-mediated neurotoxic signal via dimerization. J Pharmacol Exp Ther 306:889-902.

Hazzalin CA, Cano E, Cuenda A, Barrattt MJ, Cohen P, Mahadevan LC (1996). p38/ERK is essential for stress-induced nuclear responses: JNK/SAPKs and c-Jun/ATF-2 phosphorylation are insufficient. Curr Biol 6:1028-1031.

Helbecque N, Abderrahamani A, Meylan L, Riederer B, Mooser V, Miklossy J, Delplanque J, Boutin P, Nicod P, Haefliger JA, Cottel D, Amouyel P, Froguel P, Waeberg G (2003). Islet-brain1/c-Jun N-terminal kinase interacting protein-1 (IB1/JIP-1) promoter variant is associated with Alzheimer's disease. Mol Psychiatr 8:413-422.

Herreman A, Serneels L, Annaert W, Collen D, Schoonjans L, De Strooper B (2000). Total inactivation of γ-secretase activity in presenilin-deficient embryonic stem cells. Nat Cell Biol 2:461-462.

Hoozemans JJ, Veerhuis R, Rozemuller AJ, Arendt T, Eikelenboom P (2004). Neuronal COX-2 expression and phosphorylation of pRb precede p38 MAPK activation and neurofibrillary changes in AD temporal cortex. Neurobiol Dis 15:492-499.

Huang Y, Liu F, Grundke-Iqbal I, Iqbal K, Gong CX (2005). NF precursor, p105, and NFκB inhibitor, IκBγ, are both elevated in Alzheimer disease brain. Neurosci Lett 373:115-118.

Hunter T, Karin M (1992). The regulation of transcription by phosphorylation. Cell 70:375-387.

Hutton M, Lewis J, Dickson D, Yen SH, McGowan E (2001). Analysis of tauopathies with transgenic mice. Trends Mol Med 7:467-470.

Hwang DY, Cho JS, Lee SH, Chae KR, Lim HJ, Min SH, Seo SJ, Song YS, Song CW, Paik SG, Sheen YY, Kim YK (2004). Aberrant expression of pathogenic phenotype in Alzheimer's diseased transgenic mice carrying NSE-controlled APPsw. Exp Neurol 186:20-32.

Ip YT, Davis RJ (1998). Signal transduction by the c-Jun-N-terminal kinase (JNK): from inflammation to development. Curr Opin Cell Biol 10:205-219.

Irizarry MC, McNamara M, Fedorchak K, Hsiao K, Hyman BT (1997a). APPSw transgenic mice develop age-related Aβ deposits and neuropil abnormalities, but not [?] neuronal loss in CA1. J Neuropathol Exp Neurol 56:965-973.

Irizarry MC, Soriano F, McNamara M, Page KJ, Schenk D, Games D, Hyman BT (1997b). Aβ deposition is associated with neuropil changes, but not with overt neuronal loss in the human amyloid precursor protein V717F (PDAPP) transgenic mouse. J Neurosci 17:7053-7059.

Jang JH, Surth YJ (2002). β-amyloid induces oxidative DNA damage and cell death through activation of c-Jun N terminal kinase. Ann NY Acad Sci 973:228-236.

Jenkins SM, Zinnerman M, Garner C, Johnson GV (2000). Modulation of *tau* phosphorylation and intracellular localization by cellular stress. Biochem J 345 part 2: 263-270.

Johnson GV, Bailey CD (2003). The p38 MAP kinase signaling pathway in Alzheimer's disease. Exp Neurol 183:263-268.

Kallunki T, Deng T, Hibi M, Karin M (1996). c-Jun can recruit JNK to phosphorylate dimerization partners via specific docking interactions. Cell 87:1-20.

Kallunki T, Su B, Tsigelny Isluss HK, Dérijard B, Moore G, Davis R, Karin M (1994). JNK2 contains a specific-determining region responsible for efficient c-Jun binding and phosphorylation. Genes Dev 8:2996-3007.

Kaltschmidt B, Uherek, Volk B, Baeuerle PA, Kaltschmidt C (1997). Transcription factor NF-κB is activated in primary neurons by amyloid β peptides and in neurons surrounding early plaques from patients with Alzheimer disease. Proc Natl Acad Sci USA 94:2642-2647.

Karin M (1994). Signal transduction from the cell surface to the nucleus through the phosphorylation of transcription factors. Curr Opin Cell Biol 6:415-424.

Karin M (1995). The regulation of AP-1 activity by mitogen-activated protein kinases. J Biol Chem 270:16483-16486.

Kassed CA, Butler TL, Navidomskis MT, Gordon MN, Morgan D, Pennypacker KR (2003). Mice expressing human mutant presenilin-1 exhibit decreased activation of NFκB p50 in hippocampal neurons after injury. Brain Res Mol Brain Res 110:152-157.

Kihiko ME, Tucker HM, Rydel RE, Estus S (1999). c-Jun contributes to β-induced neuronal apoptosis but is not necessary for β-amyloid induced c-jun induction. J Neurochem 73:2609-2612.

King GD, Scott Turner R (2004). Adaptor protein interactions: modulators of amyloid precursor protein metabolism and Alzheimer's disease risk? Exp Neurol 185:208-219.

Kitamura Y, Shimohama S, Ota T, Matsuoka Y, Nomura Y, Taniguchi T (1997). Alteration of transcription factor NFκB and STAT1 in Alzheimer's disease brains. Neurosci Lett 237:17-20.

Lahiri DK (2004). Apolipoprotein E as a target for developing new therapeutics for Alzheimer's disease based on studies from protein, RNA, and regulatory region of the gene. J Mol Neurosci 23:225-233.

Lee KW, Lee SH, Kim H, Song JS, Yang SD, Paik SG, Han PL (2004). Progressive cognitive impairment and anxiety induction in the absence of plaque deposition in C57BL/6 inbred mice expressing transgenic amyloid precursor protein. J Neurosci Res 76:572-580.

Lewis J, Dickson DV (2003). Transgenic animal models of tauopathies. In: Neurodegeneration: The Molecular Pathology of Dementia and Movement Disorders (Dickson D, ed.), pp. 150-154. Basel: ISN Neuropath Press.

Lim F, Hernandez F, Lucas JJ, Gómez-Ramos P, Morán MA, Avila J (2001). FTDP-17 mutations in *tau* transgenic mice provoke lysosomal abnormalities and *tau* filaments in the forebrain. Mol Cell Neurosci 18:702-714.

Lin A, Minden A, Martinetto H, Claret FX, Lange-Carter C, Mercurio F, Johnson GL, Karin M (1995). Identification of a dual specificity kinase that activates the Jun kinase and p38-Mpk2. Science 268:289-290.

MacGibbon GA, Lawlor PA, Walton M, Sirimanne E, Faull RL, Synek B, Mee E, Connor B, Dragunow M (1997). Expression of Fos, Jun, and Krox family proteins in Alzheimer's disease. Exp Neurol 147:316-332.

Marcus DL, Strafaci JA, Miller DC, Masia S, Thomas CG, Rosman J, Hussain S, Freedman ML (1998). Quantitative neuronal c-fos and c-jun expression in Alzheimer's disease. Neurobiol Aging 19:393-400.

Markesbery WR, Carney JM (1999). Oxidative alterations in Alzheimer's disease. Brain Pathol 9:133-146.

Marques CA, Keil U, Bonert A, Steiner B, Haass C, Muller WE, Eckert A (2003). Neurotoxic mechanisms caused by Alzheimer's disease-linked Swedish amyloid precursor protein mutation: oxidative stress, caspases, and the JNK pathway. J Biol Chem 278:28294-28302.

Matsuda S, Yasukawa T, Homma Y, Ito Y, Niikura T, Hiraki T, Hirai S, Ohno S, Kita Y, Kawasumi M, Kouyama K, Yamamoto T, Kyriakis JM, Nishimoto I (2001). c-Jun N-terminal kinase (JNK)-interacting protein-1b/islet-brain-1 scaffolds Alzheimer's amyloid precursor protein with JNK. J Neurosci 21:6597-6607.

McGowan E, Pickord F, Dickson DW (2003). Alzheimer animal models: models of Aβ deposition in transgenic mice. In: Neurogeneration: The Molecular Pathology of Dementia and Movement Disorders (Dickson D, ed.), pp. 74-79. Basel: ISN Neuropath Press.

Mielke K, Herdegen T (2000). JNK and p38 stress kinases. Degenerative effectors of signal-transduction cascades in the nervous system. Progr Neurobiol 61:45-60.

Minden A, Karin M (1997). Regulation and function of the JNK subgroup of MAP kinases. Biochem Biophys Acta 1333:F85-F104.

Morishima Y, Gotoh Y, Zieg J, Barrett T, Takano H, Flavell R, Davis RJ, Shirasaki Y, Greenberg ME (2001). β-amyloid induces neuronal apoptosis via a mechanism that involves the c-Jun N-terminal kinase pathway and the induction of Fas ligand. J Neurosci 21:7551-7560.

Morris SW, St Clair DM (1994). Eliminating c-fos as a candidate gene for early-onset familial Alzheimer's disease. Neurology 44:1762-1764.

Munujos P, Vendrell M, Ferrer I (1993). Proto-oncogene c-fos induction in thiamine-deficient encephalopathy. Protective effects of nicardipine on pyrithiamine-induced lesions. J Neurol Sci 118:175-180.

Nakabeppu Y, Ryder K, Nathans D (1988). DNA binding activities of three murine Jun proteins: stimulation by Fos. Cell 55:907-915.

Nicoll JAR, Wilkinson D, Holmes C, Steart O, Markham H, Weller RO (2003). Neuropathology of human Alzheimer disease after immunization with amyloid-β peptide: a case report. Nat Med 9:4448-4452.

Niikura T, Yamada M, Chiba T, Aiso S, Matsuoka M, Nishimoto I (2004). Characterization of V642I-AβPP-induced cytotoxicity in primary neurons. J Neurosci Res 77:54-62.

Okazawa H, Estus S (2002). The JNK/c-jun cascade and Alzheimer's disease. Am J Alzheimer Dis Other Demen 17:79-88.

Otth C, Mendoza-Naranjo A, Mujica L, Zambrano A, Concha II, Maccioni RB (2003). Modulation of JNK and p38 pathways by cdk5 protein kinase in a transgenic mouse model of Alzheimer's disease. NeuroReport 14:2403-2409.

Pamplona R, Dalfó E, Ayala V, Bellmunt MJ, Ferrer I, Portero-Otin M (2005). Proteins in human brain cortex are modified by oxidation, glycoxidation and lipoxidation: effects of Alzheimer's disease and identification of lipoxidation targets. J Biol Chem 280:21522-21530.

Papolla MA, Omar RA, Kim KS, Robalds NK (1992). Immunohistochemical evidence of oxidative stress in Alzheimer's disease. Am J Pathol 140:621-628.

Pastorcic M, Das HK (2003). Ets transcription factors ER81 and Elk1 regulate the transcription of the human presenilin 1 gene promoter. Brain Res Mol Brain Res 113:57-66.

Pei JJ, Braak E, Braak H, Grundque-Iqbal K, Winblad W, Cowburn RF (2001). Localization of active forms of c-Jun kinase (JNK) and p38 kinase in Alzheimer's disease brains at different stages of neurofibrillary degeneration. J Alzheimer's Dis 3:41-48.

Perez M, Ribe E, Rubio A, Lim F, Moran MA, Gomez-Ramos P, Ferrer I, Gomez Isla MT, Avila J (2005). Characterization of a double (amyloid precursor protein-tau) transgenic: tau phosphorylation and aggregation. Neuroscience 130:339-347.

Perry G, Srinivas R, Nunomura A, Smith MA (2003). The role of oxidative mechanisms in neurode-generative diseases. In: Neurodegeneration: The Molecular Pathology of Dementia and Movement Disorders (Dickson D, ed.), pp. 8-10. Basel: ISN Neuropath Press.

Pocock JM, Liddle AC (2001). Microglial signaling cascades in neurodegenerative disease. Progr Brain Res 132:555-565.

Pozas E, Aguado F, Ferrer I (1999). Localization and expression of Jun-like immunoreactivity in apoptotic neurons induced by colchicines administration in vivo and in vitro depends on the antisera used. Acta Neuropathol 98:119-128.

Pozas E, Ballabriga J, Planas AM, Ferrer I (1997). Kainic acid-induced excitotoxicity is associated with a complex c-Fos and c-Jun response which does not preclude either cell death or survival. J Neurobiol 33:232-246.

Puig B, Gómez-Isla T, Ribe E, Cuadrado M, Ferrer I (2004). Expression of stress-activated kinases c-Jun N-terminal kinase (SAPK/JNK-P) and p38 kinase (p38-P) links oxidative stress and tau hyperphosphorylation in neurites surrounding Aβ plaques in APP Tg2576 mice. Neuropathol Appl Neurobiol 30:491-502.

Puig B, Viñals F, Ferrer I (2004). Active stress kinase p38 enhances and perpetuates abnormal tau phosphorylation and deposition in Pick's disease. Acta Neuropathol 107:185-189.

Quitschke W, Golgaber D (1992). The amyloid protein precursor promoter. A region essential for transcriptional activity contains a nuclear factor binding domain. J Biol Chem 267:17362-17368.

Raingeaud J, Gupta S, Rogers J, Dickens M, Han J, Ulevitch RJ, Davis RJ (1995). Pro-inflammatory cytokines and environmental stress causes p38 MAP kinase activation by dual phosphorylation on tyrosine and threonine. J Biol Chem 270:7420-7426.

Raingeaud J, Whitmarsh AJ, Barrett T, Dérijard B, Davis RJ (1996). MKK3-and MKK6-regulated gene expression is mediated by p38 mitogen-activated protein kinase signal transduction pathway. Mol Cell Biol 16:1247-1255.

Rauscher FJ, Sambucetti LC, Curran T Distel RJ, Spegelman BM (1988). Common DNA binding site for Fos protein complexes and transcription factor AP-1. Cell 52:471-480.

Reynolds CH, Betts JC, Blackstock WP, Nebreda AR, Anderton BH (2000). Phosphorylation sites on tau identified by nanoelectrospray mass spectrometry: differences in vitro between the mitogen-activated protein kinases ERK2, c-Jun N-terminal kinase and p38, and glycogen synthase kinase-3β. J Neurochem 74:1587-1595.

Reynolds CH, Nebreda AR, Gibb GM, Utton MA, Anderton BH (1997a). Reactivating kinase/p38 phosphorylates tau protein in vitro. J Neurochem 69:191-198.

Reynolds CH, Utton MA, Gibb GM, Yates A, Anderton BH (1997b). Stress-activated protein kinase/c-Jun N-terminal kinase phosphorylates tau protein. J Neurochem 68:1736-1744.

Ribera J, Ayala V, Esquerda JE (2002). c-Jun-like immunoreactivity in apoptosis is the result of a cross-reaction with neoantigenic sites exposed by caspase-3-mediated proteolysis. J Histochem Cytochem 50:961-972.

Robinson MJ, Cobb MH (1997). Mitogen-activated protein kinase pathways. Curr Opin Cell Biol 9:180-186.

Rogaev EI, Lukiw WJ, Vaula G, Haines JL, Rogaeva EA, Tsuda T, Alexandrova N, Liang Y, Mortilla M, Amaducci L (1993). Analysis of the c-Fos gene on chromosome 14 and the promoter of the amyloid precursor protein gene in familial Alzheimer's disease. Neurology 43:2275-2279.

Sahara N, Vega IE, Ishizawa T, Lewis J, McGowan E, Hutton M, Dickson D, Yen SH (2004). Phospho-rylated p38MAPK specific antibodies cross-react with sarkosyl-insoluble hyperphosphorylated *tau* proteins. J Neurochem 90:829-838.

Salbaum JM, Weidemann A, Masters CL, Beyreuther K (1989). The promoter of Alzheimer's disease amyloid A4 precursor gene. Progr Clin Biol Res 317:277-283.

Santacruz K, Lewis J, Spires T, Paulson J, Kotilinek L, Ingelsson M, Guimaraes A, DeTure M, Ramsden M, McGowan E, Forster C, Yue M, Orne J, Janus C, Mariash A, Kuskowski M, Hyman B, Hutton M, Ashe KH (2005). Tau suppression in a neurodegenerative mouse model improves memory function. Science 309:476-481.

Sanz O, Estrada A, Ferrer I, Planas AM (1997). Differential cellular distribution and dynamics of HSP-70, cyclooxygenase-2, and c-Fos in the rat brain after transient focal ischemia or kainic acid. Neuroscience 80:221-232.

Savage MJ, Lin YG, Ciallella JR, Flood DG, Scott RW (2002). Activation of c-Jun N-terminal kinase and p38 in an Alzheimer's disease model is associated with amyloid deposition. J Neurosci 22:3376-3385.

Scheinfeld MH, Roncarati R, Vito P, Lopez PA, Abdellah M, D'Adamio L (2002). Jun NH2-terminal kinase (JNK) interacting protein 1 (JIP1) binds the cytoplasmic domain of the Alzheimer's β-amyloid precursor protein (APP). J Biol Chem 277:3767-3775.

Selznick LA, Holtnman DM, Han BH, Gökden M, Srinavasan AN, Jonson EM, Roth KA (1999). *In situ* immunodetection of neuronal caspase-3 activation in Alzheimer's disease. J Neuropathol Exp Neurol 58:1020-1026.

Shaywitz AJ, Greenberg ME (1999). CREB: a stimulus-induced transcription factor activated by a diverse array of extracellular signals. Annu Rev Biochem 68:821-861.

Shoji M, Iwakami N, Takeuchi S, Waragai M, Suzuki M, Kanazawa I, Lippa CF, Ono S, Okazawa H (2000). JNK activation is associated with intracellular β-amyloid accumulation. Brain Res Mol Brain Res 85:221-233.

Smith CD, Carney JM, Starke-Reed PE, Oliver CN, Stadtman ER, Floyd RA, Markesbery WR (1991). Excess brain protein oxidation and enzyme dysfunction in normal aging and in Alzheimer's disease. Proc Natl Acad Sci USA 88:10540-10543.

Smith MA, Richey PL, Taneda S, Kutty RK, Sayre LM, Monnier VM, Perry G (1994). Advanced Maillard end products, free radicals, and protein oxidation in Alzheimer's disease. Ann NY Acad Sci 738:447-454.

Soriano MA, Ferrer I, Rodriguez-Farre E, Planas AM (1995). Expression of c-fos and inducible hsp-70 mRNA following a transient episode of focal ischemia that had non-lethal effects on the rat brain. Brain Res 670:317-320.

Sun A, Liu M, Nguyen XV, Bing G (2003). P38 MAP kinase is activated at early stages in Alzheimer's disease brain. Exp Neurol 183:394-405.

Swatton JE, Sellers LA, Faull RL, Holland A, Iritani S, Bahn S (2004). Increased MAP kinase activity in Alzheimer's and Down syndrome but not in schizophrenia human brain. Eur J Neurosci 19:2711-2719.

Tamagno E, Robino G, Obbili A, Bardini P, Aragno M, Parola M, Danni O (2003). H_2O_2 and 4-hydroxynonenal mediate amyloid β-induced neuronal apoptosis by activating JNKs and p38MAPK. Exp Neurol 180:144-155.

Tan Y, Rouse JR, Zhang A, Cariati S, Boccia C, Cohen P, Comb MJ (1996). FGF and stress regulated CREB and ATF-1 via a pathway involving p38 MAP kinase and MAPKAP kinase-2. EMBO J 15:4629-4642.

Tanaka S, Takehashi M, Matoh N, Iida S, Suzuki T, Futaki S, Hamada H, Masliah E, Sugiura Y, Ueda K (2002). Generation of reactive oxygen species and activation of NFκB by non-Aβ component of Alzheimer's disease amyloid. J Neurochem 82:305-315.

Taru H, Iijima K, Hase M, Kirino Y, Yagi Y, Suzuki T (2002). Interactions of Alzheimer's amyloid precursor family proteins with scaffold proteins of the JNK signaling cascade. J Biol Chem 277:20070-20078.

Trejo J, Massamiri T, Deng T, Dewji NN, Bayney RM, Brown JH (1994). A direct role for protein kinase C and the transcription factor Jun/AP-1 in the regulation of the Alzheimer's β-amyloid precursor protein gene. J Biol Chem 269:1682-1690.

Troy CM, Rabacchi SA, Xu Z, Maroney AC, Connors TJ, Shelanski ML, Greene LA (2001). β-amyloid-induced neuronal apoptosis requires c-Jun N-terminal kinase activation. J Neurochem 77:157-164.

van Dam H, Duyndam M, Rottier R, Bosch A, De Vries-Smits L, Herrlich P, Zantema A, Angel P, van der Eb AJ (1993). Heterodimer formation of c-Jun and ATF-2 is responsible for induction of c-Jun by the 243 amino acid adenovirus EIA protein. EMBO J 12:479-487.

van Leuven F (2000). Single and multiple transgenic mice as models for Alzheimer's disease. Progr Neurobiol 61:305-312.

Vélez-Pardo C, Ospina GG, Jiménez del Rio M (2002). Aβ [25-35] peptides and iron promote apoptosis in lymphocytes by an oxidative stress mechanism: involvement of H2O2, caspase-3, NFκB, p53 and c-Jun. Neurotoxicology 23:351-365.

Wei W, Norton DD, Wang X, Kusiak JW (2002). Aβ17-42 in Alzheimer's disease activates JNK and caspase-8 leading to neuronal apoptosis. Brain 125:2036-2043.

Whitmarsh AJ, Davis RJ (1996). Transcription factor AP-1 regulation by mitogen-activated protein kinase signal transduction pathways. J Mol Med 74:589-607.

Yamada T, Yoshiyama Y, Kawaguchi N (1997). Expression of activating transcription factor-2 (ATF-2), one of the cyclic AMP response element (CRE) binding proteins, in Alzheimer disease and non-neurological brain tissues. Brain Res 749:329-334.

Yamamoto Y, Gaynor RB (2001). Role of NFκB pathway in the pathogenesis of human disease states. Curr Mol Med 1:287-296.

Yamamoto-Sasaki M, Ozawa H, Saito T, Rosler M, Riederer P (1999). Impaired phosphorylation of cyclic AMP response element binding protein in the hippocampus of dementia of Alzheimer type. Brain Res 824:300-303.

Yang D, Tournier CM, Wysk M, Lu HT, Xu J, Davis RJ, Flavell RA (1997). Targeted disruption of the MKK4 gene causes embryonic death, inhibition of c-Jun NH2-terminal kinase activation, and defects in AP-1 transcriptional activity. Proc Natl Acad Sci USA 94:3004-3009.

Yoshiyama Y, Arai K, Hattori T (2001). Enhanced expression of Iκb with neurofibrillary pathology in Alzheimer's disease. Neuroreport 12:2641-2645.

Zhang P, Hirsch EC, Damier P, Duyckaerts C, Javoy-Agid F (1992). c-fos protein-like immunoreactivity: distribution in the human brain and over-expression in the hippocampus of patients with Alzheimer's disease. Neuroscience 46:9-21.

Zhang Z, Nadeau P, Song W, Donoviel D, Yuan M, Bernstein A, Yankner BA (2000). Presenilins are required for γ-secretase cleavage of β-APP and transmembrane cleavage of Notch-1. Nat Cell Biol 2:463-465.

Zhu X, Lee HG, Raina AK, Perry G, Smith MA (2002). The role of mitogen-activated protein kinase pathways in Alzheimer's disease. Neurosignals 11:270-281.

Zhu X, Ogawa O, Wang Y, Perry G, Smith MA (2003a). JKK1, an upstream activator of JNK/SAPK, is activated in Alzheimer's disease. J Neurochem 85:87-93.

Zhu X, Raina AK, Lee HG, Chao M, Nunomura A, Tabaton M, Petersen RB, Perry G, Smith MA (2003b). Oxidative stress and neuronal adaptation in Alzheimer disease: The role of SAPK pathways. Antioxid Redox Signal 5:571-576.

Zhu X, Raina AK, Rottkamp CA, Aliev G, Perry G, Boux H, Smith MA (2001a). Activation and redistribution of c-Jun N-terminal kinase/stress activated protein kinase in degenerating neurons in Alzheimer's disease. J Neurochem 76:435-441.

Zhu X, Rottkamp CA, Boux H, Takeda A, Perry G, Smith MA (2000). Activation of p38 kinase links *tau* phosphorylation, oxidative stress, and cell cycle-related events in Alzheimer disease. J Neuropathol Exp Neurol 59:880-888.

Zhu X, Rottkamp CA, Hartzler A, Sun Z, Takeda A, Boux H, Shimohama S, Perry G, Smith MA (2001b). Activation of MKK6, an upstream activator of p38, in Alzheimer's disease. J Neurochem 79:311-318.

14

Parkinson's Disease, the Dopamine System and Immediate Early Genes

XIAOQUN ZHANG and PER SVENNINGSSON

*Section for Molecular Neuropharmacology, Department of Physiology and Pharmacology,
Karolinska Institute, Stockholm, Sweden*

1. Introduction

Parkinson's disease is a movement disorder characterized by degeneration of dopamine-producing neurons. At early stages, pharmacological replacement treatment with dopamine receptor agonists or the dopamine precursor, L-DOPA, can cause symptomatic relief. However, ultimately these treatments are insufficient and cause, via largely unknown mechanisms, severe side-effects. Studies on immediate early genes, primarily c-fos, have been useful in elucidating the neuronal circuitries that are regulated by dopaminergic treatments in animal models of Parkinson's Disease. Detailed studies on immediate early gene expression have also identified the cellular phenotype of neurons affected by these treatments, particularly in the dopamine-enriched striatum. Administration of agonists directed at different dopamine receptor subtypes causes behavioral synergism that is reflected by a qualitatively distinct immediate early gene pattern in the striatum. Moreover, recent studies have demonstrated that certain immediate early gene changes reflect abnormal activation of signal transduction pathways in Parkinson's Disease. We review here some of the studies on the regulation of the dopamine pathway and immediate early genes in Parkinson's disease.

2. Parkinson's Disease

The cardinal features of Parkinson's disease (PD) are hypokinesia, rigidity, resting tremor and gait abnormalities. PD is also associated with numerous cognitive and emotional disturbances, of which depression is the most frequent (Leentjens, 2004; Weintraub et al., 2004). Pathologically, PD is characterized by degeneration of dopamine-producing neurons in the substantia nigra pars compacta (SNc) and accumulation of intracellular protein aggregates known as Lewy

*R. Pinaud, L.A. Tremere (Eds.), Immediate Early Genes in Sensory Processing, Cognitive Performance
and Neurological Disorders, 261–290, ©2006 Springer Science + Business Media, LLC*

bodies (Fahn, 1999; Dauer and Przedborski, 2003). The causes of the degeneration of dopamine neurons are largely unknown, but appears to involve multiple factors acting together, including genetic susceptibility, environmental exposures and aging. From a pathophysiological standpoint, mitochondrial dysfunction, oxidative stress, excitotoxicity, apoptosis, inflammation and proteasome failure have been observed in PD (Dauer and Przedborski, 2003). Over the past years, mutations in genes underlying familiar forms of PD have been identified (Vila and Przedborski, 2004). Some of these genes, including α-synuclein, parkin, DJ-1, and ubiquitin C-terminal hydrolase L1, participate in the ubiquitin-proteasome system and indicate that PD is associated with dysfunctional protein degradation. Mutations in PINK1 (PTEN-induced kinase (1)) are associated with PD (Valente et al., 2004). PINK provides a molecular link between mitochondria and the pathogenesis of PD as PINK1 is mitochondrially located and may exert a protective effect on the cell that is abrogated by the mutations, resulting in increased susceptibility to cellular stress. Mutations in LRRK2 (leucine-rich repeat kinase (2)), a gene encoding a large, multifunctional protein that includes a protein kinase domain of the MAPKKK (mitogen activated phospho kinase kinase kinase) is also associated with PD (Paisan-Ruiz et al., 2004; Zimprich et al., 2004).

Midbrain dopaminergic neurons are divided into two major systems; the mesostriatal system, which projects from the SNc to striatum (i.e. the caudate-putamen) and the mesolimbocortical system, which projects from the ventral tegmental area to the nucleus accumbens, olfactory tubercle, prefrontal cortex and amygdala (Dahlström and Fuxe, 1964). The mesostriatal system is primarily involved in the control of movements, whereas the mesolimbocortical system plays a more important role in reinforcing and reward-related behaviors (Nestler et al., 2005). It is primarily the dopaminergic neurons in the mesostriatal system that degenerate in Parkinson's disease. The degeneration of dopamine neurons in PD develops gradually and can, at early stages, be controlled by various drugs which potentiate dopaminergic neurotransmission, including 3,4-dihydroxyphenyl-L-alanine (L-DOPA), dopamine D1/D2 receptor agonists, D2 receptor agonists, MAOB (monoamine oxidase B) inhibitors and COMT (catechol-O-methyltransferase) inhibitors. However, at later stages of PD almost all patients receive treatment with L-DOPA, the amino acid precursor of dopamine. L-DOPA is converted to dopamine in the remaining dopaminergic nerve terminals in striatum. Unfortunately, long-term replacement treatment with L-DOPA to patients with PD frequently causes a state in which the responsiveness to L-DOPA decreases and/or becomes variable and associated with side-effects, such as on-off fluctuations, involuntary dyskinetic movements and hallucinations (Obeso et al., 2000; Dauer and Przedborski, 2003).

There are both toxin and genetic animal models of PD. The most common toxin models of PD are 6-hydroxydopamine (6-OHDA) injections into SNc, striatum or the median forebrain bundle (MFB) that interconnects SNc and striatum (Ungerstedt, 1968), repeated systemic injections of 1-methyl-4-phenyl-1,2,3,6-tetrahydropyridine (MPTP) (Heikkila et al., 1984) or, more recently, rotenone (Betarbet et al., 2001) and proteasome inhibitors (McNaught et al.,

2004). There is a lot of interest in developing genetic animal models of PD and it has, for example, been shown that mice that overexpress α-synuclein exhibit some features of PD (Masliah et al., 2000). However, at the moment, the model that best meets the criteria for human PD is the MPTP model. MPTP is highly lipophilic and readily crosses the blood-brain barrier. It is then converted into its active metabolite, 1-methyl-4-phenylpyridinium (MPP+) by MAO B, an enzyme involved in monoamine degradation (Przedborski et al., 2000). MPP+ is taken up by the dopamine transporter and is accumulated in mitochondria where it inhibits complex I of the electron transport chain. This reduces ATP production and causes an increase in free-radical production. Dopamine neurons in substantia nigra pars compacta are particurlarly vulnerable to the action of MPTP and the degeneration of these neurons are readily detectable using biochemistry and neuropathology (Giovanni et al., 1991). MPTP treatment results in easily detectable motor deficits, i.e. bradykinesia and rigidity. This model has a relatively short disease course of a few weeks, allowing rapid screening of therapeutic agents.

The unilateral 6-OHDA lesion model (Ungerstedt, 1968) has provided an invaluable tool for investigating the pathophysiology of dopamine denervation and for evaluating novel therapeutical options. The feasibility of new treatment strategies for PD, for example, brain transplants and gene therapy, has primarily been evaluated in this model. It is also easy to perform behavioral measurements in this model. Animals with unilateral 6-OHDA-induced dopamine denervation rotate ipsilaterally following administration of compounds that release dopamine, such as amphetamine, but contralaterally following administration of the dopamine precursor, L-DOPA, or dopaminergic agonists, such as apomorphine (Ungerstedt and Arbuthnott, 1970). The latter effect has been attributed to a supersensitivity of dopamine receptors and/or their signal transduction mechanisms on the dopamine-depleted side (Ungerstedt, 1971). Contralateral rotation in this model is predictive for the anti-Parkinsonian action of a given compound.

It has been reported that chronic, systemic inhibition of complex I by the lipophilic pesticide, rotenone, causes highly selective nigrostriatal dopaminergic degeneration that is associated behaviorally with hypokinesia and rigidity (Betarbet et al., 2001). Nigral neurons in rotenone-treated rats accumulate fibrillar cytoplasmic inclusions that contain ubiquitin and α-synuclein. However, more recent studies have found a more widespread neuronal damage following administration of rotenone (Hoglinger et al., 2003) and it is currently too early to conclude whether rotenone treatment represents a valuable animal model of PD.

As mentioned above, pharmacological dopamine replacement therapies can counteract the symptoms of early PD. However, numerous biochemical studies conducted in post-mortem striatal tissue from patients with PD and in animal models of this disease, such as 6-OHDA-lesioned rats and MPTP-treated mice and primates, have shown that prolonged treatment with L-DOPA does not adequately reverse the effects of dopamine depletion, but rather creates a new neurochemical state, which differs both from the normal and the dopamine-depleted one (Obeso et al., 2000). This state is associated with side-effects, such as on-off fluctuations, involuntary dyskinetic movements and hallucinations, and is characterized by pronounced changes not only in the dopamine system, but also in, for example, the acetylcholine, adenosine, glutamate and serotonin systems (Obeso et al., 2000). To understand the pathophysiology underlying L-DOPA-induced side effects, animal models have been developed in primates (Langston et al., 2000; Jenner, 2003) and, more recently, also in mice (Lundblad et al., 2004). These models are particularly useful for understanding L-DOPA-induced dyskinesias.

3. Basal Ganglia Organization

SNc and its major terminal area, striatum, are parts of the basal ganglia. The basal ganglia are also composed of the globus pallidus (external part in primates or GPe), the entopedoncular nucleus (internal part or GPi), the subthalamic nucleus (STN), and the substantia nigra pars reticulate (SNr), which are involved in the integration of sensorimotor, associative and limbic information to produce motor behaviors (Albin et al., 1989; Crutcher and Alexander, 1990; Gerfen and Wilson, 1996) (Fig. 14.1).

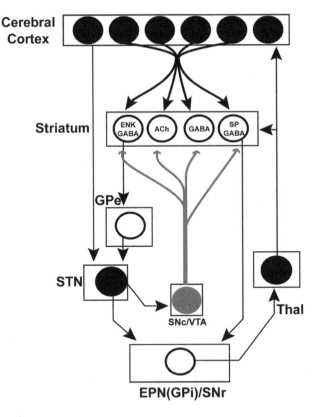

Figure 14.1. Schematic drawing of the neuronal pathways interconnecting the different subnuclei of the basal ganglia. Black circles indicate excitatory pathways. White circles indicate inhibitory pathways. Grey circles indicate modulatory pathways. Modified from Alexander and Crutcher (1990).

GABAergic medium spiny projecting neurons comprise around 95% of the striatal neurons, whereas the remaining neurons are GABAergic and cholinergic interneurons. In addition to receiving dopamine innervation from the SNc, striatum also receives dense glutamatergic inputs from cortex. The cortical inputs to the striatum are provided by excitatory glutamatergic pyramidal neurons that project from most areas of the cortex (Kitai et al., 1976; Spencer, 1976; Gerfen and Wilson, 1996). These neurons project in a topographically well-organized manner so that they define functionally distinct regions of the striatum. Inputs from sensorimotor areas innervate principally the dorsal part of the striatum, whereas inputs from prelimbic cortical areas terminate in the ventral area (Kemp and Powell, 1970; Donoghue and Herkenham, 1986; McGeorge and Fall, 1989). In primates, the caudate nucleus receives mainly prefrontal cortical inputs whereas putamen receives motor and somatosensory cortical inputs (McGeorge and Fall, 1989). In addition, excitatory projections from intralaminar nuclei of thalamus, hippocampus and basolateral parts of amygdala terminate in striatum.

Besides the striatum, the STN should be considered as an input structure to the basal ganglia (Chesselet and Delfs, 1996; Levy et al., 1997; Smith et al., 1998). There is, for example, excitatory innervation of the STN from sensorimotor cortex and parafascicular nucleus of thalamus (Canteras et al., 1990; Mouroux et al., 1993; Feger et al., 1994, 1997) and electrical stimulation of the sensorimotor cortex induces c-Fos to a similar extent in the striatum and the STN (Wan et al., 1992).

The striatal outputs are mediated through inhibitory GABAergic neurons (Yoshida and Precht, 1971; Deniau et al., 1976) which form two major output pathways, directly and indirectly, with a final common target in the SNr and the EPN. The direct pathway is monosynaptic and the indirect pathway is trisynaptic. The direct striatal pathway neurons contains high levels of substance P and dynorfin, whereas neurons in the indirect striatal pathway neurons contains enkephalin. The neurons comprising the indirect pathway project from striatum to the GPe and have another relay in the STN before terminating in EPN or SNr. The neurons in EPN and SNr are also composed of GABAergic neurons projecting to the thalamus, the pedunculopontine nucleus, and the superior colliculus (for review see Gerfen and Wilson, 1996). The direct and indirect output pathways project through the thalamus back to the cortex act in an opposing manner that is controlled by dopamine (see below).

An important role of the basal ganglia output regions in many physiological and pathophysiological processes has been established. It is, for example, known that there is a hyperactivity of the STN in PD and high frequency electrical deep brain stimulation (DBS) of the STN is now considered the most effective neurosurgical therapy for movement disorders (Limousin et al., 1998). Although the STN is the most common target for DBS in PD, stimulations of the GPi (i.e. EPN is rodents) has also been shown to be beneficial in many cases.

4. Dopamine Receptors

Dopamine exerts its action via five different dopamine receptors (Missale et al., 1998). These receptors are sub-divided into two groups based on how they regulate cAMP signaling; D_1 receptor subtypes (D_1, D_5) stimulate adenylyl cyclase and D_2 receptor subtypes (D_2, D_3, D_4) either inhibit or have no effect on adenylyl cyclase (Stoof and Kebabian, 1981). D_1 and D_2 receptors are very highly expressed in striatum. There are also moderate levels of D_3 receptors in this region, especially in its ventral part. The levels of expression of D_4 and D_5 receptors are low to moderate in striatum. The sub-cellular distributions of D_1, D_2 and D_3 receptors in striatal neurons differ. D_2 receptors are found on dopaminergic nerve terminals and postsynaptically on GABAergic medium spiny neurons, which comprise around 95% of the striatal neurons, and on cholinergic interneurons, which represent 1–2% of striatal neurons. D_1, D_3 and D_4 receptors are predominantly expressed postsynaptically on cell bodies and dendrites of GABAergic medium spiny neurons (Gerfen et al., 1990; Hersch et al., 1995; Le Moine and Bloch, 1995; Rivera et al., 2003). D_5 receptors are enriched on cholinergic interneurons (Rivera et al., 2002).

As described above, GABAergic medium spiny neurons can be divided into two equally large sub-populations, namely those terminating in SNr (striatonigral neurons) and those terminating in the GPe (striatopallidal neurons). Anatomical studies have shown that striatonigral neurons contain high levels of D_1 receptors together with the neuropeptide substance P, whereas striatopallidal neurons predominantly express D_2 receptors together with enkephalin (Gerfen et al., 1990; Hersch et al., 1995; Le Moine and Bloch, 1995). However, there is biochemical and physiological evidence supporting the idea that most neostriatal neurons possess both D_1 and D_2 receptors (Bertorello et al., 1990; Surmeier et al., 1992; Aizman et al., 2000). D_3 and D_4 receptors are expressed in both striatonigral and striatopallidal neurons (Diaz et al., 1996; Le Moine and Bloch, 1996; Rivera et al., 2003). There are no marked changes in the levels of dopamine receptors in striatum in PD. However, it has been demonstrated that repeated administration with L-DOPA causes a strong increase in the expression of D_3 receptors throughout striatum in dopamine denervated rats and monkeys (Bordet et al., 1997; Bezard et al., 2003). Interestingly, administration of a D3 receptor-selective partial agonist strongly attenuated levodopa-induced dyskinesia, but kept the therapeutic effect of levodopa, in monkeys (Bezard et al., 2003).

Given the importance of extrastriatal areas of the basal ganglia in the pathophysiology of PD, it is important to note that there is a significant dopaminergic innervation of some extrastriatal areas, including the GPe and STN. Functional D2 receptors are expressed in the GPe (Hoover and Marshall, 2004) and there are relatively high levels of D5 receptors in the STN (Svenningsson and LeMoine, 2002).

5. Dopamine Receptor-Mediated Intracellular Signaling Under Normal Conditions and in Animal Models of Parkinson's Disease. Signal Transduction Mediated via D1-like Receptors

Activation of dopamine D1 receptors leads to activation of adenylyl cyclase (Kebabian and Greengard, 1972). Two heterotrimeric G protein α subunits are capable to couple D1 receptors to adenylyl cyclase: Gαs, the ubiquitous isoform and Gαolf, initially identified in the olfactory epithelium (Jones and Reed, 1989), but also highly enriched in the basal ganglia (Drinnan et al., 1991; Hervé et al., 1993). Gαolf levels are upregulated in striatum in 6-OHDA-lesioned rats and in Parkinsonian patients suggesting that abnormal Gαolf levels could contribute to the appearance of dyskinesias (Hervé et al., 1993; Corvol et al., 2004). Gαolf as well as Gαs increase intracellular cAMP levels, which subsequently, activate protein kinase A (PKA). Dopamine and cAMP regulated phosphoprotein MW 32 kDA (DARPP-32) is a major phosphoprotein substrate for dopamine and PKA in medium spiny neurons in striatum (Walaas et al., 1983; Svenningsson et al., 2004). Deletion of DARPP-32 also results in an altered responsiveness to dopaminomimetic compounds at the biochemical, electrophysiological and behavioral levels (Svenningsson et al., 2004). Phosphorylation of DARPP-32 at Thr^{34} by D1 receptors/PKA converts it into a potent inhibitor of protein phosphatase-1 (PP-1), a major multifunctional serine/threonine protein phosphatase in the brain (Hemmings et al., 1984). It was recently observed that L-DOPA-induced dyskinesias in 6-OHDA-lesioned rats are associated with increased Thr34-DARPP-32 in a D1 receptor-mediated manner (Picconi et al., 2003). PP-1, in turn, regulates the phosphorylation state and activity of many physiological effectors, including neurotransmitter receptors, voltage-gated ion channels, an electrogenic pump and transcription factors (Svenningsson et al., 2004). In addition of activating D1/Gαolf/PKA/DARPP-32/PP-1 signaling, D1 receptors also regulate numerous other signaling pathway. It has, for example, been shown that the carboxyl terminal part of dopamine D1 receptors interact with the calcium-binding protein, calcyon (Lezcano et al., 2000). Calcyon enables D1 receptors to activate Gαq proteins. D1 receptors can shift the relative level of Gαq versus Gαolf effector coupling to stimulate robust intracellular calcium release as a result of interaction with calcyon. Moreover, several publications have recently shown that dopamine D1 receptors can activate the MAP kinase signaling cascade (e.g. Valjent et al., 2000, 2005; Gerfen et al., 2002). Using unilaterally 6-OHDA-lesioned rats, it has been observed that the regulation of MAP kinases as well as JNK kinases by D1 receptor agonists is enhanced in the direct pathway in the dopamine depleted striatum (Gerfen et al., 2002).

There is accumulating evidence that L-DOPA and dopamine receptor agonists regulate gene transcription by activating signal transduction cascades that lead to increased phosphorylation, and hence activation, of transcription factors. There is strong evidence that dopaminergic agents, via activation of D_1 receptors, stimulate phosphorylation of CREB at Ser^{133} in the normal striatum (Konradi et al.,1994; Cole et al., 1995; Simpson et al., 1995). CREB phosphorylated at

Ser[133] plays an important role in the regulation of several immediate early genes, such as different members of the Fos/Jun family (Finkbeiner et al., 1997). The regulation of Fos/Jun immediate early genes is also under the control of additional transcription factors, such as ELK. The activity of ELK is dependent upon its phosphorylation mediated via the MAP kinase signaling cascade (Xia et al., 1996; Vanhoutte et al., 1999).

Long-term changes in gene transcription may underlie the development of dyskinesias. It is therefore interesting that phosphorylation of MAPK/ELK, rather than phosphorylation of CREB, appears to be increased in PD models of dyskinesia (Andersson et al., 2001; Gerfen et al., 2002).

6. Signal Transduction Mediated via D2-like Receptors

Activation of dopamine D2 receptors leads to inhibition of adenylyl cyclase (Stoof and Kebabian, 1981). Indeed, it has been known for many years that dopamine inhibits cAMP signaling as well as Thr34-DARPP-32 phosphorylation via dopamine D2 receptors (Stoof and Kebabian, 1981; Nishi et al., 1997). However, multiple other signaling pathways are modulated directly or indirectly by activation of dopamine D2 receptors. It has, for example, been found that D2 receptor agonists activate a phospholipase C-IP3-calcineurin signaling cascade (Nishi et al., 1997; Hernandez-Lopez et al., 2000). The activation of this signaling pathway also reduces L-type Ca^{2+} currents in striatal neurons (Hernandez-Lopez et al., 2000). Since L-type Ca^{2+} channels are important for the induction of gene expression, the latter observation provides a mechanism to explain the fact that D2 receptor activation activates Ca^{2+}-dependent intracellular enzymes, but suppress gene transcription. It has also been demonstrated that activation of dopamine D2 receptors increase the MAP kinase signaling cascade (Cai et al., 2000). Interestingly, Cai and co-workers found that inhibition of MAP kinase signaling inhibits D2 agonist-induced contralateral rotation in unilaterally 6-OHDA-lesioned rats (Cai et al., 2000).

Several adaptor proteins interacting with dopamine D2 receptors have also been identified. Spinophilin has been found to bind to the third intracellular domain of the dopamine D2 receptor (Smith et al., 1999). Spinophilin is a scaffolding protein that also interacts with other GPCRs, protein phosphatase-1 and actin (Allen et al., 1997). It has recently been shown that spinophilin, antagonizes the beta-arrestin functions through blocking G protein receptor kinase 2 association with receptor-Gbetagamma complexes (Wang et al., 2004).

7. Dopaminergic Regulation of Immediate Early Gene Expression

As described in more detail in other chapters in this volume, immediate early genes (IEGs) have in common that they are rapidly (often within minutes) and strongly induced by various stimuli. The first study to demonstrate that dopaminergic drugs regulate IEG expression in the striatum was the finding that a D1 receptor agonist strongly increases c-Fos immunoreactivity in the 6-OHDA-lesioned striatum (Robertson et al., 1989). Indeed, the induction of IEGs

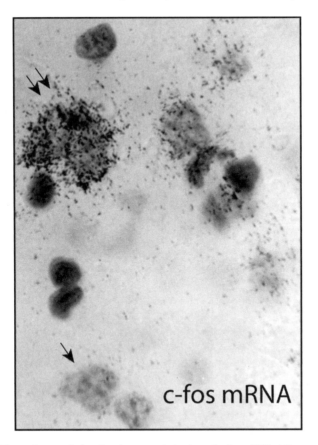

Figure 14.2. Photomicrograph showing the strong induction of c-fos mRNA (silver grains) in some striatal medium-sized neurons following treatment with L-DOPA in the 6-OHDA-lesioned striatum. Double arrows indicate neurons exhibiting c-fos induction, whereas the single arrow indicates a neuron with no induction.

can be very pronounced in striatal neurons (Fig. 14.2). Subsequent studies showed that dopaminomimetic compounds, including amphetamine and cocaine, caused a strong induction of IEGs throughout striatum (Graybiel et al., 1990). In fact, most drugs of abuse increase IEG expression in the ventral striatum (Harlan and Garcia, 1998). This induction most likely reflects the shared ability of drugs of abuse to increase dopamine levels in this region. Likewise, studies using electrical stimulation of dopaminergic fibers, leading to elevated levels of striatal dopamine, increases striatal IEG expression (Chergui et al., 1997). These stimulatory effects on IEG expression are mediated via D1 receptors. D2 receptors have an inhibitory influence on striatal IEG expression (Robertson et al., 1992; Svenningsson et al., 1999). Treatments with D2 receptor antagonists, including typical neuroleptics such as haloperidol, or acute dopamine depletion leads to induction of striatal IEG expression selectively in D2 receptor-containing neurons (Robertson et al., 1992;

Svenningsson et al., 1999). It appears that the neuronal activity of striatal medium-sized spiny projection neurons correlates with their expression of c-Fos, at least in anesthetized rats and in acute experiments (Gonon, personal communication). Indeed, electrophysiological studies have found bidirectional effects of D1 and D2 receptor stimulations in the firing of striatal neurons with D1 receptors being stimulatory and D2 receptors inhibitory (O'Donnel and Grace, 1994; Gonon, 1997; West and Grace, 2000). However, under some circumstances c-Fos expression does not correlate well with neuronal activity (Labiner et al., 1993) and it has, for example, been difficult to observe any changes in its expression in dopaminergic neurons in SNc (Dragunow and Faull, 1989).

Most IEGs, including c-*fos*, c-*jun*, *jun*D and Nurr1, have a low constitutive expression in striatum, but some, such as *jun*B, NGFI-A and NGFI-B have a rather high basal expression, making it possible to detect both reductions and inductions of their expression (reviewed in Herdegen and Leah, 1998). There is sparse knowledge on the constitutive expression of IEGs in extrastriatal basal ganglia regions. Many studies have found that several members of the Fos family, including delta-FosB, Fra1 and Fra2, are regulated in a temporally different manner when compared to the aforementioned IEGs. These late-responsive variants of Fos are induced with a slow kinetics, and, at least in the case of delta-FosB, the expression is long-lasting due to a slow degradation. It is well-established that delta-FosB is increased in response to chronic, but not acute, treatments with, for example, psychostimulants, such as cocaine and amphetamine, and neuroleptics (McClung et al., 2004). Of direct relevance to this review is also the observation that delta-FosB is increased in the striatum of animal models of Parkinson's Disease (e.g. Dragunow et al., 1991; Doucet et al., 1996; Perez-Otano et al., 1998). This induction is dependent upon a decreased D2 receptor-mediated inhibition of striatopallidal neurons. Moreover, chronic treatment with L-DOPA also increases FosB in the striatum of animal models of Parkinson's Disease (Andersson et al., 2003). This induction is mediated by increased D1 receptor-mediated stimulation of striatonigral neurons.

8. Regulation of IEGs in Animal Models of Parkinson's Disease

Numerous studies have shown that there is increased IEG levels in striatum in animal models of PD (for references, see Table 14.1). Systemic treatment with MPTP or injections with 6-OHDA into the MFB increases c-Fos and c-Fos-like proteins in striatum via disinhibition of striatopallidal neurons. Likewise, acute dopamine depletion with reserpine causes an induction of c-Fos immunoreactivity in striatopallidal neurons that can be counteracted by pretreatment with a D2 agonist.

Fewer studies have examined the regulation of IEGs in extrastriatal regions and the results show some discrepancies. Acute treatments with reserpine or intranigral 6-OHDA-lesioning increase c-fos in the entopeduncular nucleus. 6-OHDA-lesioning, but not reserpine, also increases c-fos in the SNr. On the contrary, reserpine increases c-fos expression in the GPe, whereas acute

Table 14.1. Summary of changes in c-Fos and c-Fos-like expression in the basal ganglia regions in animal models of PD. ↑ indicates up-regulation, ↓ indicates down-regulation and ↔ indicates no change

	Striatum	GPe	GPi (EPN)	STN	SNr
6-OHDA	↑[1, 2, 6, 7, 9, 10, 15, 16]	↔[15]	↑[15]	↔[11, 12]	↑[15]
MPTP	↑[3, 6, 13]				
Reserpine	↑[4, 5, 14]	↑[4]	↑[4]		↔[8]

[1]Bjelke et al. (1994); [2]Chapman et al. (1996); [3]Chen et al. (2002); [4]Cole DG et al. (1994); [5]Cooper et al. (1995); [6]Doucet et al. (1996); [7]Dragunow et al. (1991); [8]Fritschy et al. (1991); [9]Gerfen et al. (2002); [10]Jian et al. (1993); [11]Nakao et al. (1998); [12]Nielsen et al. (2003); [13]Perez-Otano et al. (1998); [14]Pollack et al. (1999); [15]Schuller et al. (2000); [16]Svenningsson et al. (1999).

6-OHDA-lesioning has no effect in this region. No changes in c-fos expression in the STN has been observed. In general, there is very limited knowledge on the neurotransmitter mechanisms that underlie IEG changes in extrastriatal regions.

9. Regulation of IEGs in Animal Models of Parkinson's Disease by Dopamine Receptor Agonists

Early studies by Dr Harold Robertson and collegues demonstrated that there is a supersensitive induction of IEGs by D1 receptor agonists in the dopamine depleted striatum (Robertson et al., 1989). Subsequent studies demonstrated that this induction occurs selectively in striatonigral neurons (Robertson et al., 1990). Many studies have thereafter confirmed and extended these initial observations on c-fos regulation by dopamine D1 receptor agonists in animal models of PD and some of them are summarized in Table 14.2. It should be emphasized that, in addition to c-fos, a large number of other IEGs are induced by this treatment. Using unbiased differential display methodology, it was found that more than thirty immediate early genes are induced following an intraperitoneal injection of a D_1 receptor agonist in the denervated side in animals with unilateral 6-OHDA-lesions of the MFB (Berke et al., 1998). Most of these genes could also be induced in normal (i.e. non-lesioned) animals by treatment with haloperidol or cocaine. Below, we will briefly review how L-DOPA and dopamine receptor agonists regulate IEG expression in various animal models of PD in striatum and in extrastriatal regions.

Acute and chronic treatment with L-DOPA increases c-fos and c-fos-like proteins in striatum in the 6-OHDA dopamine lesioned hemisphere (Fig. 14.3; for references, see Table 14.2). This induction occurs mainly in the striatonigral neurons. The acute induction by L-DOPA on c-fos as well as its prolonged effects on delta-FosB occurs predominantly in the patch compartant of the striatum (Cenci et al., 1999; Saka et al., 1999). Such up-regulation has been shown to correlate with the incidence of stereotypies and dyskinesias (Cenci et al., 1999; Graybiel et al., 2000). The patch/matrix organization has been related to limbic and non-limbic functions: the striosomes receive inputs from limbic areas (amygdala, prelimbic cortex) as well as the ventral tier of the SNc, whereas

Table 14.2. Summary of changes in c-Fos and c-Fos-like expression in the basal ganglia regions in animal models of PD following treatment with L-DOPA and dopamine receptor agonists. ↑ indicates up-regulation, ↓ indicates down-regulation and ↔ indicates no change

	Striatum	GPe	GPi (EPN)	STN	SNr
L-DOPA	↑1, 4, 6, 10, 15, 16, 26, 31, 32, 33, 34	↑32, 34		↑18	
Apomorphine (D1/D2)	↑2, 3, 5, 12, 19, 21, 29	↑3, 17, 19		↑17	
D1 agonist	↑5, 7, 8, 9, 13, 19, 20, 23, 25, 26, 27, 30, 35, 36 ↔35	↑28		↑27, 28	
D2 agonist	↔5, 20 ↓11 ↑22	↑14, 19, 20, 24, 28			

[1] Andersson et al. (2003); [2] Bronstein et al. (1994); [3] Cenci et al. (1992); [4] Cenci et al. (1999); [5] Cole et al. (1992); [6] Cole et al. (1993); [7] Ding et al. (2002); [8] Guo et al. (1998); [9] Gerfen et al. (2002); [10] Ishida et al. (1996); [11] Keefe et al. (1995); [12] Labandeira-Garcia et al. (1996); [13] LaHoste et al. (1993); [14] Marshall et al. (1993); [15] Morelli et al. (1994); [16] Mura et al. (1995); [17] Nakao et al. (1998); [18] Nielsen et al. (2003); [19] Paul et al. (1995); [20] Paul et al. (1992); [21] Pennypacker et al. (1992); [22] Pollack et al. (1997); [23] Robertson et al. (1994); [24] Robertson et al. (1990); [25] Robertson et al. (1992); [26] Robertson et al. (1989); [27] Ruskin et al. (1999); [28] Ruskin et al. (1995); [29] Saka et al. (1999); [30] Sandstrom et al. (1996); [31] Svenningsson et al. (2002); [32] Svenningsson et al. (2000); [33] Westin et al. (2001); [34] Xu et al. (2003); [35] Ziolkowska et al. (1995); [36] Zivin et al. (1996).

c-fos mRNA

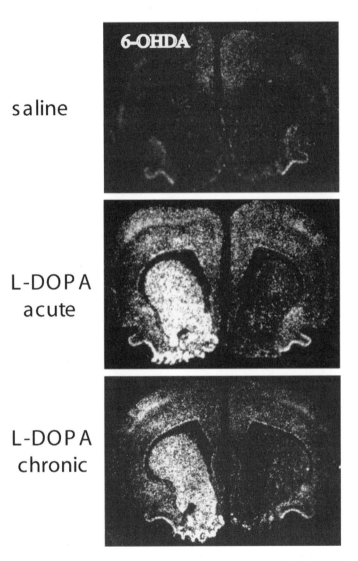

saline

L-DOPA
acute

L-DOPA
chronic

Figure 14.3. Dark-field photomicrographs showing a strong induction of c-fos mRNA in the 6-OHDA-lesioned striatum following acute or subchronic treatment with L-DOPA.

the matrix receives inputs from sensorimotor cortical areas, thalamus and the dorsal tier of the SNc. In addition, Gerfen (1989) has described the somatotopic organization of the corticostriatal projections related to the lamination of the cortex. Many other neurochemical markers, for instance cholinergic and opioid markers, are differentially distributed and probably differentially regulated in these two compartments (Herkenham and Pert, 1981; Graybiel, 1990).

Similarly, the dopamine D1/2 agonist, apomorphine, and selective D1 agonists also induce c-fos in striatum in the 6-OHDA dopamine lesioned hemisphere (Table 14.2). Like the situation in the normal striatum, there are several lines of evidence showing that dopamine D1 receptor agonists exerts a stimulatory effect on IEG expression in striatonigral neurons in animal models of PD (Robertson et al., 1990; Gerfen et al., 1995). Studies with selective D1 receptor agonists have also been performed in MPTP- and reserpine-treated animals. As expected, a D1 agonist increases c-fos expression in reserpinized mice (LaHoste et al., 1993). Surprisingly, no effect of a D1 agonist on c-fos expression was found in MPTP-treated animals (Ziolkowska et al., 1995). As the latter study used a partial D1 agonist that does not penetrate the blood-brain barrier efficiently, additional studies with brain penetrating, full agonists are warranted to better understand the role of D1 receptors on IEG expression in the MPTP model of PD.

The role of D2 agonists on IEG expression in the striatum has been much studied and appears to be rather complex. Co-treatment with D2 agonists potentiate the effects of D1 agonists on increasing IEG expression in the 6-OHDA-lesioned striatum (Paul et al., 1992; LaHoste et al., 1993; Gerfen et al., 1995). Furthermore, co-administration of D1 and D2 agonists causes a more heterogenous patchy pattern of IEG expression in striatum than a D1 agonist alone. The heterogenous pattern appears to correlate with the synergistic action of D1 and D2 agonists on various behaviors. D2 agonists, when given alone, have no effect or an inhibitory action on IEG expression. However, it should be mentioned that in a series of publications, Pollock and co-workers have found that treatment with a D2 agonist actually increases c-fos expression in the 6-OHDA denervated striatum in apomorphine-primed animals (Pollock et al., 1997, 2005).

A differential regulation of IEGs by dopamine agonists has been found in the GPe. Acute as well as chronic treatments with L-DOPA increases c-fos expression in the region. Similarly, apomorphine, selective D1 agonists and selective D2 agonists increase c-fos expression in GPe. This pallidal c-fos expression is thought to be secondary to the actions of D2/3 and D1/5 receptor agonists on striatopallidal and subthalamopallidal neurons, respectively (Ruskin and Marshall, 1995). Like the situation in the striatum, concomitant administration of selective D1 and D2 agonists causes a synergistic induction of IEG expression (Ruskin and Marshall, 1995). At the moment, few studies have examined the effects of dopaminergic agonists on IEG expression in GPe in animal models of PD other than the 6-OHDA model.

There is still no consensus on how D1/5 receptor agonists and D2/3 receptor agonists regulate c-fos expression in the STN. Using 6-OHDA lesioned rats, Marshall et al. (1993) as well as Ruskin et al. (1999) found that systemic administration of D1/5 receptor agonists causes an induction of c-fos expression in the STN. In contrast, Hassani and Feger (1999) found no effect on subthalamic c-fos expression by a local application of a D1/5 receptor agonist. However, they found an increase of c-fos expression following local application of a D2/3 receptor agonist. The action(s) whereby dopamine agonists regulate c-fos expression in subthalamic neurons could involve direct actions on local dopamine receptors as well as trans-synaptic modulation of neural pathways, such as the striato-pallido-subthalamic pathway. According to the classical model of basal ganglia organization (Albin et al., 1989; Alexander and Crutcher, 1990), the major input to the STN arises from GABAergic neurons in the GPe. It is well-established that D2/3 receptor agonists increases c-fos expression in the GPe and that this induction can be potentiated by a D1/5 receptor agonist. If the regulation of c-fos expression in the STN by dopamine agonists was entirely dependent upon the activity of the inhibitory inputs from the GPe, it would be predicted that treatment with a D1/5 receptor agonist alone would cause a weaker induction of IEGs as compared a treatment with a D1/5 receptor agonist in combination with a D2/3 receptor agonist. However, a D1/5 receptor agonist increased c-fos expression in the STN regardless whether it was administered alone or in combination with a D2/3 receptor agonist (Marshall et al., 1993). This result shows that, in addition to inputs from the GPe, other mechanisms are critically involved in regulating D1/5 agonist-mediated c-fos expression in the STN. One possibility is that systemic administration of D1/5 agonists stimulates excitatory inputs to the STN. An involvement of such a mechanism is likely since D1 and D5 receptors have been localized in the two major excitatory inputs to this nucleus, the cortex and the parafascicular nucleus of thalamus (Meador-Woodruff et al., 1992; Gaspar and Le Moine, 1995). Another possibility is that D1/5 agonists act directly on local D5 receptors in the STN to increase c-fos expression (Svenningsson and LeMoine, 2002). Few studies have examined the regulation of IEGs in the EPN and the SNr in response to dopaminergic agonists in PD models.

10. Regulation of IEGs in Animal Models of PD by Neurotransmitters Interacting with Dopamine

While PD is undoubtedly a disorder with a primary pathology of dopamine neuronal loss, the loss of dopamine and subsequent dopamine replacement therapy leads to imbalances in many non-dopaminergic transmitter systems, including the acetylcholine, glutamate, adenosine and serotonin systems. There is accumulating evidence that these neurotransmitters also contribute to various aspects of the pathophysiology of PD and we will briefly review how drugs acting via these neurotransmitters interact with the dopamine system to regulate IEGs in animal models of PD.

10.1. Muscarinic Receptor Antagonists are Effective Against PD

Functions of medium spiny neurons in the neostriatum are regulated by cholinergic interneurons in physiological and pathophysiological conditions. In a simplified model, the activity of medium spiny neurons is determined by the balance of inhibitory dopaminergic inputs and excitatory cholinergic inputs. In PD, the loss of dopaminergic neurons leads to the increased excitation of medium spiny neurons. Muscarinic M1, M3 and M5 receptors couple to phospholipase C (PLC) and eventually promote the mobilization of intracellular Ca^{2+}, whereas M2 and M4 receptors negatively couple to adenylyl cyclase. These muscarinic receptors are expressed in various neuronal subtypes of the striatum, and exert pre- and post-synaptic effects on medium spiny neurons. Due to lack of subtype-selective agonists or antagonists, the functional role of individual muscarinic receptor subtypes in the regulation of striatal dopaminergic transmission has remained unclear. Nonetheless, clinical studies have shown that parkinsonian symptoms can be counteracted with muscarinic receptor antagonists (Lang et al., 2002). Along this line, several studies have shown that the non-selective muscarinic receptor antagonist, scopolamine, potentiates the effect of D1 agonists on increasing IEG expression in the 6-OHDA-depleted striatum (Morelli et al., 1994; Sandstrom et al., 1996; Gerfen et al., 2002). By itself, scopolamine has a weak effect on IEG expression (Sandstrom et al., 1996).

10.2. Adenosine A2A Receptor Antagonists—A Novel Treatment for Parkinson's Disease

There are four cloned adenosine receptors. One of them, the A2A receptor, is highly enriched in striatum. In fact, detailed anatomical studies have shown that this receptor is only expressed in striatopallidal neurons, where they are co-localized with dopamine D2 receptors (Schiffmann et al., 1991; Fink et al., 1993; Svenningsson et al., 1997). Adenosine A2A receptors are positively coupled to adenylyl cyclase (Fredholm, 1977; Premont et al., 1977).

Over the past years, there has been considerable interest in developing A2A receptor antagonists for the symptomatic treatment of PD. Indeed, selective A2A receptor antagonists potentiate contralateral rotation induced by L-DOPA and dopaminergic agonists in rats with unilateral 6-OHDA destruction of dopamine neurons (Jiang et al., 1993; Pinna et al., 1996, Fenu et al., 1997). Furthermore, it has recently been shown that oral administration of a selective A2A receptor antagonist can counteract haloperidol-catalepsy in mice and reverse motor disabilities in MPTP-treated mice and primates in a dose-dependent manner, without inducing dyskinesia or nausea (Kanda et al., 1998). In general, there is no development of tolerance to the beneficial effects of A2A receptor antagonists following long-term treatment.

In normal animals, systemic administration of an A2A receptor agonist induces Fos-like immunoreactivity more robustly in the ventromedial than in the dorsolateral part of striatum (Morelli et al., 1994). However, in animals with unilateral 6-OHDA lesions of the median forebrain bundle, even low doses of an A2A receptor agonist induce Fos-like immunoreactivity in the dorsal part of caudate-putamen on the denervated side (Morelli et al., 1994). This induction of Fos-like immunoreactivity could be inhibited by pretreatment with a D2 agonist and partially inhibited by a muscarinic receptor antagonist (Morelli et al., 1994). Furthermore, the induction of Fos-like immunoreactivity by a D2 agonist in the GPe of the 6-OHDA-lesioned side could be counteracted by pretreatment with an A2A agonist (Morelli et al., 1994). Accordingly, an A2A receptor antagonist induces c-Fos in a few scattered cells in both the ipsi- and contralateral sides of the GPe in unilaterally 6-OHDA lesioned rats (Pollack and Fink, 1996). In accordance with the fact that selective A2A receptor antagonists potentiate contralateral rotation induced by L-DOPA and dopaminergic agonists in the 6-OHDA model of PD, such antagonists also potentiate IEG responses by L-DOPA in the striatum and the GPe (Fenu et al., 1997).

10.3. Dysfunction of the Serotonin System in PD, a Possible Explanation for the high Degree of Comorbidity Between PD and Depression?

There is a considerble overlap in dopaminergic and serotonergic innervation in the brain (Reader and Dewar, 1999). Interactions between serotonin and dopamine may provide a possible biochemical explanation for the high incidence of co-morbidity between PD and depression. Indeed, PD is traditionally viewed as a movement disorder and it is often forgotten that this disease is also accompanied by numerous cognitive and emotional disturbances. In recent reports, it has been estimated that around 50% of patients suffering from PD also are depressed (Oertel et al., 2001).

There is a moderate serotonergic innervation of the striatum. Of the 15 cloned serotonin receptors, several (i.e. 5-HT_{1B}, 5-HT_{1F}, 5-HT_{2A}, 5-HT_{2C}, 5-HT_3, 5-HT_4 and 5-HT_6) have been found on medium spiny neurons in striatum. 5-HT_{2C} receptors are very highly expressed in the subthalamic nucleus. Systemic admin-istration of D-CPP, a non-selective 5-HT receptor agonist with some preference for 5-HT2C receptors, increases c-Fos expression in the STN and the EPN in the 6-OHDA-depleted hemisphere (De Deurwaerdere and Chesselet, 2000). In the same study, it was also observed that D-CPP increases c-fos expression in the intact striatum. However, this increase was reduced in the 6-OHDA-lesioned striatum.

10.4. Ionotropic/Metabotropic Glutamate Receptors, Novel Targets to Develop anti-Parkinsonian Drugs

The basal ganglia receive a very strong glutamatergic innervation that is the major excitatory drive of striatal neurons. Glutamate elicits its responses in the CNS by activating two families of receptors; ligand-gated cation channels termed ionotropic glutamate receptors (iGluRs) and G-protein-coupled receptors termed metabotropic glutamate receptors (mGluRs). iGluRs can be divided into three major classes; NMDA, AMPA and kainate receptors. Systemic administration of NMDA receptor antagonists causes a broad spectrum of psychotropic actions which includes an increased locomotor activity. The latter action of NMDA recep-tor antagonism is thought to underlie some of the beneficial actions of amantadine in PD. The NMDA receptor antagonist, MK-801, counteracts reserpine-induced c-Fos expression in striatopallidal neurons (Pollack et al., 1999) and intrastriatal administration of an NMDA receptor antagonist counteract striatal c-fos induction that occurs in response to acute 6-OHDA-lesioning (Schuller and Marshall,

2000). Moreover, antagonism of NMDA receptors appears to counteract effects of dopaminergic agonists on IEG regulation. Using the 6-OHDA-lesion model of PD, it has been shown that systemic pre-treatment with MK-801 counteracts c-fos induction by L-DOPA (Morelli et al., 1994), apomorphine (Bronstein et al., 1994), selective D1 agonists (Paul et al., 1992) and selective D2 agonists (Pollack et al., 2005). Likewise, intrastriatal infusion of MK-801 or (+/−)-3-(2-carboxypiperazin-4-yl)-propyl-1-phosphonic acid caused a dose-dependent attenuation of D1 agonist-induced c-fos mRNA expression (Keefe and Gerfen, 1996).

Intrastriatal injections of AMPA receptor antagonists can counteract striatal c-fos induction that occurs in response to acute 6-OHDA-lesioning (Schuller and Marshall, 2000). In contrast, intrastriatal infusion of an AMPA/kainite receptor antagonist did not affect D1 receptor-mediated c-fos induction (Keefe and Gerfen, 1999). Furthermore, pre-treatment with an AMPA receptor antagonist had no effect on D2 agonist-induced c-Fos (Pollack et al., 2005).

Given the importance of glutamate in the basal ganglia and the wide-spread distribution of metabotropic glutamate receptors in basal ganglia regions, it is anticipated that these receptors will have strong influence on the physiology and pathophysiology of basal ganglia function. To date, eight mGluR subtypes (designated mGluR1 to mGluR8) have been cloned from mammalian brain (Fagni et al., 2000). These mGluRs are classified into three main groups on the basis of sequence homology, coupling to second messenger systems, and selectivity for various agonists. Group I mGluRs include mGluR1 and mGluR5, which are linked to phospholipase C and couple primarily to increases in phosphoinositide hydrolysis. Group II mGluRs include mGluR2 and mGluR3, which couple to adenylyl-cyclase inhibition. Group III mGluRs also couple to adenylyl-cyclase inhibition, and include mGluR4, mGluR6, mGluR7 and mGluR8. It has very recently been shown that stimulation of group I and II striatal mGluRs seems to play a role in diminution of parkinsonian symptoms and parkinsonian-like muscle rigidity (Kearney et al., 1997, 2000; Wolfarth et al., 2000). Despite this, few studies have examined the role of metabotropic glutamate receptors on IEG expression in animal models of PD. Nevertheless, local application of group I and III mGluR agonists increase c-fos expression in the striatum in a 6-OHDA dopamine-depleted hemisphere (Kearney et al., 1998). However, none of these agonists induced c-fos in acutely dopamine depleted animals. Intra-subthalamic injections of group II and III mGluR agonists increase c-fos expression in the GPe, EPN, STN and SNr in the intact animal (Kearney et al., 2000).

11. Future Perspectives

More studies are needed to identify IEG changes unique to PD, especially in extrastriatal basal ganglia regions.

12. Identification and Targeting of Signal Transduction Pathways Terminating in IEG Changes that Are Activated Solely in Parkinson's Disease

To better understand the pathophysiological changes occurring in PD, it is important to define intracellular signalling pathways that are dysfunctional in this condition. This holds true both in the case of non-medicated and medicated patients as it may lead to improved treatments to slow down the disease progression and to decrease the incidence of L-DOPA-induced side-effects. In particular, there is a need to find better pharmacological treatments for advanced PD as numerous biochemical studies conducted in post-mortem striatal tissue from patients with PD and in animal models of this disease, such as 6-OHDA-lesioned rats and MPTP-treated mice/primates, have shown that prolonged treatment with L-DOPA does not adequately reverse the effects of dopamine depletion, but rather creates a new neurochemical state, which differs both from the normal and the dopamine-depleted one. This state is associated with side-effects, such as on-off fluctuations, involuntary dyskinetic movements and hallucinations.

For example, it has recently been shown that the JNK/c-jun signal transduction pathway is activated by dopamine only in animal models of Parkinson's Disease. As mentioned above, dopaminomimetic compounds, such as cocaine and dopamine D_1 receptor agonists, induce the expression of multiple IEGs including c-*fos*, Δ*fos*B, *fos*B and *jun*B in the striatum of normal rats and primates. Similar results have also been found in response to dopamine D_1 receptor agonists or L-DOPA treatment in dopamine depleted rats and primates. However, there is a marked difference in the striatal regulation of c-Jun N-terminal kinases and c-*jun* mRNA expression by dopaminomimetics in the normal and dopamine depleted striatum (Gerfen et al., 2002; Svenningsson et al., 2002). Neither acute nor chronic treatments with cocaine induce c-jun expression (Svenningson et al., 2002). However, acute as well as subchronic administration with L-DOPA causes a strong and coordinated induction of c-*jun* mRNA in the dopamine-depleted striatum of unilaterally 6-OHDA-lesioned rats (Fig. 14.4; Svenningsson et al., 2002). These results provide evidence that dopamine depletion leads to adaptations in signal transduction pathways that enables L-DOPA to increase c-*jun* gene transcription in striatum. Previous work has shown that c-*fos* mRNA induction by L-DOPA, cocaine and other dopaminomimetic compounds involves an activation of a D_1/cAMP/protein kinase A/CREB pathway. However, molecular biological studies have shown that the cAMP pathway does not increase c-*jun* transcription (Angel et al., 1988). Instead the transcription of c-*jun* in striatum is strongly regulated via protein kinase C (PKC)- and Jun-N-terminal kinase (JNK)-dependent mechanism (Schwarzschild et al., 1997). Based on these data it would be interesting to examine whether JNK inhibitors can decrease the development of L-DOPA-induced side-effects. Interestingly, JNK inhibitors attenuate MPTP-induced degeneration of dopamine neurons (Saparito et al., 1999). Moreover, studies in JNK-deficient mice have shown that both JNK2 and JNK3 are required for MPTP-induced c-Jun activation and dopaminergic cell degeneration (Hunot et al., 2004). Besides JNK, there is now evidence for a

c-jun mRNA

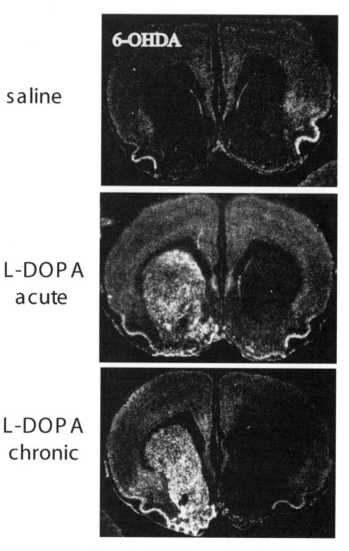

Figure 14.4. Dark-field photomicrographs showing a strong induction of c-jun mRNA in the 6-OHDA-lesioned striatum following acute or subchronic treatment with L-DOPA.

differential regulation of DARPP-32 and ERK1/2 by D1 receptor stimulation in the 6-OHDA-lesioned striatal neurons (Fig. 14.5). It is anticipated that, in addition to c-jun, multiple other IEG changes will turn out to correlate to dysfunctional regulation of signal transduction pathways occurring in PD.

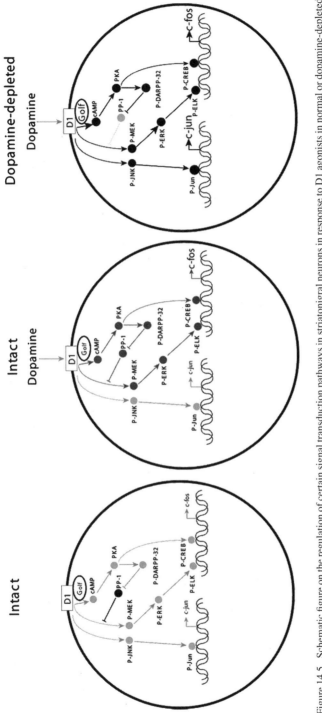

Figure 14.5. Schematic figure on the regulation of certain signal transduction pathways in striatonigral neurons in response to D1 agonists in normal or dopamine-depleted conditions. The regulations of P-JNK and P-MEK by D1 receptors involve several upstream steps that are not included. Note that the JNK/c-jun pathway is only activated by dopamine in the dopamine-depleted striatum. Light grey dots and arrows correspond to low level/activity. Strong grey dots and arrows correspond to moderate level/activity. Black dots and arrows correspond to high level/activity. Blunted arrows correspond an inhibitory action.

13. To Better Understand the Regulation and Roles of IEGs in Extrastriatal Regions

Over the past years, it has been established that adaptations in the basal ganglia output regions play major roles in many physiological and pathophysiological processes. For example, there is a hyperactivity and oscillatory firing of subthalamic neurons in PD and electrical stimulation the STN is now considered the most effective neurosurgical therapy for movement disorders (Limousin et al., 1998). Given this background, it will be important to decide whether mapping of IEG changes in extrastriatal regions can increase the understanding of the function of these regions under normal conditions and in PD. At the moment, there are relatively sparse data on this topic, in particular for the MPTP model. The existing literature also contains many results that are hard to reconcile with the current models of the basal ganglia organization. For example, all dopaminergic treatments appear to increase the expression of IEGs in the STN, despite the fact that these agents normalize the firing of subthalamic neurons in Parkinsonian patients. There is some recent evidence that IEGs other than c-fos, such cytochrome oxidase I, better reflects the neuronal activity of the STN neurons (Hirsch et al., 2000; Nielsen and Soghomonian, 2003).

It can be concluded that studies on IEGs have been very useful to localize changes in basal ganglia circuitry that occurs in PD and in response to anti-Parkinsonian therapies. At the moment, no genetic studies have associated IEG's with the neurodegenerative process of PD. However, there is accumulating evidence that the persistent changes in IEG levels following L-DOPA treatment plays an important role in the development of side effects towards this treatment, such as L-DOPA induced dyskinesias.

References

Aizman O, Brismar H, Uhlen P, Zettergren E, Levey AI, Forssberg H, Greengard P, Aperia A (2000). Anatomical and physiological evidence for colocalization of neostriatal D1 and D2 dopamine receptors. Nature Neurosci 3:226-230.

Albin RL, Young AB, Penney JB (1989). The functional anatomy of basal ganglia disorders. Trends Neurosci 12:366-375.

Alexander GE, Crutcher MD (1990). Functional architecture of basal ganglia circuits: neural substrates of parallel processing. Trends Neurosci 13:266-271.

Allen PB, Ouimet CC, Greengard P (1997). Spinophilin, a novel protein phosphatase 1 binding protein localized to dendritic spines. Proc Natl Acad Sci USA 94:9956-9961.

Andersson M, Konradi C, Cenci MA (2001). cAMP response element-binding protein is required for dopamine-dependent gene expression in the intact but not the dopamine-denervated striatum. J Neurosci 21:9930-9943.

Andersson M, Westin JE, Cenci MA (2003). Time course of striatal DeltaFosB-like immunoreactivity and prodynorphin mRNA levels after discontinuation of chronic dopaminomimetic treatment. Eur J Neurosci 17:661-666.

Angel P, Hattori K, Smeal T, Karin M (1988). The jun proto-oncogene is positively autoreulated by its product Jun/AP-1. Cell 55:875–885.

Berke JD, Paletzki RF, Aronson GJ, Hyman SE, Gerfen CR (1998). A complex program of striatal gene expression induced by dopaminergic stimulation. J Neurosci 18:5301–5310.

Bertorello AM, Hopfield JF, Aperia A, Greengard P (1990). Inhibition by dopamine of (Na^+/K^+)ATPase activity in neostriatal neurons through D1 and D2 dopamine receptor synergism. Nature 347:386-388.

Betarbet R, Sherer TB, MacKenzie G, Garcia-Osuna M, Panov AV, Greenamyre JT (2000). Chronic systemic pesticide exposure reproduces features of Parkinson's disease. Nature Neurosci 3:1301-1306.

Bezard E, Ferry S, Mach U, Stark H, Leriche L, Boraud T, Gross C, Sokoloff P (2003). Attenuation of levodopa-induced dyskinesia by normalizing dopamine D3 receptor function. Nature Med 9:762-767.

Bjelke B, Stromberg I, O'Connor WT, Andbjer B, Agnati LF, Fuxe K (1994). Evidence for volume transmission in the dopamine denervated neostriatum of the rat after a unilateral nigral 6-OHDA microinjection. Studies with systemic D-amphetamine treatment. Brain Res 662:11-24.

Bordet R, Ridray S, Carboni S, Diaz J, Sokoloff P, Schwartz JC (1997). Induction of dopamine D3 receptor expression as a mechanism of behavioral sensitization to levodopa. Proc Natl Acad Sci USA 94:3363-3367.

Bronstein DM, Ye H, Pennypacker KR, Hudson PM, Hong JS (1994). Role of a 35 kDa fos-related antigen (FRA) in the long-term induction of striatal dynorphin expression in the 6-hydroxydopamine lesioned rat. Brain Res Mol Brain Res 23:191-203.

Brotchie JM, Lee J, Venderova K. (2005). Levodopa-induced dyskinesia in Parkinson's disease. J Neural Transm 112:359-391.

Cai G, Zhen X, Uryu K, Friedman E (2000). Activation of extracellular signal-regulated protein kinases is associated with a sensitized locomotor response to D(2) dopamine receptor stimulation in unilateral 6-hydroxydopamine-lesioned rats. J Neurosci 20:1849-1857.

Canteras NS, Shammah-Lagnado SJ, Silva BA, Ricardo JA (1990). Afferent connections of the subthalamic nucleus: a combined retrograde and anterograde horseradish peroxidase study in the rat. Brain Res 513:43-59.

Cenci MA, Kalen P, Mandel RJ, Wictorin K, Bjorklund A (1992). Dopaminergic transplants normalize amphetamine- and apomorphine-induced Fos expression in the 6-hydroxydopamine-lesioned striatum. Neurosci 46:943-957.

Cenci MA, Tranberg A, Andersson M, Hilbertson A (1999). Changes in the regional and compart-mental distribution of FosB- and JunB-like immunoreactivity induced in the dopamine-denervated rat striatum by acute or chronic L-dopa treatment. Neurosci 94:515-527.

Chapman MA, Zahm DS (1996). Altered Fos-like immunoreactivity in terminal regions of the mesotelencephalic dopamine system is associated with reappearance of tyrosine hydroxylase immunoreactivity at the sites of focal 6-hydroxydopamine lesions in the nucleus accumbens. Brain Res 736:270-279.

Chen JY, Hsu PC, Hsu IL, Yeh GC (2002). Sequential up-regulation of the c-fos, c-jun and bax genes in the cortex, striatum and cerebellum induced by a single injection of a low dose of 1-methyl-4-phenyl-1,2,3,6-tetrahydropyridine (MPTP) in C57BL/6 mice. Neurosci Lett 314:49-52.

Chergui K, Svenningsson P, Nomikos GG, Gonon F, Fredholm BB, Svenson TH (1997). Increased expression of NGFI-A mRNA in the rat striatum following burst stimulation of the medial forebrain bundle. Eur J Neurosci 9:2370-2382.

Chesselet MF, Delfs JM (1996). Basal ganglia and movement disorders: an update. Trends Neurosci 19:417-422.

Cole AJ, Bhat RV, Patt C, Worley PF, Baraban JM (1992). D1 dopamine receptor activation of multiple transcription factor genes in rat striatum. J Neurochem 58:1420-1426.

Cole DG, Di Figlia M (1994). Reserpine increases Fos activity in the rat basal ganglia via a quinpirole-sensitive mechanism. Neurosci 60:115-123.

Cole DG, Growdon JH, DiFiglia M (1993). Levodopa induction of Fos immunoreactivity in rat brain following partial and complete lesions of the substantia nigra. Exp Neurol 120:223-232.

Cole RL, Konradi C, Douglass J, Hyman SE (1995). Neuronal adaptation to amphetamine and dopamine: molecular mechanisms of prodynorphin gene regulation in rat striatum. Neuron 14:813-823.

Cooper AJ, Moser B, Mitchell IJ (1995). A subset of striatopallidal neurons are Fos-immunopositive following acute monoamine depletion in the rat. Neurosci Lett 187:189-192.

Corvol JC, Muriel MP, Valjent E, Feger J, Hanoun N, Girault JA, Hirsch EC, Herve D (2004). Persistent increase in olfactory type G-protein alpha subunit levels may underlie D1 receptor functional hypersensitivity in Parkinson disease. J Neurosci 24:7007-7014.

Dahlström A, Fuxe K (1964). Evidence for the existence of monoamine-containing neurons in the central nervous system. I. Demonstration of monoamines in the cell bodies of brain stem neurons. Acta Physiol Scand 62:1-55.

Dauer W, Przedborski S (2003). Parkinson's disease: mechanisms and models. Neuron 39:889-909.

De Deurwaerdere P, Chesselet MF (2000). Nigrostriatal lesions alter oral dyskinesia and c-Fos expression induced by the serotonin agonist 1-(m-chlorophenyl)piperazine in adult rats. J Neurosci 20:5170-5178.

Deniau JM, Feger J, Le Guyader C (1976). Striatal evoked inhibition of identified nigro-thalamic neurons. Brain Res 104:152-156.

Diaz J, Levesque D, Lammers CH, Griffon N, Martres MP, Schwartz JC, Sokoloff P (1995). Phenotypical characterization of neurons expressing the dopamine D3 receptor in the brain. Neurosci 65:731-745.

Ding YM, Tang FM, Yu LP, Fu Y, Zhang GY, Jin GZ (2002). Relevance between striatal expression of Fos, proenkephalin mRNA, prodynorphin mRNA and rotation induced by 1-stepholidine in 6-hydroxydopamine-lesioned rats. Acta Pharmacol Sin 21:885-892.

Donoghue JP, Herkenham M (1986). Neostriatal projections from individual cortical fields conform to histochemically distinct striatal compartments in the rat. Brain Res 365:397-403.

Doucet JP, Nakabeppu Y, Bedard PJ, Hope BT, Nestler EJ, Jasmin BJ, Chen JS, Iadarola MJ, St-Jean M, Wigle N, Blanchet P, Grondin R, Robertson GS (1996). Chronic alterations in dopaminergic neurotransmission produce a persistent elevation of deltaFosB-like protein(s) in both the rodent and primate striatum. Eur J Neurosci 8:365-381.

Dragunow M, Faull R (1989). The use of c-fos as a metabolic marker in neuronal pathway tracing. J Neurosci Methods 29:261-265.

Dragunow M, Leah JD, Faull RL (1991). Prolonged and selective induction of Fos-related antigen(s) in striatal neurons after 6-hydroxydopamine lesions of the rat substantia nigra pars compacta. Brain Res Mol Brain Res 10:355-358.

Fagni L, Chavis P, Ango F, Bockaert J (2000). Complex interactions between mGluRs, intracellular Ca^{2+} stores and ion channels in neurons. Trends Neurosci 23:2-6.

Fahn S (1999). Parkinson's disease, the effects of levodopa, and the ELLDOPA trial. Arch Neurology 56:529-535.

Feger J, Bevan M, Crossman AR (1994). The projections from the parafascicular thalamic nucleus to the subthalamic nucleus and the striatum arise from separate neuronal populations: a comparison with the corticostriatal and corticosubthalamic efferents in a retrograde fluorescent double-labelling study. Neurosci 60:125-132.

Feger J, Hassani OK, Mouroux M (1997). The subthalamic nucleus and its connections. New electrophysiological and pharmacological data. Adv Neurol 74:31-43.

Fenu S, Pinna A, Ongini E, Morelli M (1997). Adenosine A2A receptor antagonism potentiates L-DOPA-induced turning behaviour and c-fos expression in 6-hydroxydopamine-lesioned rats. Eur J Pharmacol 321:143-147.

Fink JS, Weaver DR, Rivkees SA, Peterfreund RA, Pollack AE, Adler EM, Reppert SM (1992). Molecular cloning of the rat A_2 adenosine receptor: selective co-expression with D_2 dopamine receptors in rat striatum. Brain Res Mol Brain Res 14:186-195.

Finkbeiner S, Tavazoie SF, Maloratsky A, Jacobs KM, Harris KM, Greenberg ME (1997). CREB: a major mediator of neuronal neurotrophin responses. Neuron 19:1031-1047.

Fredholm BB (1977). Activation of adenylate cyclase from rat striatum and tuberculum olfactorium by adenosine. Med Biol 55:262-267.

Fritschy JM, Frondoza CG, Grzanna R (1991). Differential effects of reserpine on brainstem cate-cholaminergic neurons revealed by Fos protein immunohistochemistry. Brain Res 562:48-56.

Gerfen CR, Engber TM, Mahan LC, Susel Z, Chase TN, Monsma Jr FJ, Sibley DR (1990). D_1 and D_2 dopamine receptor-regulated gene expression of striatonigral and striatopallidal neurons. Science 250:1429-1432.

Gerfen CR, Keefe KA, Gauda EB (1995). D1 and D2 dopamine receptor function in the striatum: coactivation of D1- and D2-dopamine receptors on separate populations of neurons results in potentiated immediate early gene response in D1-containing neurons. J Neurosci 15:8167-8176.

Gerfen CR, Miyachi S, Paletzki R, Brown P (2002). D1 dopamine receptor supersensitivity in the dopamine-depleted striatum results from a switch in the regulation of ERK1/2/MAP kinase. J Neurosci 22:5042-5054.

Gerfen CR, Wilson CJ (1996). The basal ganglia. In: Handbook of Chemical Neuroanatomy: Integrated Systems of the CNS (Swanson LW, Björklund A, Hökfelt T, eds), pp. 371-467. Amsterdam: Elsevier Science.

Giovanni A, Sieber BA, Heikkila RE, Sonsalla PK (1991). Correlation between the neostriatal content of the 1-methyl-4-phenylpyridinium species and dopaminergic neurotoxicity following 1-methyl-4-phenyl-1,2,3,6-tetrahydropyridine administration to several strains of mice. J Pharmacol Exp Ther 257:691-697.

Gonon F (1997). Prolonged and extrasynaptic excitatory action of dopamine mediated by D_1 receptors in the rat striatum in vivo. J Neurosci 17:5972-5978.

Graybiel AM (1990). Neurotransmitters and neuromodulators in the basal ganglia. Trends Neurosci 13:244-254.

Graybiel AM, Moratalla R, Robertson HA (1990). Amphetamine and cocaine induce drug-specific activation of the c-fos gene in striosome-matrix compartments and limbic subdivisions of the striatum. Proc Natl Acad Sci USA 87:6912-6916.

Graybiel AM, Canales JJ, Capper-Loup C (2000). Levodopa-induced dyskinesias and dopamine-dependent stereotypies: a new hypothesis. Trends Neurosci 23(10 Suppl):S71-77.

Guo X, Ding YM, Hu JY, Jin GZ (1998). Involvement of dopamine D1 and D2 receptors in Fos immunoreactivity induced by stepholidine in both intact and denervated striatum of lesioned rats. Life Sci 62:2295-2302.

Harlan RE, Garcia MM (1998). Drugs of abuse and immediate-early genes in the forebrain. Mol Neurobiol 16:221-267.

Hassani OK, Feger J (1999). Effects of intrasubthalamic injection of dopamine receptor agonists on subthalamic neurons in normal and 6-hydroxydopamine-lesioned rats: an electrophysiological and c-Fos study. Neurosci 92:533-543.

Heikkila RE, Hess A, Duvoisin RC (1984). Dopaminergic neurotoxicity of 1-methyl-4-phenyl-1,2,5,6-tetrahydropyridine in mice. Science 224:1451-1453.

Hemmings Jr HC, Greengard P, Tung HY, Cohen P (1984). DARPP-32, a dopamine-regulated neuronal phosphoprotein, is a potent inhibitor of protein phosphatase-1. Nature 310:503-505.

Herdegen T, Leah JD (1998). Inducible and constitutive transcription factors in the mammalian nervous system: control of gene expression by Jun, Fos and Krox, and CREB/ATF proteins. Brain Res Brain Res Rev 28:370-490.

Herkenham M, Pert CB (1981). Mosaic distribution of opiate receptors, parafascicular projections and acetylcholinesterase in rat striatum. Nature 291:415-418.

Hernandez-Lopez S, Tkatch T, Perez-Garci E, Galarraga E, Bargas J, Hamm H, Surmeier DJ (2000). D2 dopamine receptors in striatal medium spiny neurons reduce L-type Ca^{2+} currents and excitability via a novel PLC[beta]1-IP3-calcineurin-signaling cascade. J Neurosci 20:8987-8995.

Hersch SM, Ciliax BJ, Gutekunst CA, Rees HD, Heilman CJ, Yung KL, Bolam JP, Ince E, Yi H, Levey AI (1995). Electron microscopic analysis of D_1 and D_2 dopamine receptor proteins in the dorsal striatum and their synaptic relationships with motor corticostriatal afferents. J Neurosci 15:5222-5237.

Herve D, Levi-Strauss M, Marey-Semper I, Verney C, Tassin JP, Glowinski J, Girault JA (1993). G(olf) and Gs in rat basal ganglia: possible involvement of G(olf) in the coupling of dopamine D1 receptor with adenylyl cyclase. J Neurosci 13:2237-2248.

Hirsch EC, Perier C, Orieux G, Francois C, Feger J, Yelnik J, Vila M, Levy R, Tolosa ES, Marin C, Trinidad Herrero M, Obeso JA, Agid Y (2000). Metabolic effects of nigrostriatal denervation in basal ganglia. Trends Neurosci 23:S78-85.

Hoglinger GU, Feger J, Prigent A, Michel PP, Parain K, Champy P, Ruberg M, Oertel WH, Hirsch EC (2003). Chronic systemic complex I inhibition induces a hypokinetic multisystem degeneration in rats. J Neurochem 84:491-502.

Hoover BR, Marshall JF (2004). Molecular, chemical, and anatomical characterization of globus pallidus dopamine D2 receptor mRNA-containing neurons. Synapse 52:100-113.

Hunot S, Vila M, Teismann P, Davis RJ, Hirsch EC, Przedborski S, Rakic P, Flavell RA (2004). JNK-mediated induction of cyclooxygenase 2 is required for neurodegeneration in a mouse model of Parkinson's disease. Proc Natl Acad Sci USA 101:665-670.

Ishida Y, Kuwahara I, Todaka K, Hashiguchi H, Nishimori T, Mitsuyama Y (1996). Dopaminergic transplants suppress L-DOPA-induced Fos expression in the dopamine-depleted striatum in a rat model of Parkinson's disease. Brain Res 727:205-211.

Jenner P (2003). The MPTP-treated primate as a model of motor complications in PD: primate model of motor complications. Neurology 61:S4-11.

Jiang H, Jackson-Lewis V, Muthane U, Dollison A, Ferreira M, Espinosa A, Parsons B, Przedborski S (1993). Adenosine receptor antagonists potentiate dopamine receptor agonist-induced rotational behavior in 6-hydroxydopamine-lesioned rats. Brain Res 613:347-351.

Jian M, Staines WA, Iadarola MJ, Robertson GS (1993). Destruction of the nigrostriatal pathway increases Fos-like immunoreactivity predominantly in striatopallidal neurons. Brain Res Mol Brain Res 19:156-160.

Jones DT, Reed RR (1989). Golf: an olfactory neuron specific-G protein involved in odorant signal transduction. Science 244:790-795.

Kanda T, Jackson MJ, Smith LA, Pearce RK, Nakamura J, Kase H, Kuwana Y, Jenner P (1998). Adenosine A_{2A} antagonist: a novel antiparkinsonian agent that does not provoke dyskinesia in parkinsonian monkeys. Ann Neurol 43:507-513.

Kearney JA, Frey KA, Albin RL (1997). Metabotropic glutamate agonist-induced rotation: a pharmacological, FOS immunohistochemical, and [^{14}C]-2-deoxyglucose autoradiographic study. J Neurosci 17:4415-4425.

Kearney JA, Becker JB, Frey KA, Albin RL (1998). The role of nigrostriatal dopamine in metabotropic glutamate agonist-induced rotation. Neurosci 87:881-891.

Kearney JA, Albin RL (2000). Intrasubthalamic nucleus metabotropic glutamate receptor activation: a behavioral, Fos immunohistochemical and [14C]2-deoxyglucose autoradiography study. Neurosci 95:409-416.

Kebabian JW, Greengard P (1971). Dopamine-sensitive adenyl cyclase: possible role in synaptic transmission. Science 174:1346-1349.

Keefe KA, Gerfen CR (1996). D1-D2 dopamine receptor synergy in striatum: effects of intrastriatal infusions of dopamine agonists and antagonists on immediate early gene expression. Neurosci 66:903-913.

Keefe KA, Gerfen CR (1996). D1 dopamine receptor-mediated induction of zif268 and c-fos in the dopamine-depleted striatum: differential regulation and independence from NMDA receptors. J Comp Neurol 367:165-176.

Keefe KA, Gerfen CR (1999). Local infusion of the $(+/-)$-alpha-amino-3-hydroxy-5-methyl-isoxazole-4-propionate/kainate receptor antagonist 6-cyano-7-nitroquinoxaline-2,3-dione does not block D1 dopamine receptor-mediated increases in immediate early gene expression in the dopamine-depleted striatum. Neurosci 89:491-504.

Kemp JM, Powell TPS (1970). The cortico-striate projection in the monkey. Brain 93:525-546.

Kitai ST, Kocsis JD, Wood J (1976). Origin and characteristics of the cortico-caudate afferents: an anatomical and electrophysiological study. Brain Res 118:137-141.

Konradi C, Cole RL, Heckers S, Hyman SE (1994). Amphetamine regulates gene expression in rat striatum via transcription factor CREB. J Neurosci 14:5623-5634.

Labandeira-Garcia JL, Rozas G, Lopez-Martin E, Liste I, Guerra MJ (1996). Time course of striatal changes induced by 6-hydroxydopamine lesion of the nigrostriatal pathway, as studied by combined evaluation of rotational behaviour and striatal Fos expression. Exp Brain Res 108:69-84.

Labiner DM, Butler LS, Cao Z, Hosford DA, Shin C, McNamara JO (1993). Induction of c-fos mRNA by kindled seizures: complex relationship with neuronal burst firing. J Neurosci 13:744-751.

LaHoste GJ, Yu J, Marshall JF (1993). Striatal Fos expression is indicative of dopamine D1/D2 synergism and receptor supersensitivity. Proc Natl Acad Sci USA 90:7451-7455.

Langston JW, Quik M, Petzinger G, Jakowec M, Di Monte DA (2000). Investigating levodopa-induced dyskinesias in the parkinsonian primate. Annals of Neurology 47:S79-89.

Le Moine C, Bloch B (1995). D_1 and D_2 dopamine receptor gene expression in the rat striatum: sensitive cRNA probes demonstrate prominent segregation of D_1 and D_2 mRNAs in distinct neuronal populations of the dorsal and ventral striatum. J Comp Neurol 355:418-426.

Le Moine C, Bloch B (1996). Expression of the D3 dopamine receptor in peptidergic neurons of the nucleus accumbens: comparison with the D1 and D2 dopamine receptors. Neurosci 73:131-143.

Le Moine C, Svenningsson P, Fredholm BB, Bloch B (1997). Dopamine-adenosine interactions in the striatum and the globus pallidus.inhibition of striatopallidal neurons through either D_2 or A_{2A} receptors enhance D_1 receptor-mediated effects on c-fos expression. J Neurosci 17:8038-8048.

Lang AE, Lees A. (2002). Anticholinergic therapies in the treatment of Parkinson's disease. Management of Parkinson's disease: an evidence-based review. Movement Disorders 17:S7-S12.

Leentjens AF (2004). Depression in Parkinson's disease: conceptual issues and clinical challenges. J Geriatric Psychiatry & Neurology 17:120-126.

Levy R, Hazrati LN, Herrero MT, Vila M, Hassani OK, Mouroux M, Ruberg M, Asensi H, Agid Y, Feger J, Obeso JA, Parent A, Hirsch EC (1997). Re-evaluation of the functional anatomy of the basal ganglia in normal and Parkinsonian states. Neurosci 76:335-343.

Lezcano N, Mrzljak L, Eubanks S, Levenson R, Goldman-Rakic P, Bergson C (2000). Dual signaling regulated by calcyon, a D1 dopamine receptor interacting protein. Science 287:1660-1664.

Limousin P, Krack P, Pollak P, Benazzouz A, Ardouin C, Hoffmann D, Benabid AL (1998). Electrical stimulation of the subthalamic nucleus in advanced Parkinson's disease. N Engl J Med 339:1105-1111.

Lundblad M, Picconi B, Lindgren H, Cenci MA (2004). A model of L-DOPA-induced dyskinesia in 6-hydroxydopamine lesioned mice: relation to motor and cellular parameters of nigrostriatal function. Neurobiol Dis 16:110-123.

Marshall JF, Cole BN, LaHoste GJ (1993). Dopamine D2 receptor control of pallidal fos expression: comparisons between intact and 6-hydroxydopamine-treated hemispheres. Brain Res 632:308-313.

Masliah E, Rockenstein E, Veinbergs I, Mallory M, Hashimoto M, Takeda A, Sagara Y, Sisk A, Mucke L (2000). Dopaminergic loss and inclusion body formation in alpha-synuclein mice: implications for neurodegenerative disorders. Science 287:1265-1269.

McClung CA, Ulery PG, Perrotti LI, Zachariou V, Berton O, Nestler EJ (2004). DeltaFosB: a molecular switch for long-term adaptation in the brain. Brain Res Mol Brain Res 132:146-154.

McGeorge AJ, Faull RL (1989). The organization of the projection from the cerebral cortex to the striatum in the rat. Neurosci 29:503-537.

McNaught KS, Perl DP, Brownell AL, Olanow CW (2004). Systemic exposure to proteasome inhibitors causes a progressive model of Parkinson's disease. Ann Neurol 56:149-162.

Meador-Woodruff JH, Mansour A, Grandy DK, Damask SP, Civelli O, Watson SJ (1992). Distribution of D5 dopamine receptor mRNA in rat brain. Neurosci Lett 145:209-212.

Missale C, Nash SR, Robinson SW, Jaber M, Caron MG (1998). Dopamine receptors: from structure to function. Physiol Rev 78:189-225.

Morelli M, Pinna A, Fenu S, Carta A, Cozzolino A, Di Chiara G (1994). Differential effect of MK 801 and scopolamine on c-fos expression induced by L-dopa in the striatum of 6-hydroxydopamine lesioned rats. Synapse 18:288-293.

Mouroux M, Feger J (1993). Evidence that the parafascicular projection to the subthalamic nucleus is glutamatergic. Neuroreport 4:613-615.

Mura A, Jackson D, Manley MS, Young SJ, Groves PM (1995). Aromatic L-amino acid decarboxylase immunoreactive cells in the rat striatum: a possible site for the conversion of exogenous L-DOPA to dopamine. Brain Res 704:51-60.

Nakao N, Ogura M, Nakai K, Itakura T (1998). Intrastriatal mesencephalic grafts affect neuronal activity in basal ganglia nuclei and their target structures in a rat model of Parkinson's disease. J Neurosci 18:1806-1817.

Nestler EJ (2005). Is there a common molecular pathway for addiction? Nat Neurosci 8:1445-1449.

Nielsen KM, Soghomonian JJ (2003). Dual effects of intermittent or continuous L-DOPA administration on gene expression in the globus pallidus and subthalamic nucleus of adult rats with a unilateral 6-OHDA lesion. Synapse 49:246-260.

Nishi A, Snyder GL, Greengard P (1997). Bidirectional regulation of DARPP-32 phosphorylation by dopamine. J Neurosci 17:8147-8155.

Obeso JA, Rodriguez-Oroz MC, Rodriguez M, Lanciego JL, Artieda J, Gonzalo N, Olanow CW (2000). Pathophysiology of the basal ganglia in Parkinson's disease. Trends Neurosci 23:S8-19.

O'Donnel P, Grace AA (1994). Tonic D_2-mediated attenuation of cortical excitation in nucleus accumbens neurons recorded in vitro. Brain Res 634:105-112.

Oertel WH, Hoglinger GU, Caraceni T, Girotti F, Eichhorn T, Spottke AE, Krieg JC, Poeve W (2001). Depression in Parkinson's disease. An update. Adv Neurology 86:373-383.

Paisan-Ruiz C, Jain S, Evans EW, et al. (2004). Cloning of the gene containing mutations that cause PARK8-linked Parkinson's disease. Neuron 44:595–600.

Paul ML, Currie RW, Robertson HA (1995). Priming of a D1 dopamine receptor behavioural response is dissociated from striatal immediate-early gene activity. Neurosci 66:347-359.

Paul ML, Graybiel AM, David JC, Robertson HA (1992). D1-like and D2-like dopamine receptors synergistically activate rotation and c-fos expression in the dopamine-depleted striatum in a rat model of Parkinson's disease. J Neurosci 12:3729-3742.

Pennypacker KR, Zhang WQ, Ye H, Hong JS (1992). Apomorphine induction of AP-1 DNA binding in the rat striatum after dopamine depletion. Brain Res Mol Brain Res 15:151-155.

Perez-Otano I, Mandelzys A, Morgan JI (1998). MPTP-Parkinsonism is accompanied by persistent expression of a delta-FosB-like protein in dopaminergic pathways. Brain Res Mol Brain Res 53:41-52.

Picconi B, Centonze D, Hakansson K, Bernardi G, Greengard P, Fisone G, Cenci MA, Calabresi P (2003). Loss of bidirectional striatal synaptic plasticity in L-DOPA-induced dyskinesia. Nature Neurosci 6:501-506.

Pinna A, Di Chiara G, Wardas J, Morelli M (1996). Blockade of A_{2A} adenosine receptors positively modulates turning behaviour and c-fos expression induced by D_1 agonists in dopamine-denervated rats. Eur J Neurosci 8:1176-1181.

Pollack AE, Bird JL, Lambert EB, Florin ZP, Castellar VL (1999). Role of NMDA glutamate receptors in regulating D2 dopamine-dependent Fos induction in the rat striatopallidal pathway. Brain Res 818:543-547.

Pollack AE, Fink JS (1996). Synergistic interaction between an adenosine antagonist and a D_1 dopamine agonist on rotational behavior and striatal c-Fos induction in 6-hydroxydopamine-lesioned rats. Brain Res 743:124-130.

Pollack AE, St Martin JL, Macpherson AT (2005). Role of NMDA and AMPA glutamate receptors in the induction and the expression of dopamine-mediated sensitization in 6-hydroxydopamine-lesioned rats. Synapse 56:45-53.

Pollack AE, Turgeon SM, Fink JS (1997). Apomorphine priming alters the response of striatal outflow pathways to D2 agonist stimulation in 6-hydroxydopamine-lesioned rats. Neurosci 79:79-93.

Premont J, Perez M, Bockaert J (1977). Adenosine-sensitive adenylate cyclase in rat striatal homogenates and its relationship to dopamine- and Ca^{2+}-sensitive adenylate cyclases. Mol Pharmacol 13:662-670.

Przedborski S, Jackson-Lewis V, Naini AB, Jakowec M, Petzinger G, Miller R, Akram M (2001). The parkinsonian toxin 1-methyl-4-phenyl-1,2,3,6-tetrahydropyridine (MPTP): a technical review of its utility and safety. J Neurochem 76:1265-1274.

Reader TA, Dewar KM (1999). Effects of denervation and hyperinnervation on dopamine and serotonin systems in the rat neostriatum: implications for human Parkinson's disease. Neurochem Int 4:1-21.

Rivera A, Alberti I, Martin AB, Narvaez JA, de la Calle A, Moratalla R (2002). Molecular phenotype of rat striatal neurons expressing the dopamine D5 receptor subtype. Eur J Neurosci 16:2049-2058.

Rivera A, Trias S, Penafiel A, Angel Narvaez J, Diaz-Cabiale Z, Moratalla R, de la Calle A (2003). Expression of D4 dopamine receptors in striatonigral and striatopallidal neurons in the rat striatum. Brain Res 989:35-41.

Robertson GS, Staines WA (1994). D1 dopamine receptor agonist-induced Fos-like immunoreactivity occurs in basal forebrain and mesopontine tegmentum cholinergic neurons and striatal neurons immunoreactive for neuropeptide Y. Neurosci 59:375-387.

Robertson GS, Vincent SR, Fibiger HC (1990). Striatonigral projection neurons contain D1 dopamine receptor-activated c-fos. Brain Res 523:288-290.

Robertson GS, Vincent SR, Fibiger HC (1992). D1 and D2 dopamine receptors differentially regulate c-fos expression in striatonigral and striatopallidal neurons. Neurosci 49:285-296.

Robertson HA, Peterson MR, Murphy K, Robertson GS (1989). D1-dopamine receptor agonists selectively activate striatal c-fos independent of rotational behaviour. Brain Res 503:346-349.

Ruskin DN, Bergstrom DA, Mastropietro CW, Twery MJ, Walters JR (1999). Dopamine agonist-mediated rotation in rats with unilateral nigrostriatal lesions is not dependent on net inhibitions of rate in basal ganglia output nuclei. Neurosci 91:935-946.

Ruskin DN, Marshall JF (1995). D1 dopamine receptors influence Fos immunoreactivity in the globus pallidus and subthalamic nucleus of intact and nigrostriatal-lesioned rats. Brain Res 703:156-164.

Saka E, Elibol B, Erdem S, Dalkara T (1999). Compartmental changes in expression of c-Fos and FosB proteins in intact and dopamine-depleted striatum after chronic apomorphine treatment. Brain Res 825:104-114.

Sandstrom MI, Sarter M, Bruno JP (1996). Interactions between D1 and muscarinic receptors in the induction of striatal c-fos in rats depleted of dopamine as neonates. Brain Res Dev Brain Res 96:148-158.

Saporito MS, Brown EM, Miller MS, Carswell S (1999). CEP-1347/KT-7515, an inhibitor of c-jun N-terminal kinase activation, attenuates the 1-methyl-4-phenyl tetrahydropyridine-mediated loss of nigrostriatal neurons in vivo. J Pharmacol Exp Ther 288:421-427.

Schiffmann SN, Jacobs O, Vanderhaeghen JJ (1991). Striatal restricted adenosine A_2 receptor (RDC8) is expressed by enkephalin but not by substance P neurons: an in situ hybridization histochemistry study. J Neurochem 57:1062-1067.

Schuller JJ, Marshall JF (2000). Acute immediate-early gene response to 6-hydroxydopamine infusions into the medial forebrain bundle. Neurosci 96:51-58.

Schwarzschild MA, Cole RL, Hyman SE (1997). Glutamate, but not dopamine, stimulates stress-activated protein kinase and AP-1-mediated transcription in striatal neurons. J Neurosci 17:3455-3466.

Simpson JN, Wang JQ, McGinty JF (1995). Repeated amphetamine administration induces a prolonged augmentation of phosphorylated cyclase element-binding protein and Fos-related antigen immunoreactivity in rat striatum. Neurosci 69:441-457.

Smith FD, Oxford GS, Milgram SL (1999). Association of the D2 dopamine receptor third cytoplasmic loop with spinophilin, a protein phosphatase-1-interacting protein. J Biol Chem 274:19894-19900.

Smith Y, Bevan MD, Shink E, Bolam JP (1998). Microcircuitry of the direct and indirect pathways of the basal ganglia. Neurosci 86:353-387.

Stoof JC, Kebabian JW (1981). Opposing roles for D-1 and D-2 dopamine receptors in efflux of cyclic AMP from rat neostriatum. Nature 294:366-368.

Surmeier DJ, Eberwine J, Wilson CJ, Cao Y, Stefani A, Kitai ST (1992). Dopamine receptor subtypes colocalize in rat striatonigral neurons. Proc Natl Acad Sci USA 89:10178-10182.

Svenningsson P, Arts J, Gunne L, Andren PE (2002). Acute and repeated treatment with L-DOPA increase c-jun expression in the 6-hydroxydopamine-lesioned forebrain of rats and common marmosets. Brain Res 955:8-15.

Svenningsson P, Fourreau L, Bloch B, Fredholm BB, Gonon F, Le Moine C (1999). Opposite tonic modulation of dopamine and adenosine on c-fos gene expression in striatopallidal neurons. Neurosci 89:827-837.

Svenningsson P, Gunne L, Andren PE (2000). L-DOPA produces strong induction of c-fos messenger RNA in dopamine-denervated cortical and striatal areas of the common marmoset. Neurosci 99:457-468.

Svenningsson P, Le Moine C (2002). Dopamine D1/5 receptor stimulation induces c-fos expression in the subthalamic nucleus: possible involvement of local D5 receptors. Eur J Neurosci 15:133-142.

Svenningsson P, Le Moine C, Kull B, Sunahara R, Bloch B, Fredholm BB (1997). Cellular expression of adenosine A_{2A} receptor messenger RNA in the rat central nervous sytem with special reference to dopamine innervated areas. Neurosci 80:1171-1185.

Svenningsson P, Nishi A, Fisone G, Girault JA, Nairn AC, Greengard P (2004). DARPP-32: an integrator of neurotransmission. Annu Rev Pharmacol Toxicol 44:269-296.

Ungerstedt U (1968). 6-Hydroxy-dopamine induced degeneration of central monoamine neurons. Eur J Pharmacol 5:107-110.

Ungerstedt U (1971). Stereotaxic mapping of the monoamine pathway in the rat brain. Acta Physiol Scand Suppl 367:1-48.

Ungerstedt U, Arbuthnott GW (1970). Quantitative recording of rotational behaviour in rats after 6-hydroxy-dopamine lesions of the nigrostriatal dopamine system. Brain Res 24:485-493.

Valente EM, Abou-Sleiman PM, Caputo V, et al. (2004). Hereditary early-onset Parkinson's disease caused by mutations in PINK1. Science 304:1158–1160.

Valjent E, Pascoli V, Svenningsson P, Paul S, Enslen H, Corvol JC, Stipanovich A, Caboche J, Lombroso PJ, Nairn AC, Greengard P, Herve D, Girault JA (2005). Regulation of a protein phosphatase cascade allows convergent dopamine and glutamate signals to activate ERK in the striatum. Proc Natl Acad Sci USA 102:491-496.

Vanhoutte P, Barnier JV, Guilbert B, Pages C, Besson MJ, Hipskind RA, Caboche J (1999). Glutamate induces phoshphorylaton of Elk-1 and CREB, along with c-fos activation, via an extracellular signal-regulated kinase-dependent pathway in brain slices. Molec Cell Biol 19:136-146.

Vila M, Przedborski S (2004). Genetic clues to the pathogenesis of Parkinson's disease. Nat Med 10(Suppl):S58-62.

Walaas SI, Aswad DW, Greengard P (1983). A dopamine- and cyclic AMP-regulated phosphoprotein enriched in dopamine-innervated brain regions. Nature 301:69-71.

Wang Q, Zhao J, Brady AE, Feng J, Allen PB, Lefkowitz RJ, Greengard P, Limbird LE (2004). Spinophilin blocks arrestin actions in vitro and in vivo at G protein-coupled receptors. Science 304:1940-1944.

Wan XS, Liang F, Moret V, Wiesendanger M, Rouiller EM (1992). Mapping of the motor pathways in rats: c-fos induction by intracortical microstimulation of the motor cortex correlated with efferent connectivity of the site of cortical stimulation. Neurosci 49:749-761.

Weintraub D, Moberg PJ, Duda JE, Katz IR, Stern MB (2004). Effect of psychiatric and other nonmotor symptoms on disability in Parkinson's disease. J American Geriatrics Soc 52:784-788.

West AR, Grace AA (2002). Opposite influences of endogenous dopamine D1 and D2 receptor activation on activity states and electrophysiological properties of striatal neurons: studies combining in vivo intracellular recordings and reverse microdialysis. J Neurosci 22:294-304.

Westin JE, Andersson M, Lundblad M, Cenci MA (2001). Persistent changes in striatal gene expression induced by long-term L-DOPA treatment in a rat model of Parkinson's disease. Eur J Neurosci 14:1171-1176.

Wolfarth S, Konieczny J, Lorenc-Koci E, Ossowska K, Pilc A (2000). The role of metabotropic glutamate receptor (mGluR) ligands in parkinsonian muscle rigidity. Amino Acids 19:95-101.

Xia Z, Dudek H, Miranti CK, Greenberg ME (1996). Calcium influx via the NMDA receptor induces immediate early gene transcription by a MAP kinase/ERK-dependent mechanism. J Neurosci 16:5425-5436.

Xu Y, Sun S, Cao X (2003). Effect of levodopa chronic administration on behavioral changes and fos expression in basal ganglia in rat model of PD. J Huazhong Univ Sci Technolog Med Sci 23:258-262.

Yoshida M, Precht W (1971). Monosynaptic inhibition of neurons of the substantia nigra by caudato-nigral fibers. Brain Res 32:225-228.

Zimprich A, Biskup S, Leitner P, et al. (2004). Mutations in LRRK2 cause autosomal-dominant parkinsonism with pleomorphic pathology. Neuron 44:601-607.

Ziolkowska B, Horn G, Kupsch A, Hollt V (1995). The expression of proenkephalin and prodynorphin genes and the induction of c-fos gene by dopaminergic drugs are not altered in the striatum of MPTP-treated mice. J Neural Transm Park Dis Dement Sect 9:151-164.

Zivin M, Sprah L, Sket D (1996). The D1 receptor-mediated effects of the ergoline derivative LEK-8829 in rats with unilateral 6-hydroxydopamine lesions. Br J Pharmacol 119:1187-1196.

Index